CHEMICAL ENGINEERING METHODS AND TECHNOLOGY

POWDER ENGINEERING, TECHNOLOGY AND APPLICATIONS

CHEMICAL ENGINEERING METHODS AND TECHNOLOGY

CHEMICAL ENGINEERING METHODS AND TECHNOLOGY

POWDER ENGINEERING, TECHNOLOGY AND APPLICATIONS

JASON M. BARKER
EDITOR

Nova Science Publishers, Inc.
New York

For permission to use material from this book please contact us:
Telephone 631-231-7269; Fax 631-231-8175
Web Site: http://www.novapublishers.com

LIBRARY OF CONGRESS CATALOGING-IN-PUBLICATION DATA

Powder engineering, technology, and applications / editor, Jason M. Barker.
p. cm.
Includes index.
ISBN 978-1-61761-216-9 (hardcover)
1. Powders. 2. Particles. I. Barker, Jason M.
TP156.P3P57 2010
620'.43--dc22

2010029850

Published by Nova Science Publishers, Inc. + New York

CONTENTS

PREFACE

Milling or physically dividing solids into fine fragments is the best-known method for the production of powders. It is found widely across industries from pharmaceuticals, nutraceuticals, agrochemicals, aerosols, personal care products, pigments, ceramics, cement andsemiconductors. This book presents current research data in the study of powder engineering including, particle engineering beyond size reduction; morphology-controlled fabrication and applications of metal oxide powders; organic solvent-free synthesis of magnetite nanocrystals with controlled particle sizes; and the process of nanomilling.

Chapter 1 - Several industrial applications, from the nuclear engineering to the food processing, consider the presence of surfaces immersed in bubbling fluidized beds. The surface-to-bed heat, mass and erosion rates significantly contribute to determine the actual efficiency as well as the operating life of fluidized bed unit, and this explains the paramount importance of experimental and modeling studies aimed to support reliable design and operation of these industrial processes.

The fluid dynamic field around an immersed surface presents different peculiarities respect to the average characteristics of the fluidized bed. These peculiarities are responsible for the actual values of the surface-to-bed mass and heat transfer. Similarly, they governed the momentum transfer, whose main aspect is the surface erosion caused by the impacts with the fluidized particles.

This chapter aims to analyze the transport phenomena among surfaces immersed in fluidized beds in light of the most recent experimental and modeling results reported in literature. The main goal of this contribution is to shed light on the characteristics of each transport phenomena and to point out the real analogies among them.

Chapter 2 - Magnetite (Fe3O4), which is an important member of spinel type ferrites, has been widely used in various industrial products such as recording materials because it has the excellent magnetic properties. Recently, bio-applications of Fe3O4 nanocrystals, e.g., magnetic resonance imaging, drug delivery, and hyperthermia, have attracted much attention because of the non-toxicity property and high chemical stability. Thus, the demand for Fe3O4 is increasing, and industrial synthesis methods which can meet the demand are needed. The synthesis processes should have the low environmental impact and must prepare the Fe3O4 nanocrystals with precisely controlled particle sizes suitable for the applications because the magnetic properties depend strongly on the particle size. Therefore, the authors have synthesized the size-controlled Fe3O4 nanocrystals using organic solvent-free conventional and novel methods: hydrothermal and mechanochemical synthesis methods. In the hydrothermal synthesis, ferrous and ferric hydroxides are coprecipitated or sequentially formed in high alkaline solutions, and then the suspension thus obtained is heated under a hydrothermal condition. In the coprecipitation operation, the molar ratio of ferrous ion to ferric ion is varied. Furthermore, the anionic species existing together with the iron ions are controlled. This synthesis method could control the particle size between 10 and 40 nm. On the other hand, the authors have developed a novel synthesis method using mechanochemical effect in order to prepare Fe3O4 nanocrystals with high crystallinity without using any environment-unfriendly additives such as toxic organic solvents and additional heating treatments. In this method, a tumbling ball mill is used as a synthesis reactor, and the suspension of ferrous and ferric hydroxides is ball-milled under a cooling condition. The solid phase reaction between ferrous and ferric hydroxides progresses owing to not the heat energy but the mechanical energy. This method provides superparamagnetic Fe3O4 nanocrystals with a size of less than 10 nm.

Chapter 3 - Nanostructured binary, ternary, and quaternary Fe-based mixtures have been prepared by mechanical alloying from pure elemental powders by the structural disintegration of coarser-grained structure through the severe plastic deformation. This later can force atoms into positions where they may not be at equilibrium and leads, therefore, to the formation of nanocrystalline solid solutions, intermetallic compounds, amorphous phases, etc. Microstructural, structural, morphological, thermal stability and magnetic properties changes during the milling process were investigated by X-ray diffraction, scanning electron microscopy, Mössbauer spectrometry, differential scanning calorimetry, and magnetic measurements. Depending on the milling conditions, the kinetics of elemental powders mixing can be described by one or two stages which are characterized by small values of Avrami parameter. Keywords: Nanostructured materials; Mechanical alloying; XRD; Mössbauer spectrometry; Magnetic measurements; Thermal stability.

Chapter 4 - Milling or physically dividing solids into fine fragments is the best-known method for the production of powders. It is found widely across industries from pharmaceuticals, nutraceuticals, agrochemicals, aerosols, personal care products, pigments, ceramics, cement to semi-conductors. Traditionally, milling (or also known as grinding) is regarded as a particle size reduction process, which increases the specific surface area of a solid phase for desired interaction with the neighboring phase. Advanced powder characterization techniques have revealed that milling performs more than just a size reduction role. The mechanical and thermal energies imposed onto a milled powder alter other physicochemical properties like morphology, crystallinity, polymorphism, glass transition temperature, surface energy, melting properties, as well as lead to particle size shift and changes in specific surface area during subsequent storage. These changes often have a pronounced effect on the product performance such as powder flow, aerosolization, dissolution, physical and chemical stability. The underlying scientific knowledge behind these changes has created avenues to engineer new applications as well as mitigate the milling-induced effects that are detrimental to product performance. This review presents recent developments in milling technologies such as nanomilling, co-milling and post-milling conditioning. Nanomilling using stirred media mills with suitable surfactants has produced stabilized nanosuspensions below the postulated grinding limit of half a micron, and thereby increases the specific surface area by up to 10-fold and the saturated solubility by 6-fold. Co-milling or simultaneous milling of two or more solid component has been effective in controlling the solid-state properties of milled powders. Milling-induced amorphization or the loss of long-range crystalline structure is often associated with stability problems as the amorphous phase is thermodynamically unstable and tends to re-crystallize to the stable crystalline form. This effect can be mitigated by co-milling with crystalline seeds. On the other hand, when an amorphous product is desired as it has advantages of higher solubility, dissolution rate and better compression characteristics over the crystalline form, the addition of polymeric additives during milling has led to stable amorphous solid dispersions. Co-milling with mesoporous materials also results in an amorphous form, which is entrapped inside nanosized channels. Co-grinding has been used to prepare co-crystals, which possess different physicochemical properties from the original components. To alleviate undesired milling-induced effects, new findings in post-milling conditioning as well as recent research to alternative processes to particle size reduction will be discussed.

Chapter 5 - The Discrete Particle Model (DPM) has been extensively used over the last decade to study the complicated flow behaviors in dense gas-solid fluidization systems relevant to a wide range of chemical and petrochemical processes. Numerical simulations of these real-life systems involve often tracking a large number of particles colliding at high frequency which results in a heavy computational burden. Thus, designing efficient collision-handling algorithms to operate collision events is paramount for its industrial use in studying flow dynamics in chemical reactors.

In this chapter the authors first discuss previous algorithms developed for that purpose and then present a novel, more efficient algorithm designed to handle collision events in the context of the hard-sphere DPM. The core idea is centered on the use of dynamic crosswise linked data structures. This type of data structures enables an efficient updating and managing of lists of collision events associated with particles. In addition, a chained hash table strategy is proposed to efficiently process particle collision sequence. For systems of N particles, this strategy allows a complexity O(1) time per collision regarding the event list handling operations.

Chapter 6 - The technological demands on powder-processing of functional materials have reached a particularly prosperous state. Extensive studies have been devoted to the shape-controlled production of micro/nano powders in the past few decades, and Cu2O micro/nano powders are of particular interest owing to their potential applications. This review features the recent developments regarding wet-chemical fabrication and applications of Cu2O micro/nano powders with well-controlled shapes. To understand how chemically derived processes control the architectures of Cu2O powders, the authors examine a range of synthetic strategies that give rise to the various final shapes. For our investigation, these shapes of Cu2O powders are broadly categorized into three themes, i.e., polyhedron, hollow structure, and dendritic morphology, and each includes several specific structures. As for these different powder-processing methods, important factors in achieving the controlled synthesis of Cu2O micro/nanopowders are discussed. Moreover, potential applications of the Cu2O powders with distinct morphologies in solar energy conversion, catalysis, and gas-sensing are reviewed. This work can contribute to the understanding of the control over shapes of Cu2O powders at micro- to nano-scale in the powder technology field.

Chapter 7 - Ferrites are important class of ferromagnetic materials. These materials can broadly be classified as soft and hard ferrites, depending upon their magnetic properties. It is very interesting to note that how slight variation of composition, particle size, processing technology etc affect the properties of these materials. Traditionally they have found wide range of applications in electronic, microwave, wave absorber devices, magnetic recording industries. But recently nanostructured ferrite materials have attracted attention of scientists due to their potential applications in nanotechnology, particularly drug delivery, smart materials etc. In this paper following topics will be discussed (i) a brief history of ferrite materials, (ii) different synthesis strategies, which are adopted by several scientist for preparation of different compositions of ferrites and advantages and disadvantages associated with these methods, (iii) properties and commercial applications of different ferrites, (iv) future scope of research.

Chapter 8 - Preparation of homogeneous dispersion of nanostructured catalytically active sites (metal species) within the porous, high surface area supports (such as SiO2 or Al2O3) has been a goal of catalyst scientists for many years. Mesoporous materials (e.g. SBA-15, MCM-41 etc) have exhibited their potential as catalysts, catalyst support, adsorbents for large organic molecules, guest-host chemical supports etc. Silver nanoparticles have myriad applications in diverse fields due to their catalytical, antibacterial, and biosensing properties. Several methodologies have been reported for preparation of nanosized Ag particles, such as colloidochemical, sonoelectrochemical, thermal decomposition, UV irradiation, γ-irradiation, protection-reduction, cation exchange, use of biosurfactants as reducing agent etc. However, most of these synthesis methods are complex in nature. Thus, development of a simple synthetic methodology for preparation of homogeneously dispersed Ag nanoparticles on mesoporous silica matrices is still a major challenge.

A novel chemical synthetic technique has been developed for the preparation of Ag nanoparticle loaded mesoporous silica, such as MCM-41. MCM-41 silicate material was synthesized by using a liquid crystal templating method. Here, in this synthesis process tetraethyl orthosilicate (TEOS) was used as an inorganic silica source and cetyltrimethyl ammonium bromide (CTAB) as pore structure directing template. Incorporation of Ag nanoparticle within hexagonal mesosructure was achieved through wet impregnation of AgNO3, followed by in situ chemical reduction of Ag+ ions by trisodium citrate. The synthesized materials were characterized by using room temperature powder XRD, BET surface area and porosity analysis using N2 sorption- desorption study, high resolution TEM. X-ray analysis and TEM micrographs confirmed the formation of Ag nanoparticles in the MCM-41 matrix. BET surface area of the synthesized materials was in the range of 1000 m2/ gm with narrow pore size distribution. A decrease in surface area and pore volume was observed due to impregnation for the Ag nanoparticle within porous silicate structure of MCM-41.

Chapter 9 – In stochastic modeling the focus is on one particle, namely on the time-evolution of the probability distribution of the particle's state (e.g. spatial position or physical or chemical state), given its present state. This in contrast to the standard modeling approach in process technology which is based on formulating and solving differential balance equations. Stochastic modeling is useful throughout process technology but is especially relevant in powder technology.

Stochastic modeling is found in the powder technology literature in various forms and guises. This chapter gives an up-to-date discussion of stochastic modeling in powder technology, showing through some chosen examples how the different forms of stochastic modeling are really variations of the same basic method, and bringing the literature into one unified notation for clarity.

After a brief introduction to the foundations of stochastic modeling and its strengths and limitations when applied within powder technology, some recent literature will be reviewed and specific examples of the main techniques will be analyzed in detail. The relation of stochastic modeling to other techniques, such as "discrete element modeling" and "population balance modeling" will be discussed.

Chapter 10 - Dry mechanical treatments of particulate materials such as mixing and grinding are frequently carried out in various industries such as foods, pharmaceuticals, cosmetics, electronics, etc., because of its simplicity and versatility. In the mechanical treatments, the physical properties and characteristics of powders treated depend strongly on both amount and history of the mechanical energy applied to the powders during treatment. For controlling their physical properties and characteristics, full understanding and precise control of the net applied mechanical energy are necessary. The author has proposed a simple method for measuring the applied energy and evaluating the energy characteristics of powder processing machines. In the proposed method, the stress-sensitive materials, the calcium carbonate powder containing the brittle agglomerates of fine pigment (iron oxide) particles and the copper powder with a spherical particle shape, are properly used as a standard material depending on the amount of the external force exerting on the powder to be treated. In the former, the applied energy can be estimated by analyzing the disintegration phenomenon of the agglomerates in the mixing process, and this method is available for powder mixers conducting relatively mild mixing such as a vessel-rotating type and an agitating type of powder mixers. On the other hand, the latter can estimate the applied energy from the plastic deformation of copper particles and is employed for grinding and compounding machines in which relatively larger mechanical energy applies to the powders. Based on the applied energy per unit time thus estimated, various types of powder mixers and mills could be characterized on a uniform scale regardless of their mechanical structure and mixing and grinding mechanisms. Subsequently, the author studied on the mechanical treatments for powder design of functional materials as examples of utilization of the applied energy, in which the surface of easily deformable particles with a spherical shape were mechanically modified without changing their original shape: e.g., the formation of polymer film on the surface of lead-free solder alloy particles for oxidation resistance. The author has proved through the application of the evaluation results of the applied energy to practical treatments that the operating conditions of treatment processes can be optimized based on the physical properties of raw materials to be treated and that advanced powder design can be achieved in simple dry mechanical processes.

Chapter 11 - Based on our experience of conventional crystal growth of AIIBVI, the authors have developed the techniques for nanocrystal production of some of these compounds.

Further innovations in the technological process appear to significantly reduce the development and production cost of the final product with respect to conventional single crystals, while maintaining a high level of physical properties, which conform to leading standards.

New technology of the vapor phase deposition was developed in our laboratory to produce CdTe nanoparticles of average diameter 8 nm. Further development of the technology allows us to produce 10-nm Cd-Zn-Te nanocrystals and to overcome difficulties in obtaining a ternary solid solution Cd1-xZnxTe (x = 0.04—0.1) with a stable chemical composition. Highly dense CdTe and Cd1-xZnxTe ceramics of high mechanical hardness and durability were produced by the room temperature process without any lubricants or binding materials.

It was found that CdTe ceramics produced from nanocrystals undergo wurtzite—sphalerite transition under pressure. The polymorphic transition from the hexagonal to the cubic phase in Cd1-xZnxTe ceramics caused by annealing is discussed.

The two-component texture composed of <220> axial and {100}<hk0> components was recorded in the as-compressed materials.

The effect of annealing on grain growth and texture in the ceramics is considered. It was found that our compacted ceramics guarantees high transmittance in a wide IR region of 6—25 µm, high specific resistivity on the order of 1010 Ωcm, and high microhardness on the order of 103 MPa.

The obtained properties make these materials promising for use in IR optics and for ionizing-radiation detectors.

In: Powder Engineering, Technology and Applications
Editor: Jason M. Barker, pp.1-41
ISBN: 978-1-61761-216-9

Chapter 1

HEAT, MASS AND EROSION RATES BETWEEN BUBBLING FLUIDIZED BEDS AND IMMERSED SURFACES

Francesco Di Natale* and Roberto Nigro
Department of Chemical Engineering, University of Naples
"Federico II", Naples, Italy

ABSTRACT

Several industrial applications, from the nuclear engineering to the food processing, consider the presence of surfaces immersed in bubbling fluidized beds. The surface-to-bed heat, mass and erosion rates significantly contribute to determine the actual efficiency as well as the operating life of fluidized bed unit, and this explains the paramount importance of experimental and modeling studies aimed to support reliable design and operation of these industrial processes.

The fluid dynamic field around an immersed surface presents different peculiarities respect to the average characteristics of the fluidized bed. These peculiarities are responsible for the actual values of the surface-to-bed mass and heat transfer. Similarly, they governed the momentum transfer, whose main aspect is the surface erosion caused by the impacts with the fluidized particles.

This chapter aims to analyze the transport phenomena among surfaces immersed in fluidized beds in light of the most recent experimental and modeling results reported in literature. The main goal of this contribution is to shed light on the characteristics of each transport phenomena and to point out the real analogies among them.

1. INTRODUCTION

Solid-gas fluidized beds are largely used for industrial processes also due to the very high heat and mass transfer rates either between the fluid and the solid phases or between the entire fluidized system and external surfaces, either peripheric walls or objects immersed in the bed.

* Corresponding author: Ph. +39-0817682246; Fax +39-0815936936, email francesco.dinatale@unina.it

In chemical reaction engineering, examples of applications based on the high surface-to-bed transfer rates are the use of cooling tube banks for temperature control or for heat recovery in catalytic reactors or combustors units, or the immersion of tubular membranes to control reactants distributions in fluidized bed reactors. Numbers of applications are also related to food process industry - where fluidized beds are used for food freezing and drying - and to the superficial treatments of mechanical parts - as for coating or dry painting, hardenings, chemical vapor depositions applications and so on. In all these cases, the immersed surface is usually an object of irregular shape that may be either free of moving in the fluidized system or kept at a fixed position by appropriate supports.

Surface erosion is related to the tangential stress exerted by the fluidized particles on the immersed object and it is the component of the overall momentum transfer of greatest operational relevance. Some superficial treatments of mechanical parts are based on the appreciable surface erosion rates occurring in fluidized bed. The examples of decoating or sandblast processes can be mentioned. Clearly, erosion also occurs for all the surfaces in fluidized beds, determining the true operating life of any of the immersed auxiliary devices, as the cooling tube banks or the membrane gas distributors.

The knowledge of surface-to-bed transfer rates is fundamental for process design and optimization and has been the subject of several works in the past sixty years. Nevertheless, at the moment, predictive models to estimate the transfer rates are currently available only for a limited range of operating conditions and rule-of-thumb approaches still remain the most diffused criteria for process design and scale-up. This is true, in particular, for bubbling fluidized beds, whose peculiar fluid dynamics represents an additional complexity.

One of the main issues in the analysis of surface-to-bed transport phenomena is the intrinsic difficulty in decoupling the effects of fluid dynamics field from those related to the very transfer mechanism. This concept can be easily understood with the following example. Let us consider measures of the heat transfer coefficient around a given exchange surface carried out at a same excess gas velocity, $U_{ex}=U-U_{mf}$, with sand particles of diameters ranging from tenth of microns up to several millimeters. The experimental results, known since the early seventies of the last century, showed that the heat transfer coefficient has a non-monotonic trend with a maximum around 0.1 mm and a minimum around 1-2 mm. This result can be inferred to several phenomena: 1. the number of particles in contact with the surface is higher if the particles are finer; 2. the surface renewal frequency is higher for finer particles; 3. the particle heat transfer rate is faster for smaller diameters. However, it is also known that: 4. the contribution to heat transfer due to the gas phase that passes through the emulsion is higher for coarser particles; 5. larger particles are better heat sinks and rapidly reduce the dispersion of heat from the exchange surface, thus enhancing the heat transfer rate. It is clear that a counterbalancing of all these phenomena determines the final result, but the distinction among their contribution starting from the experiments is not an easy task that requires studies on both the fluidized bed fluid dynamics and the mechanisms of heat transfer between the surface and the gas and particulate phases. Similar considerations can be done for the mass transfer or for the erosion rate.

Due to the complex interpretation of the experimental data and to their relevance for industrial applications, there is a broad literature dedicated to models for surface-to-bed transport phenomena, especially for the case of heat transfer. Therefore, a review of all the existing studies on this argument could be the subject of a dedicated book and cannot be summarized effectively in this chapter. In addition, good review articles and textbooks (e.g.

Lyckzowski and Bouillard, 2002, Yusuf et al., 2005, Scala, 2007, Rhodes, 2008) have been already presented in the last ten-fifteen years and can be proficiently used to find state-of- the art analyses on heat, mass and momentum transfer in fluidized beds.

The aim of this chapter, instead, is to provide a rationale of the transport phenomena in fluidized beds starting from their phenomenological evidences and from a short review of the main modeling approaches adopted to describe and predict the heat and mass transfer coefficients and the erosion rates in fluidized beds. For the sake of brevity, this chapter considers the case of surfaces kept at a fixed position in bubbling fluidized beds. The first assumption let us considering the largest amount of experimental evidences about heat transfer and erosion rates in fluidized beds and represents a more complex condition respect to the case of freely moving objects. The latter assumption allows us to cover the largest number of industrial applications related to superficial treatments and an appreciable percentage of the existing equipments for catalytic reactions and combustion applications.

In the case of an object fixed in a bubbling fluidized bed, the exchange surface is alternately immersed in the emulsion (o dense) and in the bubble phases for fraction of times that are typical indicated as t_d and t_b respectively. Under the assumption of ergodicity, the average values of these time fractions are commonly replaced by the volumetric fractions of the emulsion and the bubble phases, that are indicated as δ_d and δ_b – with the obvious result that δ_d = 1- δ_b. The common way to estimate the generic (heat, mass or erosion rate) time averaged transfer coefficient, φ, is:

$$\varphi = \varphi_d \cdot t_d + \varphi_b \cdot t_b + \varphi_r = \varphi_d \cdot (1-\delta_b) + \varphi_b \cdot \delta_b + \varphi_r . \qquad (1)$$

Under this approach, φ_d and φ_b represent the transfer coefficients when the surface is immersed in the emulsion and in the bubble phase respectively, while φ_r is the radiative contribution to the transfer phenomena, that exist, of course, only for the case of heat transfer.

The chapter is organized in four separate sections. In the first one, we analyzed the main features of the fluid dynamic field around fixed immersed objects in bubbling fluidized beds, due to its relevance for the transfer rates. In the following sections, we reported and discussed the experimental and modeling results on the emulsion, bubble and radiative transfer coefficients that allow the estimation of the surface-to-bed transfer coefficient in bubbling fluidized beds. Finally, in the last section we summarized the experimental and modeling results in order to stress the correlation among transport phenomena and fluid dynamic parameters, and to find out the analogies between the different transport mechanisms.

2. Fluid Dynamics Field around an Immersed Surface

The analysis of the fluid dynamic field in bubbling fluidized beds has been the subject of intense research dating back to the first studies on fluidization in the middle of the last century. The main objective of these studies was the description of the bubbles flow and the determination of the macroscopic fluid dynamic parameters, as the bed height, the bubble size and velocity, the overall void fraction of the expanded bed and the transition velocities between bubbling/homogeneous expansion and bubbling/slugging regimes. Of course, these studies also aimed to the comprehension of the mechanisms that govern the bubbles flow, and to provide quantitative estimations for the aforementioned macroscopic parameters. The first and the most acknowledged theory for the description of bubbling regime is the Two-Phase theory proposed by Harrison and Davidson in 1963. According to this theory, at a given superficial velocity, U, higher than the minimum fluidization velocity, U_{mf}, the bed appears to be made of two distinct phases. The first is the *emulsion*, whose properties are equal to that of the bed at minimum fluidization. This appears to be almost static and crossed upward by *bubbles* (the second phase) rising from the distributor. The gas flow rate passing through the bed as bubbles, Q_b, is equal to the difference between the actual flow rate and the flow rate at minimum fluidization $Q_b=(U-U_{mf})\cdot S$. Starting from these assumptions, and after the support of hundreds of experimental data on both cold model, pilot and industrial scale fluidized beds, several correlations have been developed to estimate the bubble size, the bubble rising velocity, the bed height, or the bubble fraction and frequency. References to the Two-Phase theory based models are reported in Kunii and Levenspiel (1991) and Rhodes (2008). The main criticism to the Two-Phase theory derive from two evidences. The first regards the value of superficial velocity for incipient bubbling. Indeed, bubbles appear at minimum fluidization only for Geldart B type particles, while for other materials they appear at a velocity, named "minimum bubbling velocity" and denoted as U_{mb}, that marks the transition among the homogeneous expansion and the bubbling regimes. The second evidence, partially related to the first one, is the occurrence of a certain expansion of the emulsion also in the bubbling regime. Nevertheless, after proper modifications, the overall significance of the Two-Phase theory is usually preserved and its practical use for the prediction of fluid dynamic parameters is largely diffused.

With reference to the effects of fluid dynamic field on surface-to-bed transfer rates, the most relevant parameters are represented by the bubble void fraction, δ_b - that substitute the relative values of t_d and t_b in Eq. (1) - and the value of the void fraction ε in the emulsion phase, that determine the physical properties of the emulsion, as well as the number of particles that contact the exchange surface.

Indeed, the void fraction and the superficial velocity, together with the physical properties of the gas and the solid phases, determine the values of the interstitial gas velocity and of the rate and the amplitude of the oscillation that characterize the motion of fluidized particles. These variables have direct effects on the interactions of the immersed surface with the particles and gas, thus determining the actual values of the transfer rates. In particular, the main parameters used to characterize the interactions between the particles and the surface are the duration of particle contacts with the surface, the frequency of replacement of particles close to the exchange surface – that is usually denoted as surface renewal frequency – and the aforementioned void fraction close to the surface. Differently, to describe the interactions between gas and surface, the values of the interstitial velocity and of the void fractions are required.

Several experimental evidences have pointed out that the region close to a surface immersed in the emulsion phase of a bubbling fluidized bed is characterized by a peculiar fluid dynamics. The clearest evidences of this phenomenon are the X-ray photos reported by Buyevich et al. in 1986 (Figure 1). These images highlight that the emulsion phase close to the surface consists in a boundary layer region of higher gas velocity and bed porosity whose characteristics are strongly dependent on the surface shape.

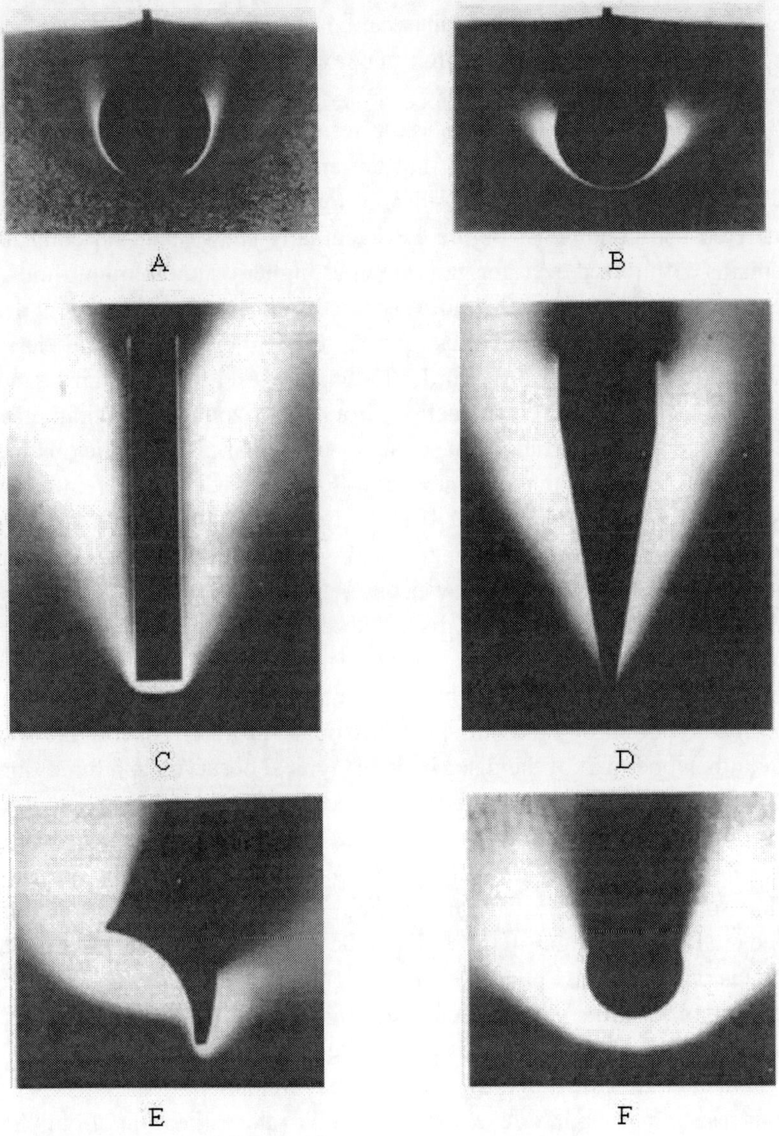

Figure 1. X ray images of bed structure close to an immersed surface (after Buyevich et al., 1986). A: Sphere at $U_{ex}/U_{mf}= 0$; B: Sphere at $U_{ex}/U_{mf}= 1$; C: Plane slab at $U_{ex}/U_{mf}= 1$; D: triangular slab at $U_{ex}/U_{mf}= 1$; E: Arc shaped surface at $U_{ex}/U_{mf}= 1$; F: Horizontal cylinder at $U_{ex}/U_{mf}= 1$

In these figures, the darker areas represent lower void fraction values, close to the minimum fluidization value, ε_{mf}, while the brighter ones correspond to higher void fractions. In all the cases, it is evident that a sort of gas cushion appears on the lower and lateral faces of the surfaces, while the upper areas are usually characterized by a void fraction close to minimum fluidization. For this reason, this last is usually denoted as *defluidized cap*. Moreover, it also appears that for slender surfaces the extension of the gas cushion is smaller than for blunter ones.

The X-ray images of Buyevich and coworkers are related to the averaged conditions around surfaces immersed in the emulsion phase only, but other evidences (e.g. Solimene et al., 2003) clearly point out that the gas cushion may eventually collapse on the upper part of the surface to give rise to indigenous bubbles.

Finally, Figure 1A shows a very important result: the gas cushion appears also at minimum fluidization velocity. This means that the emulsion phase close to the lower and lateral faces of the surface starts to be fluidized even for $U<U_{mf}$, while that close to the upper face remains at fixed bed conditions. Figure 1B eventually shows that the emulsion in this upper region remains defluidized even for gas velocities higher than minimum fluidization.

The first and - at our knowledge - the more complete collection of experimental values on surface void fraction angular profile, has been reported in 1980 by Fakhimi and Harrison for a two-dimensional fluidized bed (H=1200 mm; L=380 mm; W =11.3 mm). Capacitance probes have been used to determine the void fraction around horizontal disks immersed in the bubbling bed either singularly or arranged in square or triangular pitches. The obtained results are reported in Figure 2 for single disks immersed in fluidized beds of different particles. The angular surface void fraction profiles have been reconstructed for different gas velocities, showing an overall consistency with the images of Buyevich et al. (1986). The experimental results show that the void fraction in the lower and lateral faces of the immersed surface is higher than that in the upper part of the disk. The surface void fraction at each angular position is increased by superficial gas velocity and is higher for the larger disk.

Nevertheless, due to both the fluidized bed design and of the measurement technique, the void fraction profiles reported by Fakhimi and Harrison (1980) do not show the well known defluidized cap on the upper part of the disk (which is typical observed in a three-dimensional fluidized beds) and they present values higher than those reported by Buyevich et al. (1986) for the surface void fraction in the emulsion phase. This discrepancies however, in only apparent. In fact, it has to be recognized that the void fraction profile reconstructed from Fakhimi and Harrison analysis represents the overall, time averaged, void fraction (i.e. the ratio between the total volume of gas in the bed and the entire bed volume), that includes both the effects of emulsion and bubble phases, while the Buyevich et al. (1986) data refer to the emulsion phase only. In addition, it has to be noticed that one of the main differences between two-dimensional and three-dimensional beds rely to the characteristics of indigenous bubbles, which occur at a much larger extent in a two-dimensional system.

Recently, our research group has developed a model for the estimation of both the surface averaged and the angular profiles of surface void fraction, ε_w, close surface immersed in fluidized beds (Di Natale at al. 2008 and 2010). In particular, the surface averaged value of ε_w for vertical and horizontal cylinders in fully bubbling fluidized beds, i.e. for $U/U_{mf}>1.3$, are respectively:

$$\varepsilon_{w,avg} - \varepsilon_{mf} = 0.045 \cdot Log(Ar) \tag{2}$$

$$\varepsilon_{w,avg} - \varepsilon_{mf} = 0.050 \cdot Log(Ar) \tag{3}$$

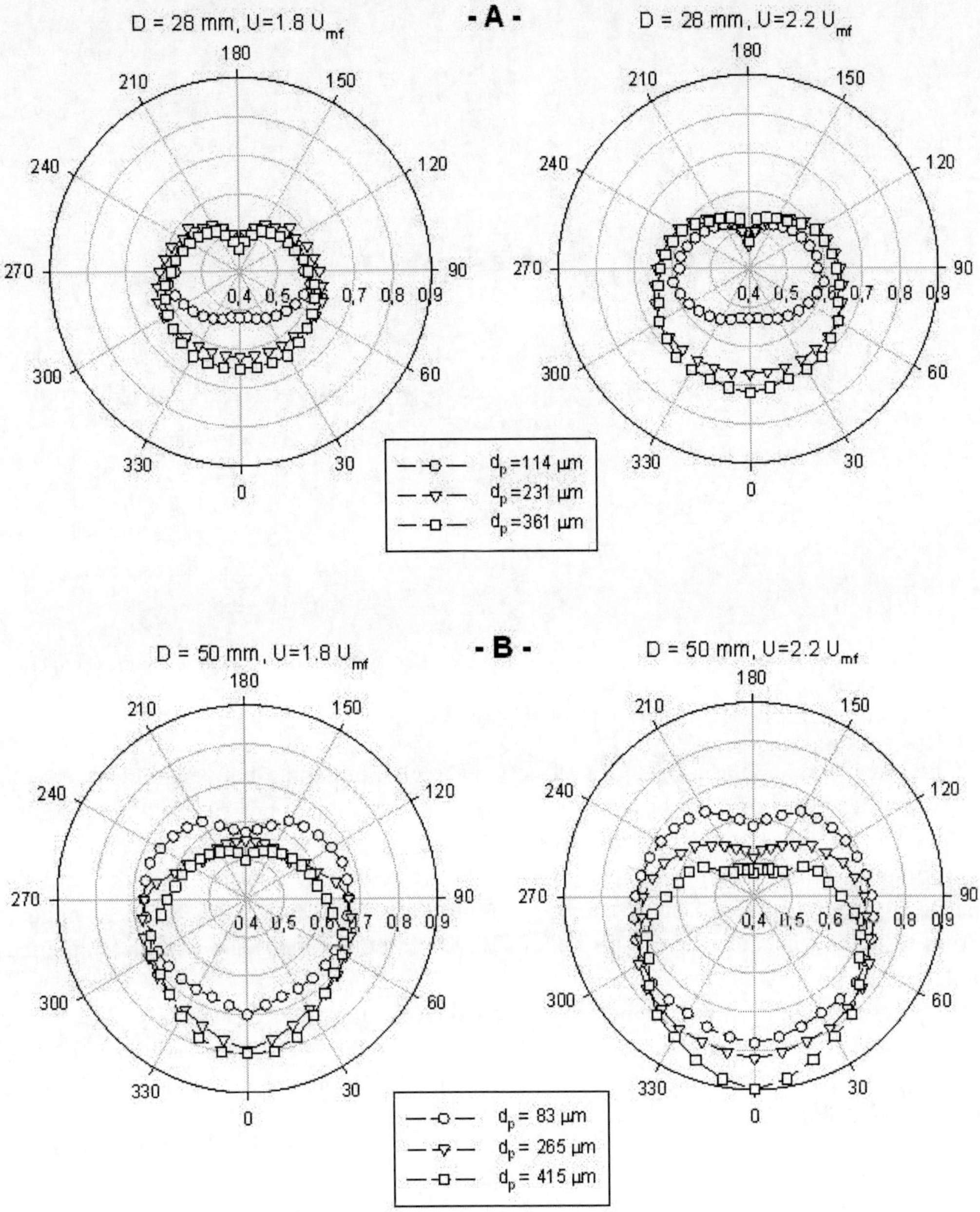

Figure 2. Time averaged values of angular surface void fraction profile close to a disk immersed in a 2D fluidized bed, parametric with the superficial velocity and the particle size. A. Disk diameter D = 28 mm; B: Disk diameter D = 50 mm. Data adapted from Fakhimi and Harrison (1980)

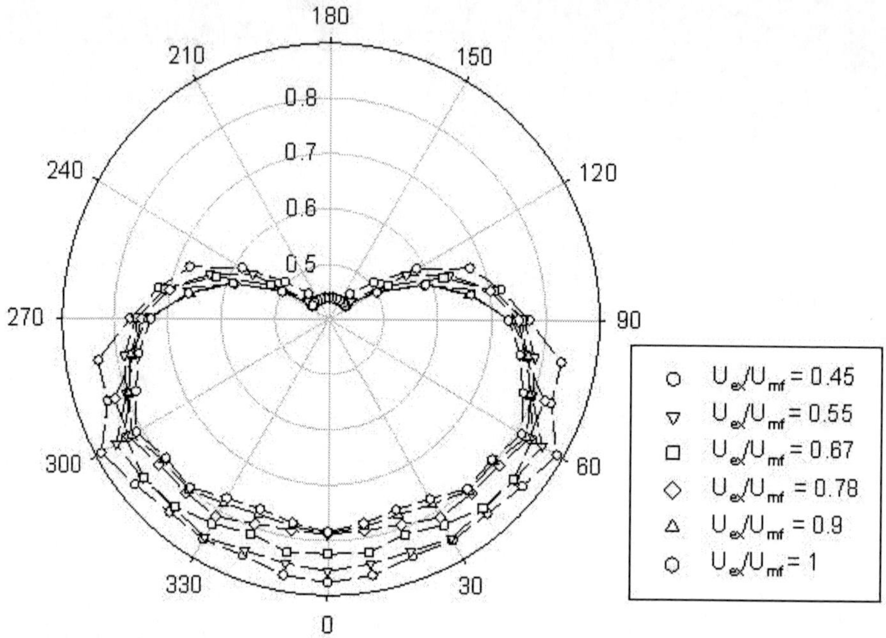

Figure 3. Surface void fraction profile in the emulsion phase as a function of the dimensionless excess gas velocity, U_{ex}/U_{mf}, for a horizontal cylinder (D=30 mm) immersed in a bubbling bed of glass beads (d_p = 500 μm) at atmospheric pressure and temperature

The calculated angular profile of surface void fraction around a horizontal cylinder with D = 30 mm for a bed of glass beads with d_p = 500 μm at ambient pressure and temperature, and for different dimensionless excess gas velocity, U_{ex}/U_{mf}, are reported in Figure 3.

The surface void fraction in the emulsion phase results to decrease slightly with the superficial velocity to approach an almost constant value for $U_{ex}/U_{mf} \geq 0.78$. These angular profiles result consistent both with the data of Buyevich et al. (1986) and with the numerical simulations of Rong et al. (1999).

In order to maintain these higher void fractions, the interstitial gas velocity close to the surface must be higher than that of the emulsion phase in the bulk of the bed. According to the well-known Richardson and Zaki law (e.g. in Rhodes, 2008), the interstitial velocity U/ε_w can be estimated as:

$$\frac{U}{\varepsilon_w} = U_t \cdot \varepsilon_w^{\,n-1} = U_{mf} \cdot \left(\frac{\varepsilon_w}{\varepsilon_{mf}}\right)^{n-1}. \tag{4}$$

In this equation, the parameter n is the Richardson and Zaki index that can be found, for example, in Rhodes et al (2008) and Di Felice and Kehlenbeck, (2000).

This higher gas velocity also leads to higher values of particle velocity near the exchange surface, that affects the values of transfer rates by determining the contact time between the fluidized particles and the surface and the amount of energy developed during particle-surface impacts.

The most common way to describe the contacts between particles in the emulsion and exchange surface is the use of a residence time distribution, E(t), that is usually described by a Gamma function with shape factor α equal to 1 or 0 (Koppel et al, 1970; Molerus et al., 1995; Wang et al., 2003):

$$E(t) = \frac{f_{avg}(U)^{\alpha+1}}{\alpha!} \cdot t^{\alpha} \cdot \exp[-f_{avg}(U) \cdot t] \tag{5}$$

where $f_{avg}(U)$ is the surface average renewal frequency, that represents the inverse of the time required to completely replace the emulsion phase in contact with the surface, t_{avg}. This value is quite different from the average duration of particle-surface contact, t_p, since, during the time t_{avg} the same particle can touch the surface several times, each of which for t_p seconds.

Indeed, the motion of particles close to the surface is related to three different phenomena: the Brownian-like microscopic motion typical of a granular flow, the mixing induced by bubbles motion, and the particle displacement caused by the peculiar fluid dynamic field in the boundary layer near the surface.

The Brownian oscillating motion of particles in the emulsion is commonly related to the concept of granular temperature and it is the leading mechanism of particle-particle interactions or for the interactions between particles and exchange surfaces of the size of millimeters, as for the tests of Pence et al. (1994) and Boerefijn et al. (1999). Typical values of the frequencies of oscillation are from 10 to 50 Hz.

The analysis of particle mixing induced by the bubble motion derives from the measures of surface averaged renewal frequency available in literature and resumed by Di Natale et al. (2008). These measures show that the surface averaged renewal frequency is a function of the physical properties of solid and gas phases, the superficial velocity and the particle diameter. For example, Figure 4 describes the results obtained by Mickley and Fairbanks (1955) on the effects of gas velocity and particle diameter on the surface renewal frequency (Figure 4A), and those of Molerus et al. (1995) on the value of the renewal frequency in a bubbling fluidized bed of 250 μm glass beads at different superficial velocities and gas pressures (Figure 4B).

Starting from these studies, Di Natale et al. (2008) provided the following expression for f_{avg}:

$$f_{avg}(U) = 2.2\frac{D^*}{d_p} \cdot \mathrm{Re}_{ex}^{\;0.287} . \tag{6}$$

with

$$\mathrm{Re}_{ex} = \frac{\rho_g \cdot d_p \cdot U_{ex}}{\mu_g} \tag{7}$$

and

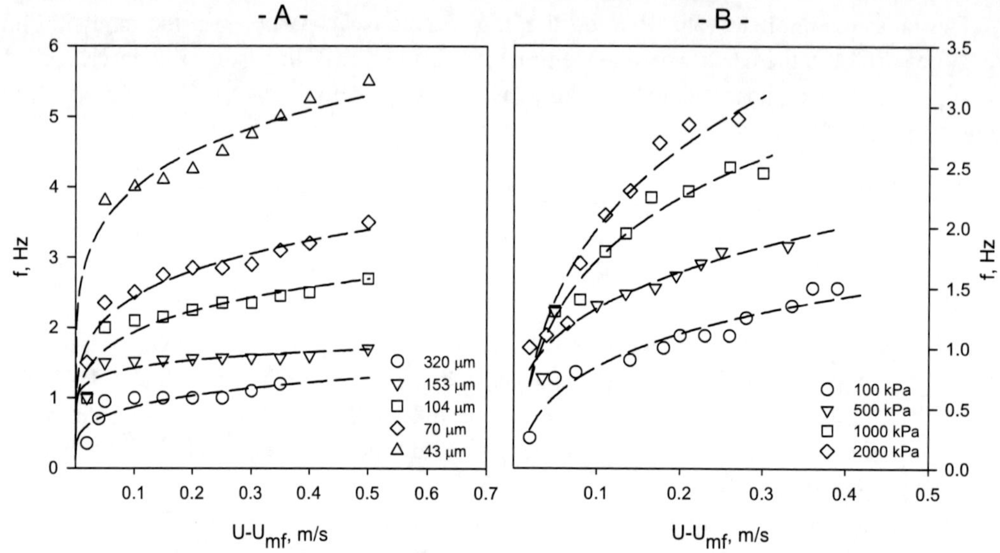

Figure 4. Surface renewal frequency in function of the excess gas velocity, pressure and particle diameter. A: Data from Mickley and Fairbanks (1955) P = 100 kPa; B: Data from Molerus et al (1995) for glass beads with d_p=250 μm

$$D^* = 56.2 \cdot \left(\frac{\mu_g^{\ 2}}{g \cdot \rho_p^{\ 2}} \right)^{1/3} \quad \text{(Cody et al, 2008)} \tag{8}$$

Usually, the values of $f_{avg}(U)$ range between 0.5 and5 Hz.

With reference to the estimation of the surface renewal frequency related to the surface boundary layer, that will be denoted as $f^{bl}(U,\theta)$, the greatest number of existing studies is related to the case of horizontal cylinder, as the experimental and numerical results of Rong et al. (1999), Kim et al. (2003), Baskakov et al. (1964), Ozkaynak and Chen (1980), Sunderesan and Clark (1995) and Wang et al. (2005).

At our knowledge, the estimation of the surface averaged value of $f^{bl}(\theta,U)$ was reported for the first time by Baskakov and coworkers (Baskakov et al., 1973). In their paper, the formula proposed to estimate the average surface renewal frequency $f_{avg,\theta}(U)$ is:

$$f^{bl}{}_{avg,\theta}(U) = \frac{1}{B} \cdot \left[\frac{U_{mf}^2 \cdot (U/U_{mf} - A)^2}{g \cdot d_p} \right]^c \cdot \left(\frac{D}{d_p} \right)^{0.225} \tag{9}$$

where D is the cylinder diameter and the parameters are A = 0.8, B = 0.485 and C = 0.143 for horizontal cylinders and A = 0.8, B = 0.44 and C = 0.140 for the vertical ones. This equation can be rearranged as follows:

$$f^{bl}{}_{avg,\theta}(U) = \frac{1}{B \cdot 56.2^{3 \cdot C}} \cdot \mathrm{Re}^{2C} \cdot \left(\frac{\rho_p}{\rho_g}\right)^{2C} \cdot \frac{D^{0.225} \cdot D*^{3C}}{d_p^{3C+0.225}} \tag{10}$$

Where:

$$\mathrm{Re}_{ex} = \frac{\rho_g \cdot d_p \cdot (U - 0.8U_{mf})}{\mu_g} \tag{11}$$

From Eq. (10), several information can be retrieved:

1. The surface renewal frequency is higher than zero at the minimum fluidization velocity, mirroring the existence of the high porosity boundary layer around the surface even at incipient fluidization, as reported in Figure 1A.
2. The surface renewal frequency is proportional to the characteristic fluid dynamic length D* defined by Cody et al. (2008).
3. the gas velocity dependence on the surface boundary layer renewal frequency is similar to that observed for the surface average renewal frequency $f_{avg}(U)$.

Of course, the overall surface renewal frequency at each given position of the exchange surface is the maximum between $f_{avg}(U)$ and $f^{bl}(U,\theta)$. This result can be easily understood by considering the example of a horizontal cylinder immersed in a bubbling or slugging fluidized bed (Figure 1F).

The angular profile of the local renewal frequency can be derived from several experimental studies. In particular, Kim et al. (2003) reported a detailed profile of the local renewal frequency around a cylinder (D = 25.4 mm) immersed in a bed of 240 μm glass beads fluidized at different gas velocities. Sunderesan and Clark (1995) instead, reported the local renewal frequency profile around a cylinder (D=50 mm) immersed in fluidized beds of different bed materials and with different gas velocities.

These results have been used by Di Natale et al. (2010) to upgrade the modified Baskakov expression of the average surface renewal frequency (Eq. 10) by adding the desired angular dependence. To this aim, the value of AU_{mf} has been substituted by the local incipient particle convection velocity, $U_{pc}(\theta)$, that represents the empty column gas velocity at which, for each given angular position, the emulsion starts to fluidize. The best fitting of experimental data has been obtained with a Gaussian function of the angle, and the final expression of the local renewal frequency is:

$$f^{bl}(U,\theta) = a \cdot \frac{D^{0.225} \cdot D*^{0.429}}{d_p^{0.654}} \cdot \mathrm{Re}^{0.286} \cdot \left(\frac{\rho_p}{\rho_g}\right)^{0.286} \cdot \exp\left[-0.5\left(\frac{\theta - \theta_0}{b}\right)^2\right] \tag{11}$$

with a = 0.184 ± 0.008, b = 55.37 ± 3.77, θ_0 = 43.07 ± 2.83. In this equation it is assumed that the angle θ=0 refers to the lowest vertical position of the cylinder. The function f(U,θ) has a maximum at θ_0, wit h a value of the order of 1.5-5 Hz, and gradually approaches 0 for θ around 150 degs, mimicking the occurrence of the defluidized cap of solid on the upper section of the cylinder (Figure 1F). The values of the local surface renewal frequencies around a horizontal cylinder (D = 30 mm) in a 500 μm glass beads bubbling fluidized bed are reported in Figure 5.

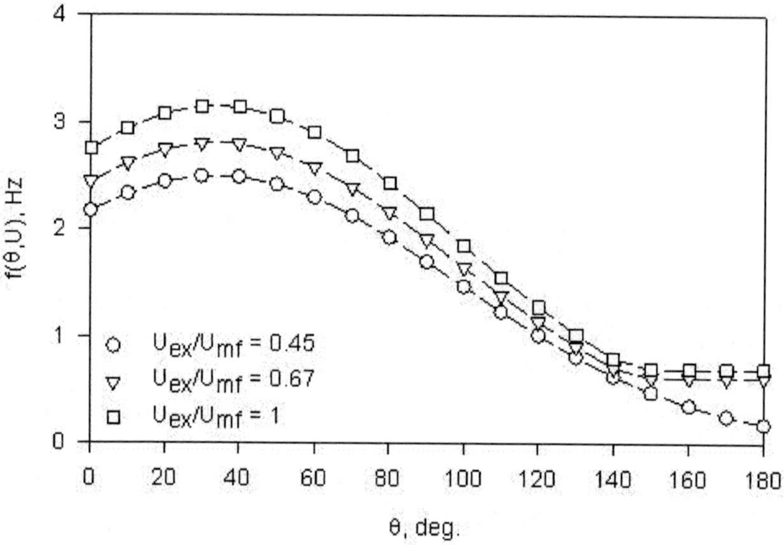

Figure 5. Surface renewal frequency, $f(U,\theta)=\max[f^{bl}(U,\theta), f_{avg}(U)]$, in function of the angular position and parametric with excess gas velocity for a horizontal cylinder immersed in a fluidized bed of glass beads with d_p = 500 μm at atmospheric pressure and temperature

Generally, for each velocity, $f^{bl}(U,\theta)$ is higher than $f_{avg}(U)$ for the angular positions from 0 to about 110 degs. On the contrary, for the upper positions of the cylinder, the particle replacement is dominated by $f_{avg}(U)$. Therefore, at low gas velocity - when the bubbles are appreciably smaller than the fluidization column diameter and their fraction in the bed is limited - the upper face of the cylinder remains undisturbed for long times. This result is consistent with the measures of heat, mass and erosion rate profiles observed in literature and reported, in the following, in Figures 7, 8 and 18.

3. Experimental Evidences on Transport Phenomena

3.1. Experimental Techniques

The study of surface-to-bed transport phenomena is strongly influenced by the techniques that can be used to measure the transfer rates – or, more commonly, the transfer coefficient φ – as a function of the different parameters that control the dynamics of the transport phenomena.

Experiments on erosion rates have been carried out since long time due to the necessity of predicting the erosion of tube banks immersed in fluidized bed combustor or heat exchangers, and minimize the risk of mechanical failures through a proper maintenance scheduling.

Usually, erosion rate tests are related to the case of tubes immersed in bubbling fluidized beds and are usually performed by measuring the loss of material after a given permanence of the tube in the bed, and by measuring an angular profile of the surface erosion through the analysis of the diameter size reduction at given angular position (e.g. Fu et al., 1991; Almsted and Wiman, 1997 and Almsted et al. 2000). Due to the long time - generally around hours - required to allow an effective measure of mass consumption and size reduction, only time-average values of erosion rates can be experimentally estimated, thus including both the effects of the emulsion and the bubble phases.

The measure of surface-to-bed mass transfer is more complex. Three different kinds of experimental techniques are the most diffused. The first considers the use of mass transfer probes made by highly sublimating solids, usually naphthalene. When immersed in the bed, the naphthalene passes in the gas phase and can be measured by an appropriate detector after a given distance from the probe, to minimize the perturbation of the fluid dynamic field near the surface. Of course, the measure can be made on-line, but it needs the naphthalene concentrations to be high enough to have a reasonable results. This can be a serious issue, since the intense solid mixing in fluidized beds spreads the gas over the entire bed section in very short times. In this case, surface averaged values of mass transfer rates can be determined. If the surface is fixed at a given position (as, for example, in the work of Garim et al., 1999), the analysis of the surface size, in terms of diameter reduction at the end of the test, allows the estimation of local values of the mass transfer rates. However, the tests have to be well planned to avoid the parallel erosion of the exchange surface. In the experiments of Garim et al. (1999), this was assured by working at small excess gas velocity and by reducing the experimental time. In order to avoid the occurrence of erosion phenomena, experiments with porous surfaces permeated by a volatile liquid have been carried out (e.g. Ziegler and Holmes, 1966). This method is far more complex even if more accurate, but it was used since now only to estimate surface averaged values of the mass transfer coefficient. Finally, another method to determine the mass transfer rate is the measure of size reduction of an evaporating exchange surface. In this case, the experiments are usually focused on objects (typically spheres) freely moving in the fluidized bed. From these studies, only surface and time averaged values of the mass transfer rates become available.

Experimental techniques to measure surface-to-bed heat transfer coefficient in bubbling fluidized beds are much more versatile and consolidated. Since the early sixties, the experimental techniques allowed an accurate estimation of surface averaged values of the heat transfer coefficients, and later several efforts to measure local heat transfer coefficient were carried out either to explain the heat transfer phenomena (e.g. George et al., 1987 and 1993; Sunderesan and Clark, 1995; Kim et al., 2003; Di Natale et al., 2010). These techniques are mainly based on a measure of surface-bed temperature difference that is assured by maintaining the surface at higher/lower temperature through a heating/cooling external source. This is usually a heater cartridge (for the case of cold model fluidized beds) or a cold-water circuit (for the case of fluidized beds at high temperature).

The main experimental limitations of these methods are usually related to the necessity of supporting electrical wires connections or water cooling piping for the heat transfer probe. Both for this limit and for the typical operational conditions of heat exchange surfaces in industrial application, experiments are usually performed for surfaces fixed at a given position within the bed. In the greatest number of studies, cylindrical exchange surfaces were considered.

However, these techniques have a unique feature: it allows an accurate estimation of the "instantaneous" heat transfer coefficient (since the typical time resolution of commercial thermocouples is lower than 0.1s). This time scale analysis allowed the realization of both local and instantaneous determinations of the heat transfer coefficient, that make possible to discriminate between the emulsion, bubbles and radiative contributions at a micro-scale level. Actually, these features explains the richness of data that can be retrieved from heat transfer experiments which eventually become the main sources of information for the investigation of transport mechanism in fluidized beds.

3.2. Experimental Evidences

Erosion rate

Experimental tests on the erosion rates are usually referred to immersed tubes and they were mainly carried out by varying particle size, d_p, excess gas velocity, $U-U_{mf}$, pressure, P, and temperature, T. In many cases, the data are collected in terms of angular erosion profiles as a function of the angle θ.

Investigations on the erosion rate are usually carried out for cylinders immersed horizontally in bubbling fluidized beds, and this eventually results the most critical condition, as shown by the tests of Fu et al. in 1991 on cylinders at different orientation (Figure 6A). The same authors also analyzed the erosion rate at different gas velocities in fixed and fluidized bed conditions (Figure 6B). The erosion rate becomes appreciable only when the bed is fluidized (at Umf =0.10 m/s *circa* in the reported tests) and it is enhanced by higher superficial velocities.

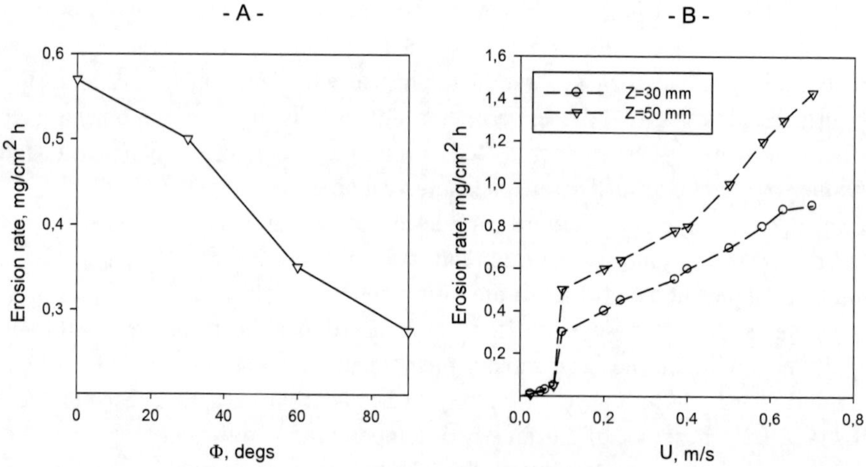

Figure 6. Surface averaged erosion rate of a tube as a function of (A) its inclination respect to the horizontal axis, Φ, at U= 0.67 m/s and (B) of the superficial velocity at Φ = 0, parametric with the distance from the distributor Z. Working conditions: Bed material: glass beads with d_p=550 μm; Ambient pressure and temperature; Wax cylinder OD =17 mm, L = 180 mm. (Data adapted from Fu et al., 1991)

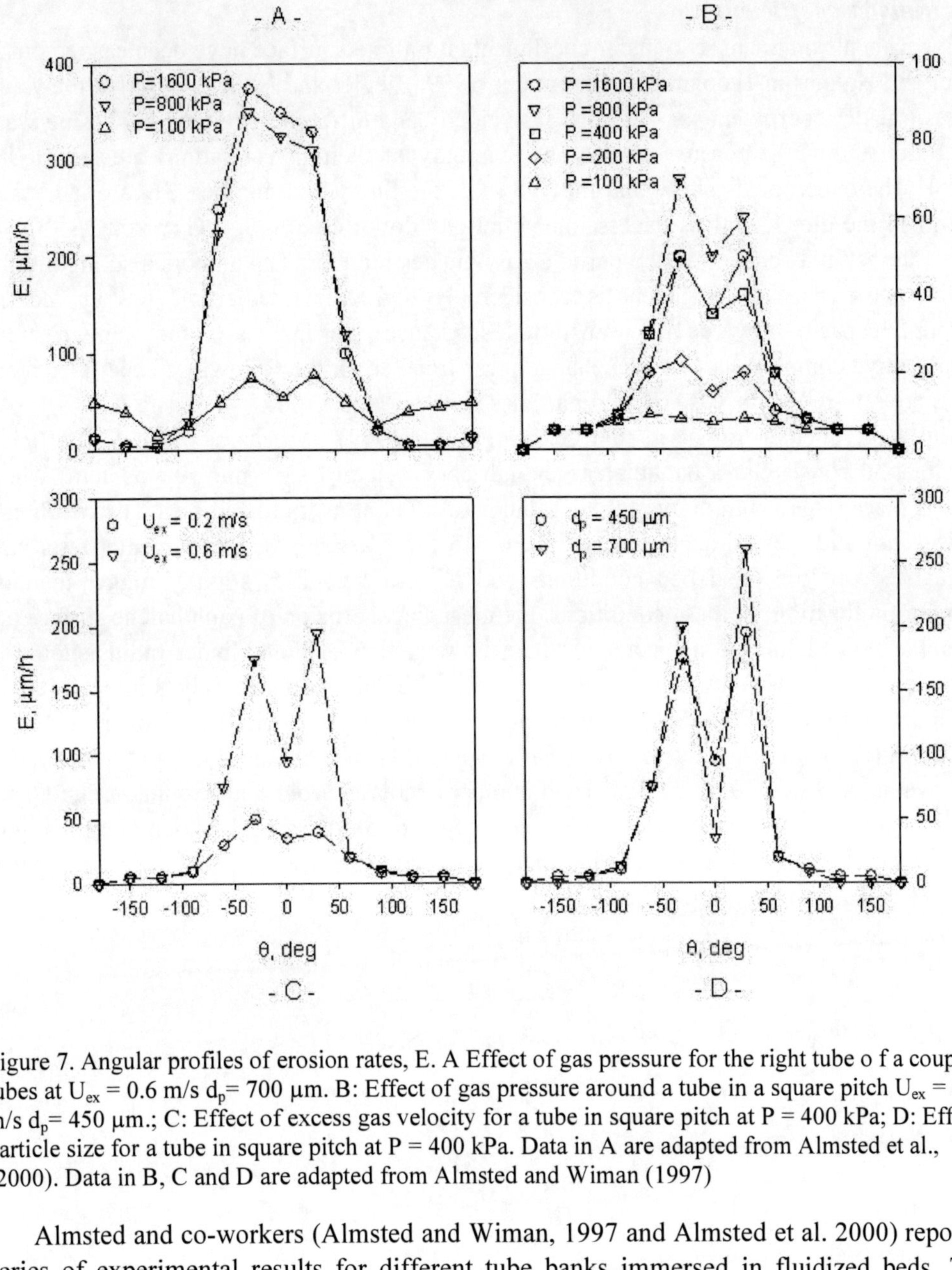

Figure 7. Angular profiles of erosion rates, E. A Effect of gas pressure for the right tube o f a couple of tubes at U_{ex} = 0.6 m/s d_p= 700 μm. B: Effect of gas pressure around a tube in a square pitch U_{ex} = 0.2 m/s d_p= 450 μm.; C: Effect of excess gas velocity for a tube in square pitch at P = 400 kPa; D: Effect of particle size for a tube in square pitch at P = 400 kPa. Data in A are adapted from Almsted et al., (2000). Data in B, C and D are adapted from Almsted and Wiman (1997)

Almsted and co-workers (Almsted and Wiman, 1997 and Almsted et al. 2000) reported a series of experimental results for different tube banks immersed in fluidized beds. These results (summarized in Figure 7) show that the gas pressure increases the erosion rate (Figure 7 A and B), but a non-monotonic trend can be observed when the cylinder is part of a tube bank. The erosion rate E increases with the excess gas velocity in agreement with Fu et al. (1991) and it is a little bit higher passing from d_p = 450 to 700 μm particles. Almsted and Wiman (1997) also provide a theoretical prediction of the effect of temperature on the erosion rate. The authors argued that the temperature effects are limited to the variation of the fluid dynamic properties of the fluidized bed and, through the use of Glicksmann (1984) scaling laws, concluded that higher temperatures lead to an increase of the pressure levels at which maximum erosion occurs for tube banks.

Mass transfer coefficient

Experiments on the mass transfer coefficients for a fixed surface have been carried out by Ziegler and Holmes in 1966 and Garim et al. in 1999. Ziegler and Holmes reported the values of mass transfer coefficient for different gas velocity and different bed materials, using water evaporation through a porous sphere. The summary of their investigations are reported in Table 1. The experiments show that the values of the Sherwood number, Sh, are increasing function of the dimensionless excess interstitial gas flow rate, $(G\text{-}G_{mf})/G_{mf}$, with $G = \rho U/\varepsilon$, and are almost independent on the particle Archimedes number. The authors also highlighted that the mass transfer coefficient is considerably higher if the particle has appreciable adsorption capacity towards the transferred substances. In this case, the same particles become active components of the exchange mass transfer. In the following, experiments and models are referred to the case of inert particles.

The experimental rig used by Garim et al. (1999) for mass transfer coefficient determination consists in a naphthalene cylindrical surface (d_s= 35 mm, L= 53 mm) whose center is placed at 75 mm above the distributor of a bubbling fluidized bed. The bed height was 150 mm and the fluidized particles were 545 μm glass beads. Experimental tests have been carried out for fixed bed conditions and for gas velocities slightly higher than the minimum fluidization. In these conditions, the mechanical erosion of naphthalene surface and the bubble flow should be neglected, so that the reduction of the cylinder radius should be directly connected to naphthalene sublimation in the emulsion phase only. Figure 8 reports the angular profiles of the Sherwood number referred to the cylinder diameter d_s, - i.e. $Sh(U,\theta) = k(U,\theta)\cdot d_s/D°$ - evaluated by Garim et al. (1999) in function of the gas velocity. The experimental data show that the Sherwood number decreases from a maximum, around $\theta = 0$ degs , to a minimum at $\theta = 180$ degs. The dependence of Sh on the gas velocity appears quite limited even if it is worth remembering that the experiments were carried out at superficial velocities close to the minimum fluidization.

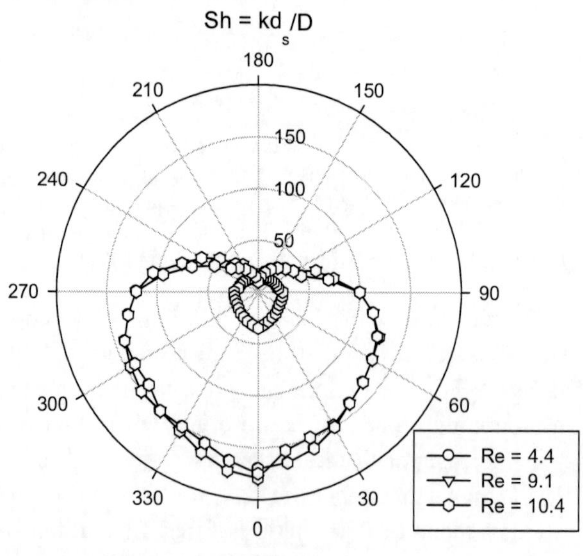

Figure 8. Local Sherwood number for a fixed cylinder (d_s=35 mm) as a function of particle Reynolds number. The cylinder is immersed in a fixed and a fluidized bed of 545 μm glass beads ($Re_{mf} \sim 9$)

Table 1. Operating conditions and results for experiments on mass transfer coefficient of a fixed spherical exchange surface. Data adapted from Ziegler and Holmes (1966)

Material	$d_{p,}$ μm	ρ_p, kg/m^3	Particle sphericiy	Ar	G_{mf}, kg/m^{2}hr	ε, -	$\frac{G-G_{,mf}}{G_{mf}}$	Sh, -
Fused Alumina A	1467	3070	0.59	330000	271	0.689	0.62	9.68
						0.733	0.93	12.55
						0.776	1.15	12.5
						0.784	1.44	14.17
						0.796	1.82	14.04
Fused Alumina B	880	3070	0.58	70700	110	0.722	3.91	13.57
						0.751	4.37	15.52
						0.763	5.18	15.92
Copper (Granules)	582	7062	0.89	47200	227	0.532	1.41	12.28
						0.632	2.17	13.16
						0.643	3.08	15.71
						0.644	3.18	15.11
Glass	1152	1946	0.88	101000	183	0.57	1.89	13.7
						0.618	2.42	11.35
						0.669	2.03	9.48
						0.678	3.19	10.63
Copper (Spheres)	1364	7062	1	608000	188	0.378	0.74	8.8
						0.495	0.95	9.06
						0.536	1.34	10.35
Nickel (Spheres)	1364	7011	1	603000	287	0.378	0.74	8.57
						0.464	1.11	9.39
						0.556	1.30	11.9
						0.603	1.61	10.72
Solder (Spheres)	1364	7000	1	602000	300	0.328	0.59	8.54
						0.495	1.15	9.34
						0.556	1.30	11.01
						0.603	1.96	10.8
						0.61	2.30	12.1

Other experimental evidences derive from studies on active particles freely moving in fluidized beds. For example, Scala (2007) estimated the effect of superficial velocity, particle and sphere size on the mass transfer coefficient of active spheres with d_s ranging between 1 and 10 mm in a bubbling fluidized bed at 450°C. The effect of superficial velocity, expressed in terms of particle Reynolds number, is reported in Figure 9. Experimental results show that the Sh number is almost constant with the superficial velocity, whatever the active particle size is. This evidence apparently disagree with the results of Ziegler and Holmes (1966), but it has to be observed that the first data referred to a fixed sphere immersed in the bed, thus accounting for the alternation of bubbles and emulsion, while, in the experiments of Scala (2007) consider that the surface is only immersed in the emulsion phase.

Starting from the independency of Sh on Re, Scala also highlighted the correlation between Sherwood number and particle Reynolds number at minimum fluidization (Fig. 10), as well as that between Sh and the size of the active particle, d_s (Fig. 11).

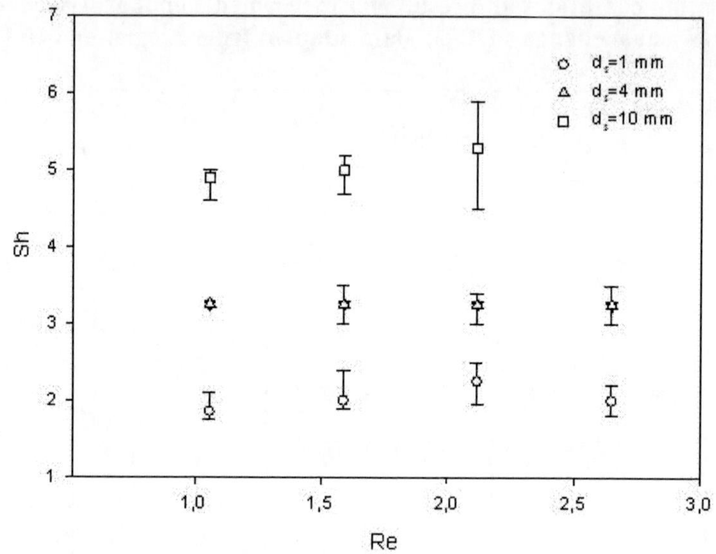

Figure 9. Sherwood number referred to the active particle size ($Sh=k \cdot d_s/D$) as a function of particle Reynolds number, ($Re= U \cdot d_p/\nu$) and for different active particle size. Freely moving active particle in fluidized bed of 500-600 μm sand; T= 450 °C; P=1 atm (adapted from Scala, 2007)

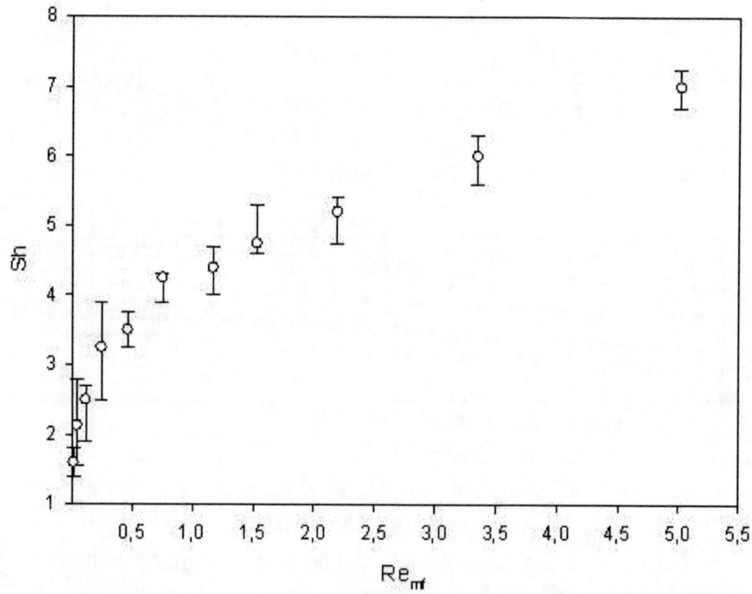

Figure 10. Sherwood number referred to the active particle size ($Sh=k \cdot d_s/D$) as a function of particle Reynolds number at the minimum fluidization, ($Re_{mf}= U_{mf} \cdot d_p/\nu$). Freely moving active particle with d_s = 4.6 mm in fluidized bed of sand; T= 450 °C; P=1 atm (adapted from Scala, 2007)

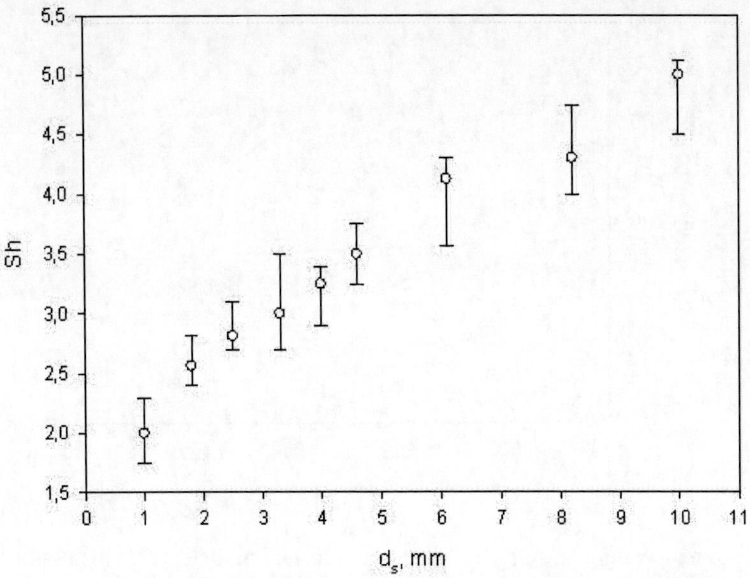

Figure 11. Sherwood number referred to the active particle size ($Sh=k{\cdot}d_s/D$) as a function of the particle size. Freely moving active particle in fluidized bed of 500-600 μm sand; T= 450 °C; P=1 atm (adapted from Scala, 2007)

Heat transfer coefficient

As said before, experiments on heat transfer are by far most diffused than those on mass and erosion rates. Experiments have been carried out to analyze the effects of many process parameters such as excess gas velocity, particle size, gas pressure and temperature, gas and particle thermal conductivity, particle density, surface size and shape.

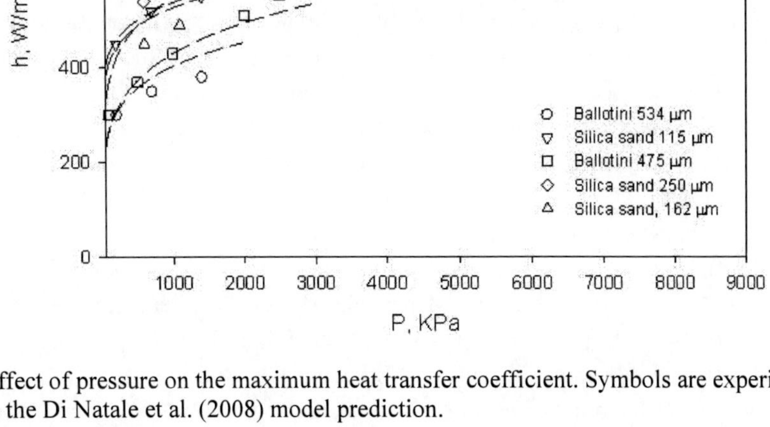

Figure 12. Effect of pressure on the maximum heat transfer coefficient. Symbols are experimental data and lines are the Di Natale et al. (2008) model prediction.

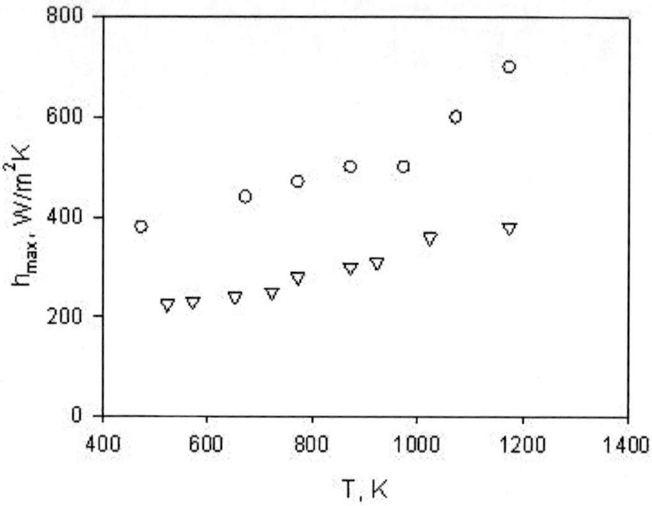

Figure 13. Effect of temperature on surface-to-bed heat transfer coefficient for silica sand 462 μm (circles) and 1276 μm (triangles). Data adapted from Botterill and Teoman (1980)

Figures 12 and 13 describes the effect of pressure and temperature on the averaged heat transfer coefficient for fully bubbling beds, as derived from several experimental studies. Both the increases of pressure and temperature lead to the increase of heat transfer coefficient. Figure 12 compares these experimental results (symbols) with the value of the heat transfer coefficient calculated by the Di Natale et al. (2008) model (lines), that allows the description of experimental data with a maximum error of estimation of the order of 7%.

Results on the maximum heat transfer coefficient as a function of temperature are reported in Figure 13. Experimental data show that higher temperatures enhance the heat transfer, with a change in the slope of $h_{max}(T)$ which appears around 1000 K for d_p = 462 μm particles, and – less evidently - at about 800 K for the 1276 μm particles. This change of slope marks the value of the transition temperature above which radiative phenomena become significant, coherently with the results of Flamant et al. (1999).

Di Natale et al. (2004) reported a series of experimental data on the surface averaged heat transfer coefficient for a vertical cylinder immersed in bubbling/slugging fluidized beds of different bed materials and gas pressure. The corresponding values of the heat transfer coefficient as a function of superficial velocity, and parametric with gas pressure is reported in Figure 14 for two different particle types. The experimental data show an abrupt increase at minimum bubbling velocity, due to the high solid mixing induced by bubbles that eventually activate the so-called particle convective heat transfer. For higher velocities, the heat transfer coefficient progressively increases until it reaches a maximum depending on bed material properties and on the diameter of the fluidization column. Figure 14 also shows that coarser particles have lower heat transfer coefficients than finer ones, and they are more affected by the pressure level.

A systematic analysis on the effect of particle diameter on heat transfer coefficient was reported by Kunii and Levenspiel (1991) that collected data for particles coarser than 50 μm obtained by different authors. These data showed that the heat transfer coefficient is maximum for particle diameter around 50 μm and minimum for very large particles. Experiments on different bed materials and particle sizes have been carried out by Di Natale et al. (2009) on three different exchange surfaces, namely a sphere and two vertical cylinders. The experimental results are reported in Figure 15 in terms of the maximum Nusselt number in a fully fluidized system ($Nu_{max} = h_{max} \cdot d_s/K_g$) as a function of the particle Archimedes number. The data shows that regardless the surface shape, Nu_{max} is the highest for Ar around 50, and it progressively decreases of about 50% for both finer and coarser particles. The effect of surface shape results in a vertical shift of each curve with maximum differences of the order of 250 for Ar~50.

The same experiments also highlight the effect of surface shape on the heat transfer. Indeed, the heat transfer coefficient is higher for slender surfaces than for blunter ones (see also Di Natale et al. 2007).

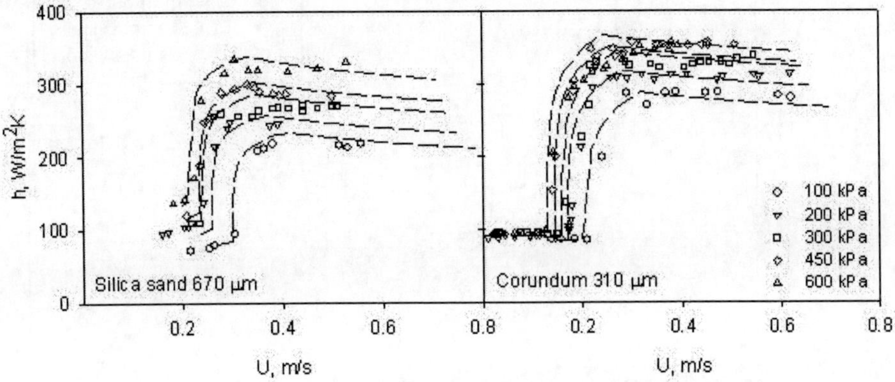

Figure 14. Effect of superficial gas velocity and gas pressure on heat transfer coefficient for a small vertical cylinder (OD=20 mm, L=20 mm) at ambient temperature (Di Natale et al. 2004). The lines represent prediction of the heat transfer coefficient with the Di Natale et al. (2008) model

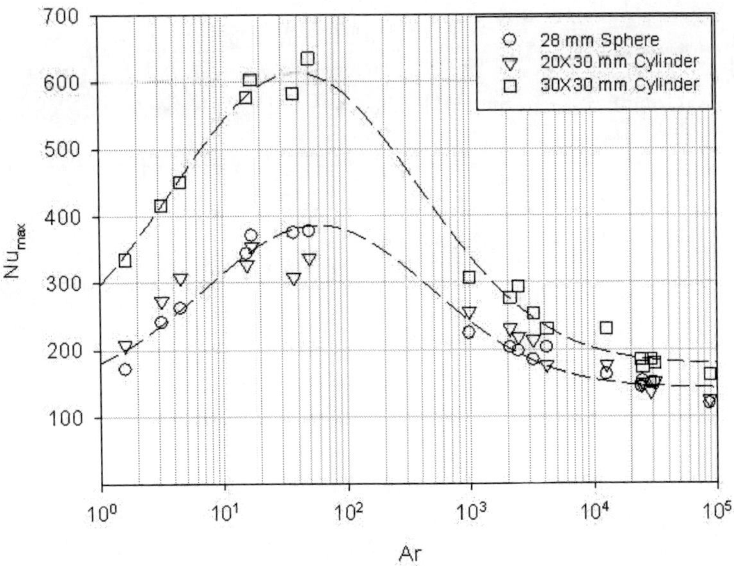

Figure 15. Surface Nusselt number versus Archimedes number for different exchange surfaces. (Di Natale et al. 2009)

In the same paper, the effect of particle thermal conductivity on heat transfer coefficient has been analyzed by plotting particle Biot numbers versus the ratio between particle and gas thermal conductivities (Figure 16). When a negligible effect of particle thermal conductivity is observed, the Biot number is inversely proportional to k_p/k_g. This condition holds for values of k_p/k_g between 30 and 1000.

Kunii and Levenspiel (1991) collected data from different authors in order to describe the effect of gas thermal conductivity on the heat transfer coefficient, showing that it monotonically increases with k_g (Figure 17).

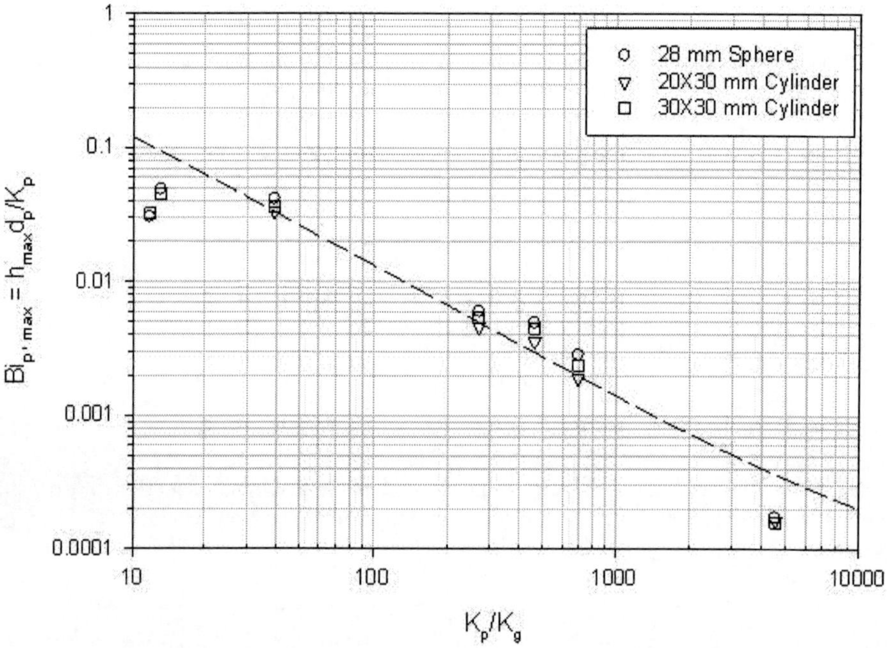

Figure 16. Particle Bi number versus thermal conductivity ratio (Di Natale et al. 2009).

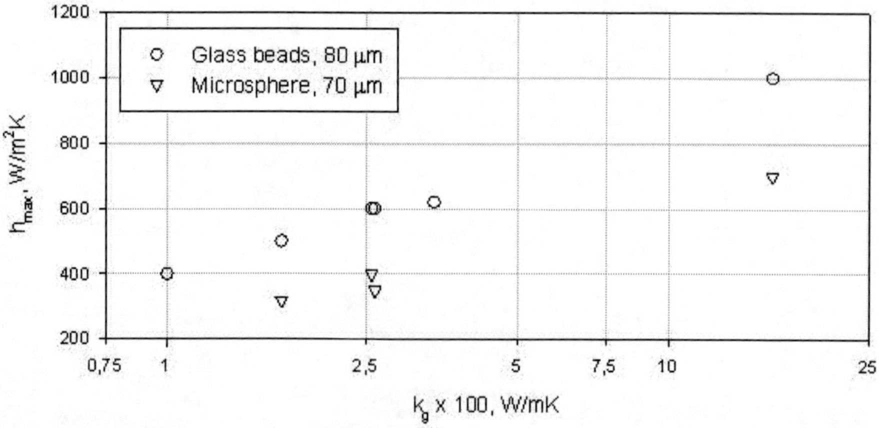

Figure 17. Effect of gas thermal conductivity on the maximum heat transfer coefficient (Kunii and Levenspiel, 1991)

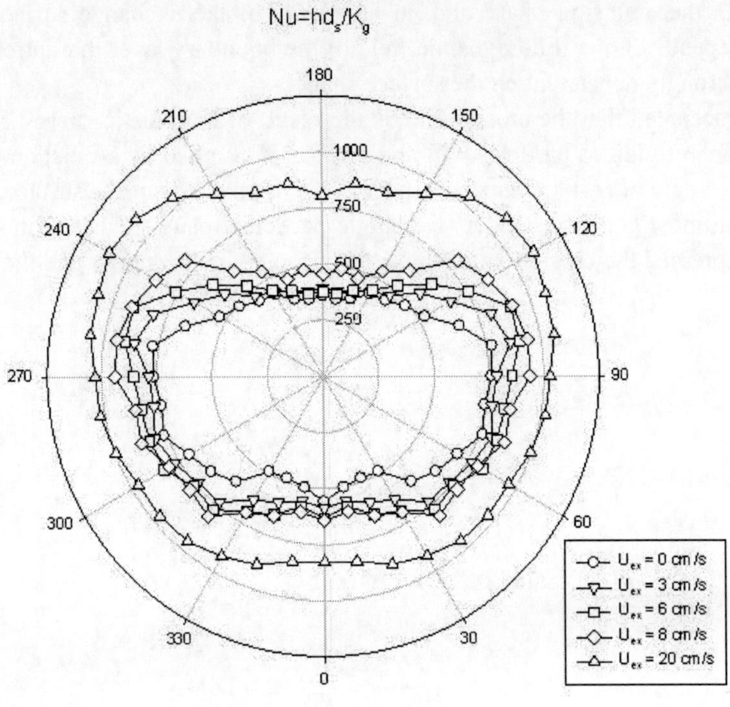

Figure 18. Angular profile of the Nusselt number as a function of excess gas velocity. Horizontal cylinder (OD = 30 mm) immersed in a fluidized bed of 500 μm glass beads and ambient pressure and temperature. Data from Di Natale et al. (2010)

Finally, Di Natale et al. (2010) estimated the local heat transfer coefficient profile close to a horizontal cylinder (OD = 30 mm) immersed in a fluidized bed of 500 μm glass beads and ambient pressure and temperature. The experimental data (reported in Figure 18 in terms of surface Nusselt number) show that the heat transfer coefficient monotonically increases with gas velocity and varies significantly around the cylinder circumference, being maximum at about 100-120 degs and minimum for θ around 180 degs.

4. Models for Transport Phenomena

4.1. Erosion Rate

Models for erosion rates are all based on the assumption that the gas has no effect on the surface erosion, except for its influence on the particle motion close to the immersed surface. This means that, by following the general approach of Eq.(1), the contribution of bubbles to the erosion coefficient E is zero as well as that pertaining to the interstitial gas. Needless to say, there are no radiative contributions.

The transport mechanism related to particle impacts with the surface can be called "particle convection" following the conventional name adopted for this phenomenon in heat transfer studies. As shown in the second paragraph, particle motion depends, on a macro-scale, on the bubble phase that mixes the solid at the length scale of the fluidization column diameter and with a characteristic frequency of about 0.5 – 3 Hz, depending on the operating conditions. In addition, the micro-scale motion of the particles is a function of the interstitial gas velocity in the emulsion phase and, in proximity of the exchange surface, the particle motion also depends on the fluid dynamic field of the boundary layer that introduces a meso-scale motion strongly dependent on the surface shape.

It is well accepted that the erosion rate is the result of the kinetic energy evolved during the impacts of the fluidized particles with the surface, multiplied by a function that takes into account for the angle of particle impact respect to the normal axis of the surface.

One of the most common models for single particle erosion is that proposed by Finnie (1960) that expressed the erosion in terms of the particle kinetic energy and the impact angle, α, as:

$$W = \frac{C}{P_H} m c_w^{\,2} f(\alpha) \tag{12}$$

where:

$$f(\alpha) = \sin(2\alpha) - 3\sin^2(\alpha) \qquad \alpha \le 18.43° \tag{13}$$

$$f(\alpha) = \frac{\cos^2(\alpha)}{3} \qquad \alpha > 18.43°$$

In this equation, C is a constant equal to 0.25, P_H is the so-called Vickers hardness of the surface, m is the particle mass and c_w is the instantaneous particle velocity near the surface (the subscript w stays for wall). The ratio between C and P_H is usually denoted as B_F.

This equation is at the basis of a rigorous approach to the estimation of the erosion rate of the entire surface based on the integration of all the contributions pertaining to the contacting particles in the unit time.

$$E = \int_0^\infty (c_w \cdot n) \cdot B_F \cdot m c_w^{\,2} \cdot f_w(r, c_w, t) \cdot dc_w \tag{14}$$

In this equation, f_w (r, c_w, t) is the particle velocity distribution.

The Finnie model can be coupled with the Kinetic Theory of Granular Flows, KTGF, to give an explicit solution of the integral (14), as reported by Lyczkowski and Bouillard (2002). Solving the integral (14) under the assumption that: i) the particle velocity c_w can be expressed in terms of its averaged value v_w and of the granular temperature Θ; ii) the impact angle can be deduced from geometric considerations; iii) the particle velocity distribution is Maxwellian:

$$f_w(r, c, t) = \frac{1}{2\pi\Theta^{1.5}} \exp\left[\frac{-c_w}{2\Theta}\right]; \tag{15}$$

the erosion rate is:

$$E = 2[1-\varepsilon_w]\rho_p B_F \times \left[\left[\frac{(2\Theta)^{3/2}}{\sqrt{\pi}} \right] F_1(\theta_c) + \frac{v_w^{\,2}}{2}\sqrt{\frac{2\Theta}{\pi}} F_1(\theta_c) + \frac{3}{2} v_w \Theta F_2(\theta_c) \right] \quad (16)$$

The parameters $F_1(\theta_c)$ and $F_2(\theta_c)$ are equal to 0.10 and 0.06 respectively.

Equation (16) describes the erosion rate as a function of three different terms, one independent from the averaged particle velocity and two directly function of it. Since the emulsion remains almost static until it is not mixed by the rising bubbles or by the exchange surface boundary layer, the first term represents the effect of the emulsion phase in quasi-static conditions while the other two are present when the particles are accelerated either by the rising bubbles (through their wake or drift fraction) or by the superficial boundary layer.

Equation (16) is not easy to handle, and for practical applications empirical or semi-empirical models are commonly adopted and have been recently resumed in the review of Lyczkowski and Bouillard (2002):

SCFMED model

$$E = E_0 \cdot (1 + 0.01167\,\mathrm{Re}) \cdot \frac{(1-\varepsilon)(\varepsilon - \varepsilon_{mf})}{\varepsilon^2} \quad (17)$$

Where:

$$E_0 = (1-e) \cdot \frac{75 \cdot \mu \cdot x_d}{(1-\varepsilon) \cdot d_p \cdot E_{sp}} \quad (18)$$

In this equation, x_d is a particle acceleration distance and E_{sp} is a specific energy of the eroding material, related to the hardness, and measured in MPa.

Yates model (1987)

$$E_0 = 0.13 \cdot (1-\varepsilon) \cdot \frac{\delta_b}{(1-\delta_b)} \cdot \rho_p \cdot g \cdot d_b^{\,4}; \quad (19)$$

where d_b is the equivalent bubble diameter.

Zhu et al ***model (1991)***

$$E_0 = C \cdot f \cdot \rho_p^{\,1.2} \cdot u_v^{\,2.1} \cdot (1-\psi); \quad (20)$$

Where f is a "void frequency" - that is similar to the aforementioned surface renewal frequency - u_v is the void velocity and ψ is the particle shape factor.

4.2. Mass Transfer

Even if based on limited experimental evidences, mass transfer coefficient is reliably modeled by following the analogy with fixed bed systems. The mass transfer coefficient between a surface and a fluidized bed of inert particles can be considered as the sum of the bubble and emulsion phase contributions (Yusuf et al., 2005) and it is entirely related to the movement of the gas phase, while particles role is restricted to their effects on gas distribution around the exchange surface. In the emulsion phase, the mass transfer coefficient is mainly proportional to the interstitial gas velocity and to the transport properties of the flowing gas (Yusuf et al., 2005; Scala, 2007; Rhodes, 2008) and typical equations are usually based on the use of classical dimensional analysis, in terms of Sherwood number versus Reynolds and Schmidt numbers. Usually, since the surface-average mass transfer coefficient in the emulsion is scarcely dependent on gas velocity, the Reynolds number is referred to the interstitial gas velocity at minimum fluidization. Starting from the analysis of several existing models and performing experimental tests on average mass transfer coefficient for spheres freely moving in bubbling fluidized beds, Scala (2007) proposed the following equation for the Sherwood number:

$$Sh = \frac{kd_s}{D} = 2 \cdot \varepsilon_{mf} + 0.7 \left(\frac{Re_{mf}}{\varepsilon_{mf}} \right)^{1/2} \cdot Sc^{1/3} \tag{21}$$

Theoretically, the mass transfer in the emulsion phase is expected to be proportional to the local interstitial gas velocity around the surface, U_w. Nevertheless, the observed overall effect on the surface-average value of the mass transfer coefficient is usually so limited that the mass transfer is often considered almost constant with gas velocity and Sherwood number is usually assumed proportional to minimum fluidization velocity, as in equation (21). However, in fluidized beds systems, velocity and void fraction are strongly correlated, since a given void fraction can be only achieved by a assuring a gas velocity high enough to keep particles at distance. As a consequence, higher interstitial gas velocity are related to higher void fractions, as expressed by the well known Richardson and Zaki equation for homogeneous bed expansion (e.g. Rhodes, 2008).

When this result is transferred to the case of a surface fixed in a fluidized bed, the estimation of the relative contribution of bubble and emulsion phase must be taken into account, and Eq.(1) usually represent the most reliable way to interpret this case. Nevertheless, different approaches have been proposed in the past. For example, in order to provide for an interpretation of their experiments, Ziegler and Holmes (1966) adopted a modification of the classical correlation of Ranz and Marshall for the Sherwood number of spheres in a homogeneous fluid. The proposed mode was:

$$Sh = Sh_e + 33.7 \cdot Y^{0.4} \tag{22}$$

Where

$$Sh_e = 2 + 0.60\left[\frac{\mathrm{Re}}{\varepsilon}\right]^{1/2} Sc^{1/3} \tag{23}$$

is the Ranz and Marshall equation referred to the interstitial gas velocity, and the second term of the sum accounts for the occurrence of local turbulence caused by particle motion. The dimensionless term Y is:

$$Y = \frac{G - G_{mf}}{(\psi \cdot d_p)^2 \cdot (\rho_p - \rho_g)\rho_g} \tag{24}$$

Agarwal et al.(1988) proposed a model for the estimation of the Sherwood number for the case of an active particle freely moving in the fluidized bed but alternatively immersed in the emulsion and the bubble phase. Indeed, their equation for the estimation of the mass transfer coefficient in the two phases can be applied to the case of a fixed surface. The proposed equations were:

$$Sh_d = K_p\left(\frac{C_{D,d}}{8}\right)\cdot\left(\frac{q}{\varepsilon}\cdot \mathrm{Re}_s\right)^{2/3} Sc^{1/3} \tag{25}$$

$$Sh_b = K_p\left(\frac{C_{D,b}}{8}\right)\cdot(3\cdot \mathrm{Re}_s)^{2/3} Sc^{1/3} \tag{26}$$

Where K_p is a parameter, $C_{D,d}$ and $C_{D,b}$ represent the active particle drag coefficient in the emulsion and the bubble phase respectively and q is the tortuosity of the emulsion phase. The Reynolds number, Re_s, is referred to the active particle size.

Recently Di Natale and Nigro (2010) calculated the Sherwood number in the emulsion phase for the local mass transfer coefficient around a horizontal cylinder reported by Garim et al. (1999). In this model, the authors started from the determination of angular surface void fraction profile obtained from heat transfer experiments and modeling studies. The local value of emulsion Sherwood number is:

$$Sh_d(U,\theta) = 0.82\left[\frac{\mathrm{Re}_w(U,\theta)}{\varepsilon_w(U,\theta)}\right]^{0.73}\cdot Sc^{1/3} \tag{27}$$

where the subscript w denotes that the surface Reynolds number and the void fraction refer to conditions near the wall of the exchange surface. Indeed, by using the Richardson and Zaki law to correlate surface Reynolds number to void fraction, it results:

$$\frac{Re_w(U,\theta)}{\varepsilon_w(U,\theta)} = Re_{p,t}\cdot\varepsilon_w(U,\theta)^{n-1}\frac{d_s}{d_p} \tag{28}$$

Therefore, since the authors found that in a bubbling fluidized bed the surface void fraction is almost constant with the gas velocity, also the Sherwood number results to be almost constant with U. This result is of fundamental relevance since it explains the independence of emulsion mass transfer coefficient from fluidization velocity observed in the former experimental studies (Scala, 2007; Rhodes, 2008).

Heat transfer

Among the transport phenomena, the models for heat transfer coefficients are by far the most advanced, since the richness of data and the reliability of the experimental techniques allowed the analysis of the heat transfer at the microscopic scale.

Nevertheless, the heat transfer is by far the most complex of the transport phenomena since both the gas and the particles directly contribute to the heat exchange. Indeed it has been well established that the heat transfer rate mainly depends on three different exchange mechanisms, namely: i) the convection/conduction with the gas phase flowing as bubbles or within the interstices of the emulsion phase, ii) the convection/conduction with the moving particulate phase and iii) the presence of radiative heat transfer phenomena at temperature higher than 800 K (Flamant et al. 1999).

Usually, to clearly point out these different mechanisms, the overall surface-to-bed heat transfer coefficient is considered to be made up of three additive components, providing an alternative way of writing Eq. (1):

$$h = (1 - \delta_b) \cdot (h_{gc} + h_{pc}) + h_r + h_b \cdot \delta_b; \tag{29}$$

where h_{gc}, h_{pc}, h_r are representative of the gas convective, the particle convective and the radiative heat transfer coefficients.

Gas convective heat transfer coefficient

Gas convection became relevant only for particle coarser than about 1 mm and, in this case, its dependence on gas velocity has to be properly assessed for a correct prediction of the heat transfer coefficient, h. Nevertheless, it is well accepted that for sand-like particles finer than 1 mm, h_{gc} is of the order of magnitude of the heat transfer coefficient at minimum fluidization velocity, when particle motion does not exists yet (Botterill and Denloye, 1963). Heat transfer coefficient at minimum fluidization velocity, h_{mf}, can be estimated by any of the models reported in Table 2.

Particle convective heat transfer coefficient

The first mechanistic model for particle convection has been proposed by Mickley and Fairbanks in 1955. This model represents a milestone in the study of heat transfer in fluidized bed since, for the first time, the intrinsic transient nature of the contact between fluidized particles and exchange surface was recognized. Indeed, Mickley and Fairbanks reported experimental tests on the time series of the heat transfer coefficient of a small heated cylinder immersed vertically in fluidized beds of different sand-like materials ranging from 50 to 300 μm. First of all, the authors showed that the heat transfer coefficient is small for beds at fixed and incipient fluidization conditions but it increases sharply in bubbling conditions. Then, they noticed that, in a bubbling bed, all the experiments followed a similar trend: the heat transfer coefficient is lower – even negligible - when the surface is surrounded by the rising bubbles, while it increases abruptly when it comes into contact with the emulsion phase. While these phenomena occur with a frequency that was directly related to that of the bubbles (around 0.5-3 Hz), the time series also showed high frequency fluctuations that were related, for the first time, to the occurrence of transient contacts between the exchange surface and the particles in the emulsion phase. Indeed, the authors found that such a contact is related to an "ensemble" of particles moving all together as a cluster that was denoted as *packet*. It was assumed that packets have the same physical properties of the emulsion in terms of density, specific heat and thermal conductivity and it remains in contact with the surface for a given time before it is replaced by another one coming from the emulsion. This time is the reciprocal of the surface renewal frequency discussed in Section 2.

It was assumed that, during its contact with the surface, the packet exchanges heat following the penetration theory for heat transfer like a homogeneous medium whose physical properties may be considered as the averaged of the solid-gas emulsion at minimum fluidization conditions. Even if the packet model is valid only for the particle convection, the experiments of Mickley and Fairbanks were conducted for fine particles for which the gas convection is almost negligible. As a consequence, the experimental heat transfer coefficient mainly coincides with the particle convective one.

The heat transfer coefficient was thus expressed as:

$$\frac{h \cdot d_p}{k_g} = \frac{1.13 \cdot d_p}{k_g} \cdot \left[\left(1-\varepsilon_{mf}\right)\cdot\left(1-\delta_b\right)\cdot f(U)\cdot \rho_p \cdot c_p \cdot k_{eff}\right]^{0.5} \tag{30}$$

Table 2. Models for heat transfer coefficient at minimum fluidization

Equations for h_{gc}	Author
$h_{mf} = 0.863 Ar^{0.39} dp^{-0.5} k_g$	Botterill and Denloye (1963)
$h_{mf} = \left(\frac{4k_{eff^0}{}^0 \rho_g c_g U_{mf}}{\pi L}\right)^{0.5}$ plane slab $h_{mf} = \left(\frac{4k_{eff^0}{}^0 \rho_g c_g U_{mf}}{\pi L}\right)^{0.5} + \frac{k_{eff}{}^0}{d_{sonda}}$ vert.cyl. $k_{eff}{}^0 = k_g + 0.1\rho_g c_g d_p U_{mf}$	Gabor (1970)
$h_{mf} = 0.009 Ar^{0.5} \Pr^{0.33} d_p k_g$	Baskakov (1972)
$h_{mf} = 0.4527 Ar^{0.2332} \Pr^{0.33} d_p{}^{-1} k_g$	Molerus (1995)

Table 3. Model equations for effective emulsion thermal conductivity, k_{eff}

Equations	References
$k_e^0 = k_g \cdot \left(\frac{k_p}{k_g}\right)^z$ $z = 0.28 - 0.757 \cdot \log(\varepsilon_{mf}) - 0.057 \cdot \log\left(\frac{k_p}{k_g}\right)$	Krupiczka (1967)
$k_e{}^\circ = 0.9065 \cdot \left(\frac{2}{k_p{}^{-1} + k_g{}^{-1}}\right) \cdot \left[\frac{k_p}{k_p - k_g} \cdot \left(\ln\left(\frac{k_p}{k_g}\right) - 1\right)\right] + 0.0935 k_g$	Swift (1966)
$k_e{}^\circ = k_p \cdot \left[\frac{2k_p + k_g + 2\varepsilon_{mf} \cdot (k_g - k_p)}{2k_p + k_g - \varepsilon_{mf} \cdot (k_g - k_p)}\right]$	Maxwell (1954)
$k_e{}^\circ = \varepsilon_{mf} k_g + (1 - \varepsilon_{mf}) \cdot k_p \cdot \left[\phi_b (k_p / k_g) + \frac{2}{3}\right]^{-1}$	Kunii and Smith (1960)
$k_e{}^\circ = k_p \cdot \frac{\varepsilon_{mf}{}^{2/3} + \frac{k_p}{k_g} \cdot \left(1 - \varepsilon_{mf}{}^{2/3}\right)}{\varepsilon_{mf}{}^{2/3} - \varepsilon_{mf} + \frac{k_p}{k_g} \cdot \left(1 + \varepsilon_{mf} - \varepsilon_{mf}{}^{2/3}\right)}$	Russell (1935)
$k_e{}^\circ = 0.9065 \cdot \left(\frac{2}{k_p{}^{-1} + k_g{}^{-1}}\right) \cdot \left[\frac{kp}{kp - kg} \cdot \left(\ln\left(\frac{k_p}{k_g}\right) - 1\right)\right] + 0.0935 k_g$	Gabor (1970)

In this equation k_{eff} is the effective thermal conductivity of the emulsion phase, that is a function of both the void fraction and the physical properties of the solid and gas phase, but also of the particle arrangement in the emulsion. As a consequence, values of k_{eff} are complex to determine by experiments and are usually obtained either from theoretical investigations, based on some assumptions on the particle arrangement, or from numerical regression of experimental data with given heat transfer models. Some of the equations used for k_{eff} are reported in Table 3.

Albeit for its historical and conceptual significance, the Mickley and Fairbanks model overestimates the heat transfer coefficient of coarser particles. To explain this phenomenon, between the 1960 and 1980 several researchers, as Gliksmann, Decker, Botterill, Denloy and Williams, studied the problem of heat transfer from a plane surface to a single sphere and showed that the temperature of particles larger than 1 mm did not change appreciably for a residence time as long as 1 s. In particular, Botterill et al. (1967) studied the case of a double layer of particles surrounding a plane slab, considering the particle residence times (i.e. the surface renewal frequencies) usually observed in fluidized beds, and showed that for particles coarser than about 400-500 μm, the heat flow can hardly pass the first layer of particles. This characteristic size is similar to the estimation given by Leva (1949) and later by Decker and Gliskmann (1981) and by Phillips (1990) on the extension of the thermal boundary layer and the Kolmogorov length scale in fluidized beds. Therefore, it was argued that once the residence time is lower than a characteristic particle heating time, τ, the heat transfer coefficient is related to the interactions between the exchange surface and the single contacting particle. In this case, the heat transfer through the gas phase that surrounds each particle plays a fundamental role, becoming an additional thermal resistance to estimate.

On the contrary, when the particle residence time is higher than the particle heating time, the heat flows through several layers of particles, according to the mechanism described by Mickley and Fairbanks. In this case, the thermal resistance of the gas layer is usually neglected but the heat transfer strongly depends on the value of the effective thermal conductivity of the emulsion, k_{eff}, that is a complex parameter to account for.

This result gave rise to two different families of heat transfer model, a first one - based on the single particle approach - used to describe the case of particles coarser than 300 μm, and a second one - based on packet or emulsion resistance approach - used for particles finer than about 500 μm.

Two main modifications have been applied to these original approaches. The first one is related to the single particle approach and consists in the introduction of a particle residence time distribution (usually a Gamma function – see Eq. 5) to integrate the contribution of each particle according to their contact time with the surface. This concept has been largely applied in the past and was adopted, for example, in the model of Ziegler et al. (1964a and b), Martin (1981), Di Natale et al. (2008). In this case, the overall heat transfer coefficient is the integral of the contribution pertaining to each particle, weighted according to their residence time distribution. For example, Di Natale et al. (2008) calculate h_{pc} from the following equation:

$$h_{pc} = \frac{\dot{Q}}{A_s\left(T_w - T_\infty\right)} = \frac{1}{A_s\left(T_w - T_\infty\right)} \int_0^{+\infty} q(t) \cdot E(t) \cdot n \, dt \tag{31}$$

Where q(t) is the single particle-surface heat flow, n is the number of contacting particles (a function of the surface void fraction) and E(t) is expressed according to Eq.(5).

Solving Eq. (31) under the assumption that particles are monosized spheres with negligible intra-particle thermal gradient, the authors obtain:

$$h_{pc} = \rho_p c_p d_p (1 - \varepsilon_w) \frac{f}{1 + f\tau} \tag{32}$$

Where τ is the characteristic particle heating time that, for spherical particles is:

$$\tau = \frac{\rho_p c_p d_p}{6} \frac{\delta}{k_g} \tag{33}$$

and δ is the gas gap surrounding the contacting particles, equal to $d_p/10$ for spherical particle and $d_p/24$ for irregular shaped ones.

The second modification is related to the attempt of extending the validity of packet resistance model to the case of coarser particles. To do so, an additional gas thermal resistance, in series with the emulsion one has been considered in order to reduce the overall heat transfer coefficient to the observed values. An example of this approach is the classical model proposed by Xavier (1977) to describe the particle heat transfer coefficient in slugging fluidized beds.

$$h_{pc} = \frac{K_{mf}\rho_{mf}c_{mf}\delta}{k_g t_c}\left[\exp\left[\left(\frac{k_g}{\delta}\right)^2 \frac{t_c}{K_{mf}\rho_{mf}c_{mf}}\right]\cdot erfc\left|\frac{k_g}{\delta}\left(\frac{t_c}{K_{mf}\rho_{mf}c_{mf}}\right)^{0.5}\right|-1\right]+2\left(\frac{K_{mf}\rho_{mf}c_{mf}}{\pi t_c}\right) \quad (34)$$

More detailed analyses of particle convective heat transfer models are available in the works of Chen et al (2005), Di Natale et al. (2008) and (2009) and in the review of Yusuf et al. (2005).

Radiative heat transfer

According to Flamant et al (1999), radiative contributions to heat transfer coefficient are likely to occur above 800 K and mainly for coarser particles. Even though this condition is largely diffused in industrial applications (e.g. fluidized bed combustors or catalytic reactors), there are no reliable models to predict the radiative heat exchange in fluidized beds. Indeed, this absence is related to the very nature of radiative heat transfer that is proportional to the difference of the fourth power of surface and bed temperatures and to the assessment of some optical properties (as emissivity, view factors etc.) of the fluidized system. However, realistic estimations of the radiation, absorption and scattering phenomena in a fluidized bed are quite difficult and require a complex mathematic solution. A simplified approach is to treat radiation phenomena as the exchange between opaque grey medium (the particles and the surface) separated by an inert medium (the gas).

According to this approach, and assuming a view factor F =1, an engineering estimation of the radiative heat transfer coefficient (Chen et al., 2005) is:

$$h_r = \frac{q_r}{T_w - T_b} = \frac{\sigma\cdot\left(T_w^{\ 4} - T_b^{\ 4}\right)}{\left(T_w - T_b\right)\cdot\left[1+\frac{1-e_b}{e_b}+\frac{1-e_w}{e_w}\right]} \quad (35)$$

Where q_r is the radiative heat flux, e_W and e_b are the emissivities of the bed and of the surface, both ranging between 0.7 and 1, and σ is the Stephan-Boltzmann constant.

Empirical and semi-empirical models for heat transfer coefficients

Even if the mechanistic models give explanations of the observed functional dependencies of the heat transfer coefficient on the different process parameters (in particular, pressure, temperature, gas velocity, particle diameters and bed material properties), for practical applications some empirical or semi-empirical models mainly based on Dimensional Analysis have been adopted. Among them, the most acknowledged are those for maximum heat transfer coefficient as the following:

Zabrodsky (1966):

$$h_{max} = 35.7 \cdot \rho_p^{0.2} k_g^{0.6} d_p^{-0.36} \tag{36}$$

Vreedemberg (1958) for horizontal cylinder:

$$Nu_{max} = \frac{h_{max} d_p}{k_g} = 0.66 C_R \cdot \left(\frac{c_g \cdot \rho_g}{k_g}\right)^{0.3} \cdot \left[\frac{(1-\varepsilon_{mf})}{\varepsilon_{mf}} \cdot \frac{\rho_p}{\rho_g} \cdot \mathrm{Re}_s\right]^{0.44} \tag{37}$$

Wender and Cooper (1958) for vertical tubes:

$$Nu_{max} = \frac{h_{max} d_p}{k_g} = 0.01844 C_R \cdot (1-\varepsilon_{mf}) \cdot \left(\frac{c_g \cdot \rho_g}{k_g}\right)^{0.43} \cdot \mathrm{Re}^{0.23} \cdot \left(\frac{c_p}{c_g}\right)^{0.8} \cdot \left(\frac{\rho_p}{\rho_g}\right)^{0.8} \tag{38}$$

Borodulya et al. (1990) for vertical cylinders:

$$Nu_{max} = \frac{h_{max} d_p}{k_g} = 0.116 Ar^{0.3} \Pr^{1/3} + 0.0175 \cdot Ar^{0.44} \Pr^{1/3} \tag{39}$$

To account for the exchange surface shape and size, Prins et al. (1989) proposed the following equation for the description of experimental data for the case of 103 μm FCC fluidized bed and spherical exchange surfaces of different size (6 - 30 mm):

$$Nu_{p,max} = \frac{h_{max} d_p}{K_g} = 4.175 \cdot Ar^n \cdot \left(\frac{D}{d_p}\right)^{-0.278} \tag{40}$$

With:

$$n = 0.087 \left(\frac{D}{d_p}\right)^{0.128} \tag{41}$$

Later, starting from experimental data on different kind of bed materials and different exchange surface shapes, Di Natale et al. (2007) proposed a generalization of Prins et al. model for the maximum surface Nusselt number:

$$Nu_{max} = \frac{h_{max} D_{eq}}{K_g} = a \cdot Ar^b \cdot \left(\frac{H}{D_{eq}}\right)^c \tag{42}$$

Where: a=837.2, b = -0.141, c = 0.572, H is the exchange surface height and D_{eq}, is the diameter of a circumference whose area, S_{eq}, is equivalent to the average cross sectional area of the exchange surface perpendicular to the main gas flow direction (the horizontal one in typical fluidized beds):

$$D_{eq} = \sqrt{\frac{4 \cdot S_{eq}}{\pi}} = \left[\frac{4}{\pi \cdot H} \int_0^H S(y)dy\right]^{0.5} \tag{43}$$

5. Toward a Generalized Model for Transport Phenomena in Fluidized Beds

Starting from the analysis of experimental and modeling studies reported in the previous sections, it is clear that each of the three transport phenomena, is characterized by a peculiar behavior that give rise to substantial differences in the functional dependence upon process parameters experimentally observed.

A synoptic table of these functional dependencies is reported in Table 4.

Table 4. Synoptic of the qualitative dependence of surface transport coefficient on the main process parameters

	Erosion rate, E	Mass transfer coefficient, k	Heat transfer coefficient, h
d_p	↑	↑	Non monotonic
Ar	↑	↑	Non monotonic
U_{ex}	↑	↑	↑
D	n.a.	↓	↓
ρ_g	↑	↑	↑
ρ_p	↑	-	↑
μ_g	↓	↓	↓
k_s	-	-	↑ (if K_p/K_g<30)
k_g	-	-	↑
c_p	-	-	↑
c_g	-	-	↑
D°	-	↑	-
T	n.a.	↑	↑
P	↑	↑	↑

Table 5. Synoptic of the qualitative dependence of the main fluid dynamic parameters on the main process parameters

	Surface renewal frequency, f	Surface void fraction, ε_W	Bubble fraction, δ_b	Interstitial gas velocity, U_W/ε_W	Particle velocity, c
d_p	↓	↑	~	↑	↑
Ar	↓	↑	~	↑	↑
U_{ex}	↑	~	↑	~	↑
D	↑	↑	↑	↑	↑
ρ_g	↑	↑	↓	↓	↑
ρ_p	↓	↑	↓	↑	↑
μ_g	↓	↓	-	↓	↓
k_s	-	-	-	-	-
k_g	-	-	-	-	-
c_p	-	-	-	-	-
c_g	-	-	-	-	-
D°	-	-	-	-	-
T	↓	↓	↑	↑	↓
P	↑	↑	↓	↓	↑

Analogously, Table 5 reports the same synoptic chart for the case of the most relevant fluid dynamic parameters.

From Table 4 it results that the overall qualitative effect of a given process parameter is similar for all the transport mechanisms, apart, obviously, for the specific physical properties, as thermal conductivity or mass diffusivity, that influences only one kind of transport phenomenon. However, the actual effect of a given parameter is quite different for each case and even if several efforts have been made in the past to find out a reliable generalized correlation for heat, mass and momentum balance in bubbling fluidized bed, starting from the powerful approach of Dimensional Analysis, all the attempts were unfruitful.

With reference to the results summarized in Table 5, it is worth noticing that the existing studies on granular flows, partially reported in Section 2, showed that both the particle and the interstitial gas velocities can be deduced once the fluidization velocity and the void fraction are known. Therefore, the fluid dynamic field around the exchange surface can be characterized by the knowledge of the surface renewal frequency, the surface void fraction and the bubble fraction only.

As said before, the first accounts for the particle replacement time near the exchange surface and thus it is related to the number of free particles, coming from the bulk of the bed, that contact with the surface for each unit time. The second parameter is related to the number of particles that are contemporary in contact with the surface and, eventually, with the value of the particle fluctuation velocity (i.e. the granular temperature) and with the value of interstitial velocity required to sustain that specific void fraction. Finally, the bubble fraction allows to weight between emulsion and bubble phase contribution and is responsible for the macro-scale mixing of particles in the fluidized bed. However, the existing correlations allow reliable determination of δ_b from the value of U_{ex} and of the bed material properties.

The mechanistic models developed in the past clearly pointed out that each transport phenomena can be decomposed in several contributions, each of which related to the fluid dynamic field in a peculiar way. In particular:

1. The erosion rate is only dependent on the particle convective contribution. It depends on the number of contacting particles per unit time and surface area (a function of the void fraction near the surface and of the surface renewal frequency) and on the tangential stress exerted by each contacting particle (directly proportional to its kinetic energy).
2. In the case of particles without any adsorption capacity toward the transported substance, the mass transfer coefficient is independent on any direct particle contribution. Indeed, it depends only on the gas phase, divided in the contribution of the interstitial gas (a function of the surface void fraction) and of the rising bubbles (a function of the overall fluid dynamics).
3. The heat transfer coefficient is the results of particle convection, gas convection, bubble flow and radiative phenomena. The particle convection is related to the number of contacting particles per unit time and surface area (a function of the void fraction near the surface and of the surface renewal frequency) and to the thermal properties of the contacting particles (thermal conductivity, specific heat, particle diameter). Traditionally, the gas convection sums up two different phenomena: the contribution of the interstitial gas (function of the local void fraction) and the heat conduction through the particles in the emulsion. As a matter of fact, it resumes the entire heat transfer phenomena of a fixed bed and, for this reason, it is usually assumed to be of the same order of magnitude of the overall heat transfer coefficient at minimum fluidization. The effect of the rising bubbles is considered analogously to the case of convection with a homogeneous fluid and it is usually neglected. Radiative mechanisms, in the end, are active only at $T>800K$ and are complex and in large part unknown functions of the particle concentration (i.e. of the surface void fraction) near the exchange surface.

Following these indications, it is possible to define three different kinds of parameters: i) the physical properties of the bed materials (ρ_g, c_g, c_s etc.); ii) the fluid dynamic parameters (resumed by ε_W, f and U_{ex}) and iii) *Intrinsic* transport parameters. This last represent those parameters that directly influence the transport phenomena between the surface and the particles and between the surface and the gas phase. Apart from those parameters already appearing in the class of physical properties, intrinsic parameters are the particle size, d_p, (that determine the particle impact energy and the particle characteristic heating time) and the instantaneous particle velocity, c, for its effect, once more, on the impact energy.

Table 6 is a revised synoptic of the transport phenomena aimed at resuming the effect of the different parameters on each transport mechanisms. These considerations point out that while a direct similitude between overall heat, mass and erosion rates are actually unreliable, several analogies among the different type of transport mechanisms can be hypothized. Needless to say, the easiest one is the analogy between bubbles phase contribution to mass and heat transfer, for which the Chilton-Colburn analogy for homogeneous fluid can be applied. However bubble contributions are usually negligible respect to the other contributions and this analogy do not provide any practical improvement for engineering estimations.

The existence of some analogies between gas convective contributions is intuitive, but its actual evidence is limited by the common meanings of these mechanisms in mass and heat transfer. In fact, while in the first case the mass transfer is only related to the interstitial gas flow, the concept of h_{gc} resume both the interstitial gas contribution and the heat conduction trough the solid phase so that it consider an active role of the particulate phase that is absent in the case of mass transfer. For this reason, while the analogy between heat and mass transfer rate between the interstitial gas and the fluidized particles can be easily assumed, there is no apparent correlation between the existing data and models for gas convective heat and mass transfer with immersed surfaces. On the contrary, a stronger similitude between the particle convection mechanisms, active for erosion and heat transfer, can be observed. This is true, in particular, when single particle heat transfer models are considered. In this case, the overall particle convective contribution for heat transfer or erosion is the results of the quantity of heat, or tangential stress, transferred by each particle that contacts the exchange surface, weighted on the number of particles contacts per unit time and surface area. These variables are mainly related to the concepts of surface void fraction and surface renewal frequency. However, albeit for this intuitive similitude, at our knowledge, there is no mechanistic model aimed at estimating the particle convective heat transfer the erosion rate such as different aspects of a unique phenomenon.

Table 6. Synoptic of the qualitative dependencies of the transfer coefficients on the physical properties, the fluid dynamic and the intrinsic transport parameters

	Particle convection		Gas convection		Bubbles flow		Radiative
	E	h_{pc}	k_d	h_{gc}	k_b	h_b	h_r
Physical parameters							
ρ_g	$\uparrow$	$\uparrow$	$\uparrow$	$\uparrow$	$\uparrow$	$\uparrow$	-
ρ_p	$\uparrow$	$\uparrow$	-	$\uparrow$	-	-	-
μ_g	$\downarrow$	$\downarrow$	$\downarrow$	$\downarrow$	$\downarrow$	$\downarrow$	-
K_s	-	$\uparrow$ if $K_p/K_g<30$	-	n.a.	-	-	-
K_g	-	$\uparrow$	-	$\uparrow$	-	$\uparrow$	-
c_p	-	$\uparrow$	-	$\uparrow$	-	-	-
c_g	-	$\uparrow$	-	$\uparrow$	-	$\uparrow$	-
D°	-	-	$\uparrow$	-	$\uparrow$	-	-
T	n.a.	$\downarrow$	$\downarrow$	$\downarrow$	$\downarrow$	$\downarrow$	$\uparrow$
P	$\uparrow$	$\uparrow$	$\uparrow$	$\uparrow$	$\uparrow$	$\uparrow$	-
Fluid dynamic parameters							
f	$\uparrow$	$\uparrow$	-	-	-	-	-
ε_w	$\downarrow$	$\downarrow$	$\uparrow$	$\uparrow$	-	-	$\uparrow$
U_{ex}	$\uparrow$	$\uparrow$	~	~	$\uparrow$	$\uparrow$	-
Intrinsic parameters							
d_p	$\uparrow$	$\downarrow$	-	-	-	-	-
c	$\uparrow$	-	-	-	-	-	-

6. CONCLUSION

This chapter is focused on the analysis of surface-to-bed transport phenomena that play a fundamental role in the fluidization technology.

The study of the main experimental and modeling results reported in the pertinent literature shows that heat, mass and erosion rates in fluidized beds are complex phenomena accounting for specific contributions of both the gas and the particulate phases. These contributions are functions of the peculiar fluid dynamic field near the exchange surface for which, at this moment, there are only a limited number of experimental and modeling results valid for specific operating conditions, usually referred to vertical or horizontal cylinders immersed in the fluidized bed. Heat transfer rates are also characterized by the occurrence of radiative phenomena that appears at high temperature.

Gas phase contribution is recognized to be related to either the interstitial gas flow though emulsion and to the gas flow through the bubbles. This contribution does not directly affect the erosion rates, but plays the same role in heat and mass transfer, due to both convective and diffusive phenomena.

Models for gas convection due to the bubble phase and to the interstitial gas may follow the same approach used to describe fixed bed systems, and can written according to the general indication of the Chilton-Colburn analogy. By this way, according to the assumptions of the generalized boundary layer model for homogeneous fluid, the gas convective contribution can be reconstructed as the result of a similar fluid dynamic effect (depending on the global fluidization regime as well as on the features of the superficial boundary layer region around the exchange surface, and usually expressed by a Reynolds number) and of the correct transport properties of the gas phase (either the gas diffusivity or the thermal diffusivities that appears in the Schmidt and Prandtl numbers) required to scale the transfer phenomena.

However, for practical purposes, the gas convective heat transfer mechanism in the emulsion phase is usually consider as the results of both the interstitial gas convection and of the heat conduction between the gas and fluidized particles. As a matter of fact, this corresponds to the overall heat transfer rate for a fixed bed, and the corresponding value of the gas convective heat transfer coefficient is roughly assumed to be of the same order of overall heat transfer coefficient at minimum fluidization. Actually, this leads to an incongruous comparison between the models for gas convective coefficient for heat and mass transfer.

Particles contribution to the surface–to-bed transport phenomena is twofold. A first effect is related to their role in dispersing heat and mass through the emulsion phase, where they act both as gas-mixers and eventually as heat (or mass) sinks. They are also responsible for the actual stress tensor of the emulsion phase, influencing the momentum transfer phenomena. On the other hand, particles play a crucial role in determining the heat and erosion rates due to their unsteady impacts with the exchange surface. In the first case, the particle is, for example, a cold sink that exchanges heat from a warm object by means of conduction and convection phenomena involving the gas gap between the same particle and the surface. In the second case, the particles are carrier of kinetic energy (proportional to their inertia) that generates tangential stresses on the impacting surface.

This contribution is usually defined as particle convection and it is the leading transfer mechanisms in erosion and heat transfer, determining, for example, the well-known rapid increase in the corresponding erosion and heat transfer coefficient, once the superficial velocity overcame the minimum fluidization conditions. Particle convection is strongly related to the superficial boundary layer fluid dynamic field, and, in particular to the values of the surface void fraction and of the surface renewal frequency. Nevertheless, although representative of the same phenomenon, heat transfer and erosion rates have been never analyzed in light of any fluid dynamic analogies and the existing model approaches are quite different from each other.

Finally, it is worth noticing that one of main reason for the lack of a mechanistic model to describe the surface-to-bed transport phenomena is the scarcity of information on the superficial boundary layer fluid dynamic field. In the last ten years, our group dedicated an intense research activity to this subject, obtaining some relevant results especially for the case cylindrical and spherical surfaces. Notwithstanding this, we envisage that even more experimental evidences and theoretical works are required to finally reach the goal of a predictive and generalized model for surface-to-bed transport phenomena in fluidized beds.

ACKNOWLEDGMENTS

The authors sincerely wish to thank Dr. Roberta Saltelli for her precious contribution in the realization of this chapter.

REFERENCES

[1] Almstedt, AE; Gustavsson, M. *Chem. Eng. Sci.*, 2000, 55, 867-879.

[2] Almstedt, AE; Wiman, J. *Chem. Eng. Sci.*, 1997, 52(16), 2677-2695.

[3] Baskakov, AP; e Suprun, VM. *Int. Chem. Eng.*, 1972, 12, 53-58.

[4] Baskakov, AP; Berg, BV; Vitt, OK; Filippovskii, NF; Kirakosyan, VA; Goldobin, Y; Yu, M; Maskaev, VK. *Powd. Techn.*, 1973, 8(5-6), 273-82.

[5] Boerefijn, R; Poletto, M; Salatino, P. *Powd. Techn.*, 1999, 102, 53-63.

[6] Borodulya, VA; Teplitsky, AP; Sorokin, VV; Markevich, II; Hassan, AF; Yeryomenko, TP. *Int. J. Heat Mass Transf.*, 1991, 34, 47-53.

[7] Botterill, JSM; Teoman, Y. *Proc. Int. Fluid. Conf.*, 1980, 93-100.

[8] Botterill, JSM; Brundrett, GW; Cain, GL; Elliott, DE. *Chem. Eng. Symp. Ser.*, 1966, 62(1), 1-6.

[9] Botterill, JSM. *AIChE Symp. Ser.*, 1973, 69(128), 26-7.

[10] Botterill, JSM.; Denloye, AO. *AIChE Symp. Ser.*, 1978, 176, 194-202.

[11] Botterill, JSM.; Williams, JR. *Trans. Inst. Chem. Eng.*, 1963, 41(5), 217-30.

[12] Buyevich, YA; Korolyov, VN; Syromyatnikov, NI. in *"Heat and Mass transfer in fixed and fluidized bed"*, W.P.M. Van Swaaj, N.H. Afgan eds., Springer & Verlag, New York, (USA) 1986.

[13] Chen, JC; Grace, JR; Golriz, MR; *Powd. Techn.*, 2005, 150(2),123-132.

[14] Cody, GD; Johri, J; Goldfarb, D. *Powd. Techn.*, 2008, 182(2), 146-170.

[15] Davidson, JF; Harrison, D. *Fluidized Particles*, Cambridge, The University Press, London, UK, 1963.
[16] Decker, NA; Glicksman, LR. *AIChE Symp. Series*, 1981, 208(77), 341-349.
[17] Decker, N; Glicksman, LR. *Int. J. Heat Mass Transf.*, 1983, 26(9), 1307-20.
[18] Di Felice, R; Kehlenbeck, R. *Chem. Eng. Techn.*, 2000, 23(12), 1123-1126.
[19] Di Natale, F; Lancia, A; Musmarra, D; Nigro, R. In: XI; Fluidization, P; Salatino, R; Chirone, U; Arena, M. Miccio, eds., *ECI Conference Series*, New York, USA, 2004, 911-919.
[20] Di Natale, F; Lancia, A; Nigro, R. *Powd. Techn.*, 2007, 174(3), 75-81.
[21] Di Natale, F; Lancia, A; Nigro, R. *Powd. Techn.*, 2008,187, 68-78.
[22] Di Natale, F; Lancia, A; Nigro, R. *Powd. Techn.*, 2009, 195(2), 135-142.
[23] Di Natale, F; Bareschino, P; Nigro, R. Heat transfer and void fraction profiles around a horizontal cylinder immersed in a bubbling fluidized bed, *International Journal of Heat and Mass Trnasfer*, 2010, in press.
[24] Fakhimi, CyS; Harrison, D. *Trans. Inst. Chem. Eng.*, 1980, 58, 125-131.
[25] Finnie, I. *Wear*, 1960, 3(2), 87-103.
[26] Flamant, G; Lu, JD; Variot, B. *Chem Eng. Sci.,* 1993, 48(13), 2493-2503.
[27] Fu, T; Lee, W; Nieh, S. *Powd. Techn.*, 1991, 67, 229-235.
[28] Gabor, JD. *Chem Eng. Sci.*, 1970, 25, 979-985.
[29] Gabor, JD. *AIChE Symp. Ser.*, 1970, 66, 76-86.
[30] Garim, MM; Freire, JT; Goldstein, RJ. *Powd. Techn.*, 1999, 101(2), 134-141.
[31] George, AH; Int. *J. Heat Mass Transf.*, 1993, 36(2), 337-345.
[32] George, AH. *Int. J. Heat Mass Transf.*, 1987, 30(4), 763-769.
[33] Glicksmann, LR; *Chem. Eng. Sci.*,1984, 39,1561-1379.
[34] Kim, SW; Ahn, JY; Kim, SD; Lee, D. Hyun, Int. *J. Heat Mass Transf.*, 2003, 46(3), 399-409.
[35] Koppel, LB; Patel, RD; Holmes, JT. *AIChE J.*, 1970, 16(3), 456-64.
[36] Krupiczka, R. *Int. Chem. Eng.*, 1967, 7, 122-128.
[37] Kunii, D; Levenspiel, O. *Fluidization Engineering,* 2nd ed., Butterworth-Heinemann, Newton (MA), 1991.
[38] Kunii, D; Smith, JM; *AIChE J.*, 1960, 6, 71-78.
[39] Leva, M; Weintraub, M; Grummer, M. *Chem. Eng. Progress*, 1949, 45, 563-72.
[40] Lyczkowski, RW; Bouillard, J. *X. Progress En. Comb.Sci.*, 2002, 28, 543–602.
[41] Martin, H. *Chem. Eng. Comun.*, 1981, 13(1) 1-16
[42] Maxwell, JC. *A Treatise on Electricity and Magnetism*, third ed., Dover Publications Inc., New York, 1954.
[43] Mickley, HS; Fairbanks, CA. *AIChE J.*, 1955, 1, 374-84.
[44] Molerus, O; Burshka, A; Dietz, S. *Chem Eng. Sci.*, 1995, 50(5), 871-877.
[45] Molerus, O; Mattmann, W. *Chem. Eng. Techn.*, 1992, 15(3), 139-50.
[46] Ozkaynak, TF; Chen, JC. *AICh E J.*, 1980, 26(4), 544-50.
[47] Pence, DV; Beasley, DE; Figliola, RS. *J. Heat Trans*, 1994, 116(4), 929-37.
[48] Phillips, MC. *Chem. Eng. Sci.*, 1990, 45(8), 2195-2202.
[49] Prins, W; Harmsen, GJ; de Jong, P; van Swaaij, WPM. In: VI; Fluidization, JR; Grace, LW; Shmeilt, MA. Bergougnou, eds., *Engineering foundations*, Alberta,Canada, 1989, 677-684.

[50] Rhodes, M. *Introduction to Particle Technology*, John Wiley and Sons, Chichester, UK, 2008.
[51] Rong, D; Mikami, T; Horio, M. *Chem. Eng. Sci*., 1999, 54, 5737-5754.
[52] Russell, HW. *J. Am. Ceramic Soc*., 1935, 18(1), 1-5.
[53] Scala, F. *Chem. Eng. Sci*., 2007, 62(16), 4159-4176.
[54] Solimene, R; Marzocchella, A; Salatino, P. *Powd. Techn*., 2003, 133(1-3), 79-90.
[55] Sunderesan, SR; Clark, NN. *Int. J. of Multiphase Flow*, 1995, 21(6), 1003-1024.
[56] Swift, DL. *Int. J. Heat Mass Transf*., 1966, 9, 1061.
[57] Vreedenberg, HA. *Chem. Eng. Sci*., 1958, 9(1), 52-60.
[58] Wang, L; Wu, P; Ni, X. *Powd. Techn*., 2005, 149 (2-3), 127-138.
[59] Wang, XS; Rhodes, MJ. *Chem Eng.Sci*., 2003, 58(2), 387-395.
[60] Wender, L; Cooper, GT. *AIChE J*., 1958, 4(1), 15-23.
[61] Xavier, AM. *Heat transfer between a fluidized bed and a surface*, Ph.D. Thesis, University of Cambridge, U.K. 1977.
[62] Yates, JG. *Chem. Eng. Sci*., 1987, 42(2), 379-80.
[63] Yusuf, R; Melaaen, MC; Mathiesen, V., *Chem. Eng. & Techn*., 2005, 28(1), 13-24.
[64] Zabrodsky, SS. *Hydrodynamics and heat transfer in fluidized beds,* M.I.T. Press, Cambridge, MA, 1966
[65] Zhu, J; Lim, CJ; Grace, FR; Lund, JA. *Chem Eng. Sci,* 1991, 46(4),1151–6.
[66] Ziegler, EN; Koppel, LB; Brazelton, WT. *Ind. Eng. Chem. Fund*., 1964, 3(2) 324-330.
[67] Ziegler, EN; Brazelton, WT. *Ind. Eng. Chem. Fund*., 1964, 3(2), 94-98.
[68] Ziegler, EN; Holmes, JT. *Chem. Eng. Sci*., 1996, 21, 117-122.

In: Powder Engineering, Technology and Applications ISBN: 978-1-61761-216-9
Editor: Jason M. Barker, pp.43-80

Chapter 2

ORGANIC SOLVENT-FREE SYNTHESIS OF MAGNETITE NANOCRYSTALS WITH CONTROLLED PARTICLE SIZES

Tomohiro Iwasaki
Department of Chemical Engineering, Osaka Prefecture University,
Sakai, Osaka, Japan

ABSTRACT

Magnetite (Fe_3O_4), which is an important member of spinel type ferrites, has been widely used in various industrial products such as recording materials because it has the excellent magnetic properties. Recently, bio-applications of Fe_3O_4 nanocrystals, e.g., magnetic resonance imaging, drug delivery, and hyperthermia, have attracted much attention because of the non-toxicity property and high chemical stability. Thus, the demand for Fe_3O_4 is increasing, and industrial synthesis methods which can meet the demand are needed. The synthesis processes should have the low environmental impact and must prepare the Fe_3O_4 nanocrystals with precisely controlled particle sizes suitable for the applications because the magnetic properties depend strongly on the particle size. Therefore, we have synthesized the size-controlled Fe_3O_4 nanocrystals using organic solvent-free conventional and novel methods: hydrothermal and mechanochemical synthesis methods. In the hydrothermal synthesis, ferrous and ferric hydroxides are coprecipitated or sequentially formed in high alkaline solutions, and then the suspension thus obtained is heated under a hydrothermal condition. In the coprecipitation operation, the molar ratio of ferrous ion to ferric ion is varied. Furthermore, the anionic species existing together with the iron ions are controlled. This synthesis method could control the particle size between 10 and 40 nm. On the other hand, we have developed a novel synthesis method using mechanochemical effect in order to prepare Fe_3O_4 nanocrystals with high crystallinity without using any environment-unfriendly additives such as toxic organic solvents and additional heating treatments. In this method, a tumbling ball mill is used as a synthesis reactor, and the suspension of ferrous and ferric hydroxides is ball-milled under a cooling condition. The solid phase reaction between ferrous and ferric hydroxides progresses owing to not the heat energy but the mechanical energy. This method provides superparamagnetic Fe_3O_4 nanocrystals with a size of less than 10 nm.

INTRODUCTION

Iron oxides powders have been used as color pigments for a long time, and the industrial uses, in particular, the applications to magnetic materials using magnetite (Fe_3O_4) and maghemite (γ-Fe_2O_3) such as recording media have rapidly expanded since the beginning of the 20th century because of the excellent magnetic properties [1]. At present, magnetic iron oxides are also utilized for printing and electrophotography such as copying toner [2] and carrier powders [3]. Recently, the non-toxicity property and high chemical stability of Fe_3O_4 have attracted much attention, and many biomedical applications of Fe_3O_4 nanoparticles have been studied actively: e.g., sensors [4–6], magnetic resonance imaging (MRI) [7–9], drug delivery system (DDS) [10, 11], and cancer therapy [12–14].

In order that Fe_3O_4 nanoparticles exhibit required magnetic efficiency in biomedical applications, Fe_3O_4 nanoparticles should have the controlled size suitable for the applications because the magnetic properties depend strongly on the particle size. For example, Fe_3O_4 nanoparticles of about 10 to 20 nm are used for DDS and cancer therapy, and below 10 nm of superparamagnetic Fe_3O_4 nanoparticles with good dispersibility are needed for MRI. For preparing the size-controlled Fe_3O_4 nanoparticles, various synthesis methods using solid, liquid and gaseous phases reactions have been developed: e.g., thermal decomposition of iron complex in organic solvent [15], pulsed laser ablation (PLA) [16], mechanochemical processing [17], and micro-emulsion [18, 19]. Many synthesis methods developed to control the particle size so far have serious problems that the synthesis process is relatively complicated and organic solvents which are unsuitable for biomedical applications are often used.

Among the preparation methods, coprecipitation of ferrous and ferric hydroxides in organic solvent-free water system is useful because of its simplicity of the process and relatively low environmental impact [20–25]. However, this method has a disadvantage that obtained Fe_3O_4 nanoparticles tend to have relatively low crystallinity and poor magnetic properties [26–28]. Therefore, following hydrothermal treatment of the suspension of precursor, ferrous hydroxide-goethite composites, formed by the coprecipitation is effective to improve easily the crystallinity [17]. During the hydrothermal treatment, the Fe_3O_4 nanoparticles have a tendency to aggregate by heating. Thus, surfactant is often added to the suspension for preventing the aggregation [29, 30]. For medical applications, however, it is undesirable to use any surfactants because surfactants are unsuitable for human body, and if used, sufficient separation and washing treatments to remove surfactants are needed, leading to a complicated preparation process. Accordingly, a preparation method without using any surfactants is required in the hydrothermal synthesis of Fe_3O_4 nanoparticles.

In this chapter, several hydrothermal synthesis methods we developed to control the size of Fe_3O_4 nanoparticles without using any surfactants are present. In addition, we propose a novel preparation process in which formation and crystallization of Fe_3O_4 nanoparticles are promoted by means of not heat energy but mechanical energy.

Size Control in Hydrothermal Synthesis

Effect of Ferrous/Ferric Ions Molar Ratio on Reaction Mechanism for Hydrothermal Synthesis of Magnetite Nanoparticles

The reaction mechanism in the hydrothermal synthesis of Fe_3O_4 is as follows. When an alkaline solution is added to an acidic solution containing ferric and ferrous ions, ferrous hydroxide ($Fe(OH)_2$) and goethite (α-FeOOH) which is transformed from ferric hydroxide ($Fe(OH)_3$) are coprecipitated according to Eqs. (1) to (3).

$$Fe^{2+} + 2OH^- \rightarrow Fe(OH)_2 \quad (1)$$

$$Fe^{3+} + 3OH^- \rightarrow Fe(OH)_3 \quad (2)$$

$$Fe(OH)_3 \rightarrow \alpha\text{-}FeOOH + H_2O \quad (3)$$

At the coprecipitation, $Fe(OH)_2$ grows on α-FeOOH, resulting in formation of the $Fe(OH)_2$-α-FeOOH composites. And then, a part of them is converted into homogeneous Fe_3O_4 according to the following solid phase reaction [31].

$$Fe(OH)_2 + 2\alpha\text{-}FeOOH \rightarrow Fe_3O_4 + 2H_2O \quad (4)$$

The reaction rate changes remarkably depending on the molar ratio of Fe^{2+}/Fe^{3+} [32] and the presence of oxidizing [33] and reducing agents [34]. Accordingly, for the formation of Fe_3O_4 nanoparticles with controlled sizes, the initial Fe^{2+}/Fe^{3+} molar ratio of 0.5 corresponding to the stoichiometric ratio in the Fe_3O_4 formation reaction is needed and no use of oxidizing and reducing agents is preferred. It is well known that this reaction is relatively slow at room temperature and that the reaction rate can be increased by heating. Therefore, effects of the Fe^{2+}/Fe^{3+} molar ratio on the reaction mechanism were examined.

All chemicals used in the experiments were of analytical reagent grade and were used without further purification. A proper amount of ferrous sulfate heptahydrate ($FeSO_4 \cdot 7H_2O$) and ferric chloride hexahydrate ($FeCl_3 \cdot 6H_2O$) were dissolved in 40 ml of deionized and deoxygenated water in a stainless steel autoclave. After several minutes, 20 ml of 1.0 kmol/m^3 sodium hydroxide (NaOH) solution was added into the solution at a constant addition rate under vigorous stirring using a magnetic stirrer in an argon atmosphere for oxidation prevent of Fe^{2+}. The addition rate of NaOH solution was set to be 3.0 ml/min for forming the precipitates homogeneously.

The initial Fe^{2+}/Fe^{3+} molar ratio β in the solution was varied in the range of $\beta = 0.25$ to 2.0 by adjusting the amount of $FeSO_4$ under the constant amount of $FeCl_3$ (= 2 mmol). After air in the autoclave was replaced with argon, the sealed autoclave was maintained at 120°C for a given heating time in an electric oven and then cooled down to room temperature. The heating time was varied as 0 (i.e., without heating), 4 and 20 h. The obtained black precipitate was washed with deionized water and decanted after centrifugation at the centrifugal acceleration of 1500 G. After the washing operation was repeated 3 times, the sample was dried overnight in air at 80°C.

The powder X-ray diffraction (XRD) pattern of samples was measured with CuKα radiation ranging from 2θ = 10 to 80° at a scanning rate of 1.0°/min using a Rigaku RINT-1500 powder X-ray diffractometer. Figure 1 shows the XRD patterns of samples obtained after heating for a given time under various β. The morphology of prepared samples was observed with a field emission scanning electron microscope (FE-SEM, JEOL JSM-6700F). Figures 2 and 3 indicate the SEM images of samples prepared under representative molar ratios of β = 0.5 and 2.0 as an example, respectively. In addition, the average crystallite size was calculated from the full-width at half-maximum (FWHM) of the Fe_3O_4 (311) diffraction peak at $2\theta \approx 35.5°$ using the Scherrer's formula. Figure 4 shows the change in the average crystallite size with the heating time. As can be seen in these figures, in the samples prepared at $\beta \leq 0.5$ without heating, the typical peak of α-FeOOH phase was seen at $2\theta \approx 21.2°$ in the XRD patterns, and the needle-shaped particles corresponding to α-FeOOH were observed (see Figure 2a) because of incomplete reaction. At β = 0.5, α-FeOOH was consumed as the reaction progressed by heating and then homogeneous Fe_3O_4 phase with a relatively high crystallinity could be obtained. However, when β = 0.25, α-FeOOH still remained after heating for 20 h because of the lack of $Fe(OH)_2$. No peaks were observed in the XRD patterns at 2θ = 14.9, 23.7 and 26.0°, which are typical peaks of maghemite (γ-Fe_2O_3) corresponding to (103) and (110) lines, (106) and (203) lines, and (213) and (116) lines, respectively. In addition, the lattice constant calculated from the XRD patterns of the samples of β = 0.5 to 2.0 after heating for 20 h was between 8.37 and 8.42 Å, and close to that of Fe_3O_4 (8.396 Å) compared to that of γ-Fe_2O_3 (8.345 Å). These also support that the samples of $\beta \geq 0.5$ after heating were Fe_3O_4 phase.

As can be seen in Figure 2, at β = 0.5 the particle size after heating for 20 h was almost the same (about 30 nm) as that before heating whereas the average crystallite size increased with increasing in the heating time as shown in Figure 4. This implies that the reaction progresses rapidly by heating and the crystal nucleation (i.e., formation of precursors) of Fe_3O_4 mainly occurs, resulting in inhibition of the crystal growth, and that the coalescence of crystallites in the inside of a particle increases the crystallinity.

On the other hand, when $\beta \geq 1.0$ in which the amount of $Fe(OH)_2$ in the starting solution exceeded the stoichiometric ratio, α-FeOOH was consumed almost completely before heating as shown in Figures 1a and 3a. Also, as can be seen in Figures 2 and 3a, the particle size before heating was almost the same as that at β = 0.5. The results reveal that the reaction rate increases even at room temperature in the presence of surplus $Fe(OH)_2$. In the heating period, the average crystallite size increased as the heating time elapsed similar to the case of β = 0.5. However, the particle size and crystallinity increased greatly with increasing in the heating time. This tendency was remarkable as β increased, and especially at higher β for longer heating times, the nanoparticles with a shape close to a regular octahedron, which is the crystal form of Fe_3O_4 [35], were obtained. The results suggest that the crystallization of Fe_3O_4 nanoparticles progresses in the heating period according to the different mechanisms depending on the initial $Fe(OH)_2$ content; the growth of crystallites remarkably occurs at higher β in the heating period, caused by the Schikorr reaction in which $Fe(OH)_2$ is transformed into Fe_3O_4 at higher temperatures according to the following reaction [36, 37].

$$3Fe(OH)_2 \rightarrow Fe_3O_4 + H_2 + 2H_2O \quad (5)$$

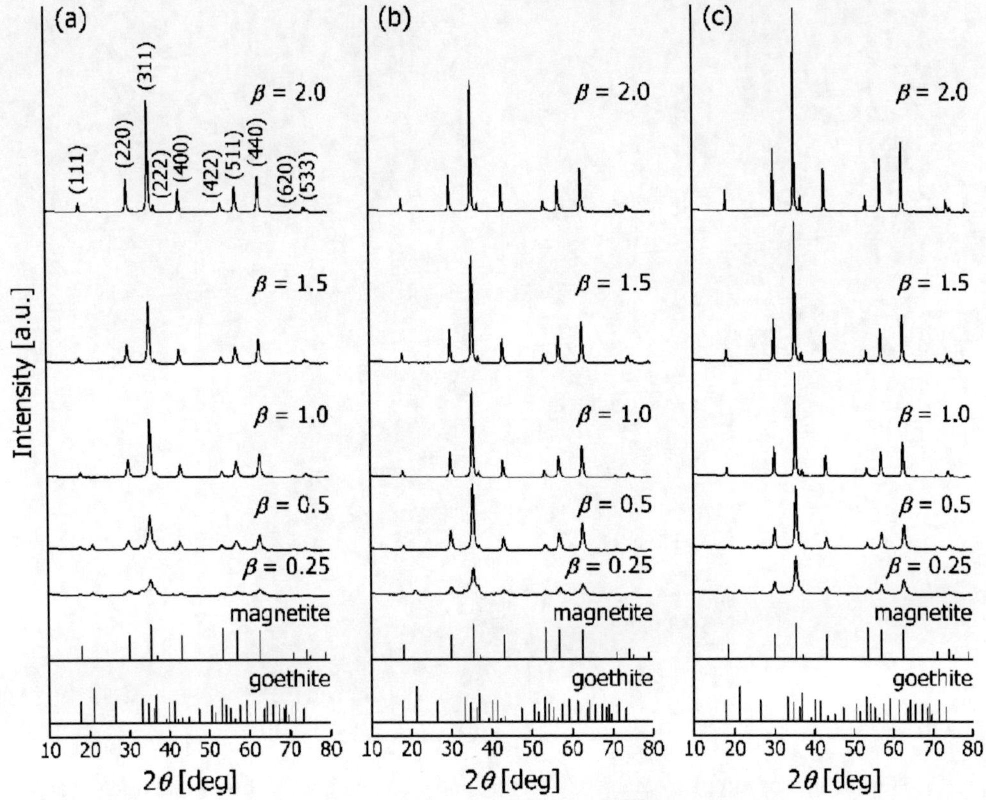

Figure 1. XRD patterns of samples prepared under various Fe^{2+}/Fe^{3+} molar ratios β (a) before and after heating for (b) 4 h and (c) 20 h. Solid bars indicate the standard patterns for magnetite (ASTM card #11-614) and goethite (ASTM card #3-249)

The rate of the Schikorr reaction is very slow as compared with that of the reaction expressed by Eq. (4). Therefore, in the presence of excess $Fe(OH)_2$ the crystal nucleation was completed rapidly at room temperature according to Eq. (4), and then using surplus $Fe(OH)_2$ the crystallites in the Fe_3O_4 nanoparticles grew up slowly under hydrothermal conditions according to Eq. (5) rather than their coalescence.

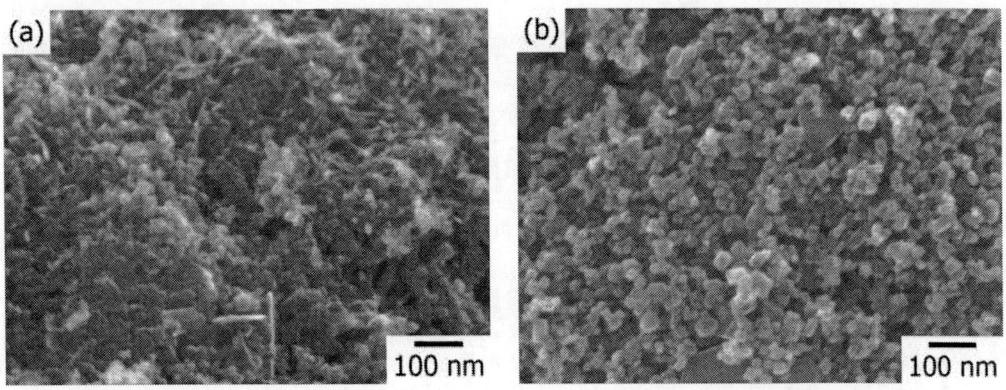

Figure 2. SEM images of samples obtained (a) before and (b) after heating for 20 h at $\beta = 0.5$

Figure 3. SEM images of samples obtained (a) before and after heating for (b) 4 h and (c) 20 h at β = 2.0

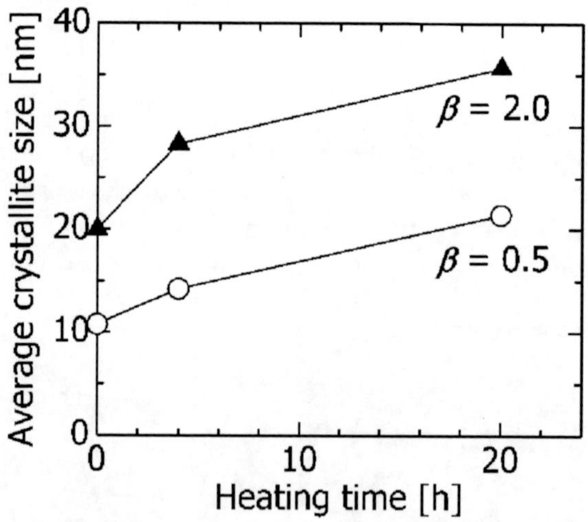

Figure 4. Change in average crystallite size of samples with heating time

The magnetic property of samples was analyzed using a superconducting quantum interference device (SQUID) magnetometer (Quantum Design model MPMS) at a magnetic field of 100 Oe in the temperature rage between 5 and 400 K for both field-cooling (FC) and zero-field-cooling (ZFC). The hysteretic cycles were also recorded at room temperature in the rage of magnetic field between –10000 and 10000 Oe. Figure 5 shows the magnetization-temperature (i.e., ZFC/FC) curves and the magnetization-magnetic field hysteresis of samples obtained after heating for 20 h. The Verwey transition and hysteresis curves were observed in both samples. The prepared Fe_3O_4 nanoparticles exhibited a ferromagnetic behavior with a high saturation magnetization of about 80 emu/g, which was higher than that of γ-Fe_2O_3 (56 emu/g) and, however, a little bit lower than that of the corresponding bulk (92 emu/g) because the particle size was smaller than that of bulk [38]; this result was similar to those reported by Yu et al. [35] and Zaitsev et al. [39]. Accordingly, it has been clarified that the Fe_3O_4 nanoparticles consist of almost the same magnetic phase [40] and have the similar magnetite properties regardless of β whereas the particle size and crystallinity change depending on β.

In summary, the hydrothermal synthesis of Fe_3O_4 nanoparticles were carried out without adding any oxidizing and reducing agents under various initial Fe^{2+}/Fe^{3+} molar ratios in order to clarify effects of the molar ratio on the reaction mechanism. When the molar ratio was 0.5 corresponding to the stoichiometric ratio in the formation reaction of Fe_3O_4 from $Fe(OH)_2$ and α-FeOOH, the nucleation of Fe_3O_4 crystals progressed rapidly in an initial stage of the hydrothermal synthesis, resulting in formation of the Fe_3O_4 nanoparticles having a smaller size and lower crystallinity. On the other hand, when the molar ratio was more than 1.0, the particle size and crystallinity increased with increasing in the molar ratio because the crystallites of Fe_3O_4 nanoparticles grew up slowly using surplus $Fe(OH)_2$ under hydrothermal conditions according to the Schikorr reaction. The Fe_3O_4 nanoparticles prepared under various molar ratios had good magnetic properties regardless of the molar ratio.

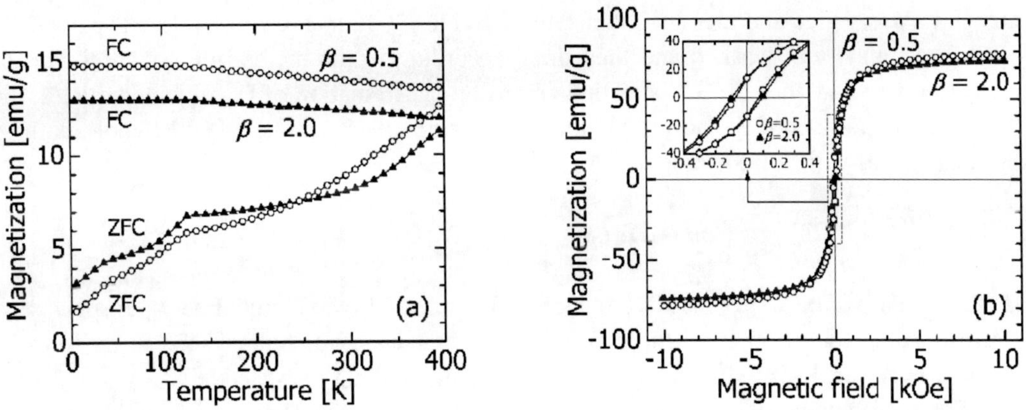

Figure 5. Magnetic properties of samples obtained after heating for 20 h: (a) magnetization-temperature curves and (b) magnetization-magnetic field curves

Size Control of Magnetite Nanoparticles in Hydrothermal Synthesis by Coexistence of Lactate and Sulfate Ions

In normal hydrothermal synthesis technique, a mixed aqueous solution of ferrous and ferric ions is used as the starting solution, and precipitates of $Fe(OH)_2$ and α-FeOOH are formed as the precursor via coprecipitation in an alkaline solution [31]. As mentioned above, the size of Fe_3O_4 nanoparticles varied depending on the Fe^{2+}/Fe^{3+} molar ratio. This implies that the particle size can also depend on the size of precursor $Fe(OH)_2$ and α-FeOOH particles. Therefore, in order to control the size of obtained Fe_3O_4 nanoparticles, the formation process of $Fe(OH)_2$ and α-FeOOH precipitates should be controlled. Several studies on control of the crystal growth of α-FeOOH were reported. For controlling the shape of α-FeOOH crystal, Kandori et al. [41, 42] used silicate, phosphoric and citrate ions as the anion coexisting with ferric ion in the solution. The anions formed iron complex with ferric ion and tended to adsorb onto the surface of formed α-FeOOH crystals. As a result, the formation and crystal growth of α-FeOOH were remarkably influenced by the anions. Ishikawa et al. [43] reported that coexisting carboxylate ions such as formate and oxalate ions have the effect to inhibit the crystal growth of α-FeOOH. These results suggest that the coexisting anion can contribute effectively to control the size of Fe_3O_4 nanoparticles. Therefore, lactate ion, which is carboxylate ion forming relatively stable complex with ferric ion, was used as the coexisting anion in the hydrothermal synthesis of Fe_3O_4 nanoparticles. The size control using the effects of coexisting anion was investigated, and the effects of lactate ion on the formation mechanism were analyzed.

In all the experiments, $FeCl_3{\cdot}6H_2O$ was used as the ferric ion source. As the ferrous ion source, the mixture of $FeSO_4{\cdot}7H_2O$ and ferrous lactate trihydrate ($Fe(lac)_2{\cdot}3H_2O$) was used, and the total molar amount of ferrous ion in the mixture was kept at 1 mmol. 2 mmol of $FeCl_3{\cdot}6H_2O$ and 1 mmol of the mixture of ferrous ion salts were dissolved in 40 ml of deionized and deoxygenated water in a stainless autoclave, in which the Fe^{2+}/Fe^{3+} molar ratio was the chemical stoichiometric ratio (= 0.5) in the formation reaction of Fe_3O_4. After aging for 10 min, 20 ml of 1.0 kmol/m^3 NaOH solution was dropped into the solution at the rate of 3.0 ml/min under vigorous stirring using a magnetic stirrer in an argon atmosphere. The autoclave was sealed and then maintained at 120°C for 20 h in an electric oven. The obtained black precipitate was washed and then dried according to the method described earlier. In order to investigate the role of coexisting anions on the formation of Fe_3O_4 nanoparticles, the initial mixing ratio m of $Fe(lac)_2$ to the total molar amount of ferrous ion (= 1 mmol), defined by Eq. (6), was varied in the range of $m = 0$ to 1.

$$m = [Fe(lac)_2]/\{[Fe(lac)_2]+[FeSO_4]\} \qquad (6)$$

$[Fe(lac)_2]$ and $[FeSO_4]$ represent the molar amount of $Fe(lac)_2$ and $FeSO_4$, respectively. Corresponding to $m = 0$ to 1, the concentrations of lactate and sulfate ions in the starting solution were varied from 0 to 33.3 mM and from 16.7 to 0 mM, respectively.

The morphology of samples was observed with the FE-SEM and a field emission transmission electron microscope (FE-TEM; EM-002BF, TOPCON). Figure 6 shows the SEM images of samples prepared by using either $FeSO_4$ or $Fe(lac)_2$ as the ferrous ion source, i.e. under the conditions of $m = 0$ and 1, respectively. TEM image of the sample of $m = 1$ is also shown in Figure 7. The size of Fe_3O_4 nanoparticles at $m = 0$ was about 20–40 nm. At $m =$ 1, however, the Fe_3O_4 nanoparticles with a size below 10 nm were obtained. The XRD patterns of samples obtained under various values of m are represented in Figure 8. The diffraction intensity peaks indicating the formation of other compounds were not observed regardless of m. In addition, the lattice constant was also determined from several diffraction angles showing high intensity peaks. The lattice constant of samples of $m = 0$ to 1 was between 8.38 and 8.42 Å, which was close to that of Fe_3O_4 (8.396 Å) rather than that of γ-Fe_2O_3 (8.345 Å). The results support that the samples of $m = 0$ to 1 had a single Fe_3O_4 phase.

The average crystallite size was calculated from the FWHM of the Fe_3O_4 (311) diffraction peak at $2\theta \approx 35.5°$ using the Scherrer's formula. The hydrodynamic particle size distribution of samples dispersed in deionized water was measured by dynamic light scattering (DLS-700, Otsuka Electronics) and the median size (number basis) was determined. Figure 9 indicates the average crystallite size and median particle size of samples. Both the crystallite size and the median size decreased with increasing in *m*. This tendency was remarkable within a range of lower *m*. The result reveals that coexistence of a small amount of lactate ion greatly decreased them and the size of obtained Fe_3O_4 nanoparticles was remarkably influenced by formation of the precursor α-FeOOH and $Fe(OH)_2$ precipitates. This suggests that the molar amount of lactate ion can be the operating factor to control the size of Fe_3O_4 nanoparticles.

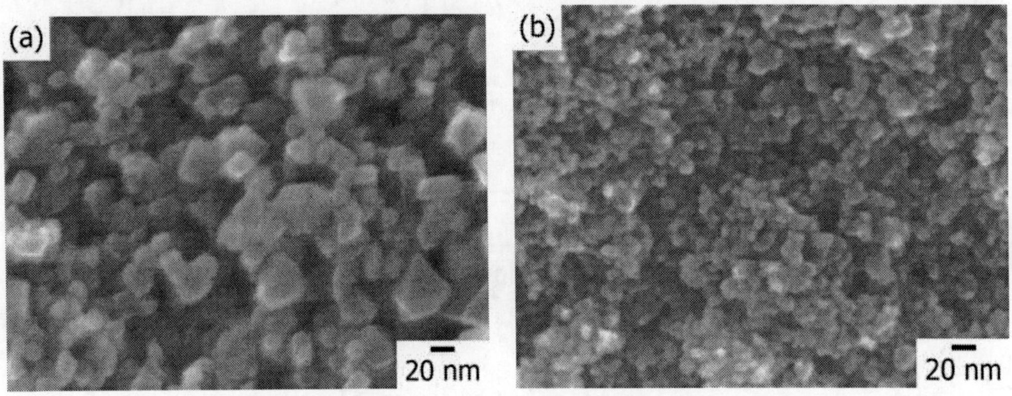

Figure 6. SEM images of samples prepared at m = (a) 0 and (b) 1

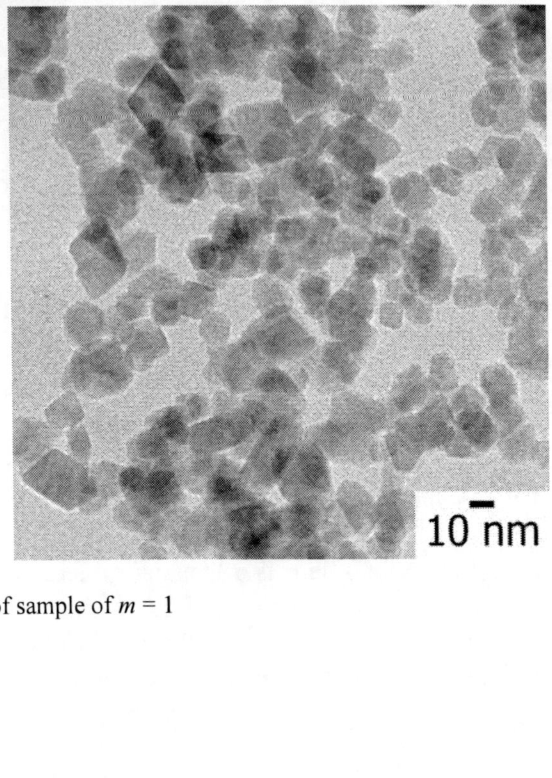

Figure 7. TEM image of sample of m = 1

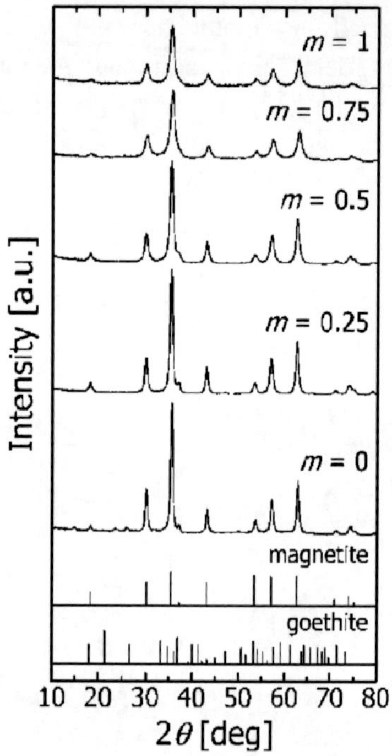

Figure 8. XRD patterns of samples prepared under various addition amounts of $Fe(lac)_2$

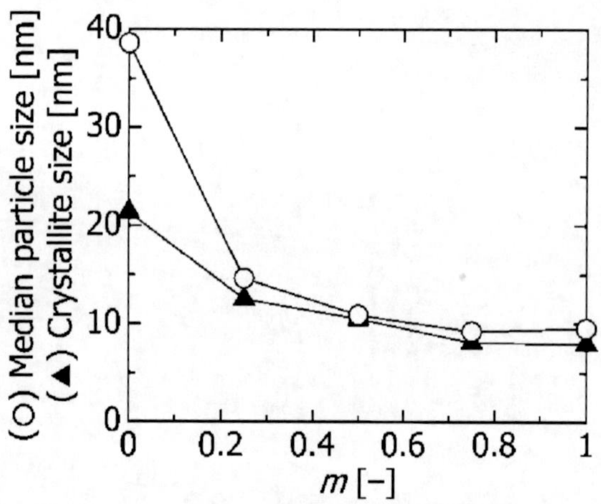

Figure 9. Effect of addition amount of $Fe(lac)_2$ on average crystallite size and median particle size of samples

For evaluating the relative stability of ferric complexes with the anions, the UV analysis [44] for the 0.2 mM $FeCl_3$ solutions containing 0.2 mM of either Na_2SO_4 or Na(lac) was carried out at room temperature using a spectrophotometer (Ubest V-530, JASCO). Figure 10 represents the UV spectra of ferric ion solutions containing sulfate, chloride and lactate ions. The UV spectrum of ferric ion solution containing both sulfate and chloride ions had a strong absorption peak at a wavelength of approximately 290 nm, which was similar to that of the ferric chloride solution without adding any sodium salts. When lactate ion coexisted with chloride ion, however, the UV spectrum greatly changed, and the strong absorption peak was observed at a wavelength of 347 nm because ferric ion tends to form relatively stable trinuclear complexes with lactate ion [45]. It was confirmed that formation of the ferric complex with chloride ion was not influenced by sulfate ion and lactate ion formed more easily the complex with ferric ion than chloride ion. This decreased ionic strength of the solution and promoted that the formation of crystalline nuclei occurred frequently as compared with the particle growth, resulting in the formation of finer precipitates.

In order to investigate the dispersion stability of Fe_3O_4 nanoparticles in the solution containing the anions, a small amount of Fe_3O_4 nanoparticles (prepared by the hydrothermal synthesis using 1 mmol of $FeCl_2$ and 2 mmol of $FeCl_3$) were dispersed in the solution containing a proper amount of sodium salts of lactate and sulfate ions. The addition molar amount of sodium salts was expressed in terms of m' defined by Eq. (7) and varied in the range of $m' = 0$ to 1 corresponding to $m = 0$ to 1.

$$m' = [\mathrm{lac}^-]/\{[\mathrm{lac}^-]+2[\mathrm{SO_4}^{2-}]\} \qquad (7)$$

$[\mathrm{lac}^-]$ and $[\mathrm{SO_4}^{2-}]$ represent the concentrations of lactate and sulfate ions in the solution, respectively. The zeta potential was measured with a zeta potential analyzer (Zetasizer Nano ZS, Malvern Instruments). Figure 11 shows change in the zeta potential of Fe_3O_4 nanoparticles with m'. The zeta potential was shifted to a more negative range with increasing in the lactate ion concentration. It was found that the presence of lactate ion was effective to decrease the surface potential of Fe_3O_4 nanoparticles and to increase the dispersion stability. This result implies that the aggregation of Fe_3O_4 particles was inhibited during both the precipitate formation and the hydrothermal treatment, resulting in the formation of finer Fe_3O_4. As can be seen in Figure 9, the crystallite size almost agreed with the median size at $m \geq 0.5$. In addition, typical HR-TEM image of the sample of $m = 1$ is shown in Figure 12 as an example. These results suggest that the Fe_3O_4 nanoparticles having a single-crystal structure (i.e., high crystallinity) were formed at larger m due to inhibiting the aggregation. On the contrary, the crystallite size at $m = 0$ was 20 nm, whereas the median size was over 30 nm. This result indicates that at $m = 0$ the crystal growth occurred accompanied with aggregation of crystallites; on the other hand, each Fe_3O_4 nanoparticle obtained at $m \geq 0.5$ was almost constituted by single crystallite. Therefore, ultrafine Fe_3O_4 nanoparticles were formed under the conditions that crystallites kept good dispersion inhibiting the aggregation.

Figure 13 shows magnetization-magnetic field hysteresis of the samples of $m = 0$ and 1 measured at room temperature. The coercitivity and saturation magnetization of samples obtained at various values of m are also indicated in Figure 14. The obtained Fe_3O_4 nanoparticles were found to exhibit a ferromagnetic behavior. The coercitivity decreased with increasing in m, and the samples at larger m showed superparamagnetism. The saturation magnetization also decreased to 64.1 emu/g at $m = 1$ which was higher than that of γ-Fe_2O_3 (56 emu/g); however, it was lower than that of the corresponding bulk (92 emu/g) due to their smaller size [38].

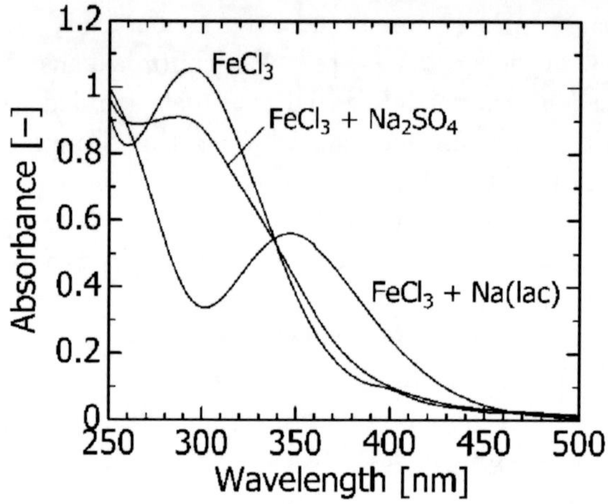

Figure 10. UV spectra of solutions of $FeCl_3$, $FeCl_3$+Na_2SO_4 and $FeCl_3$+Na(lac)

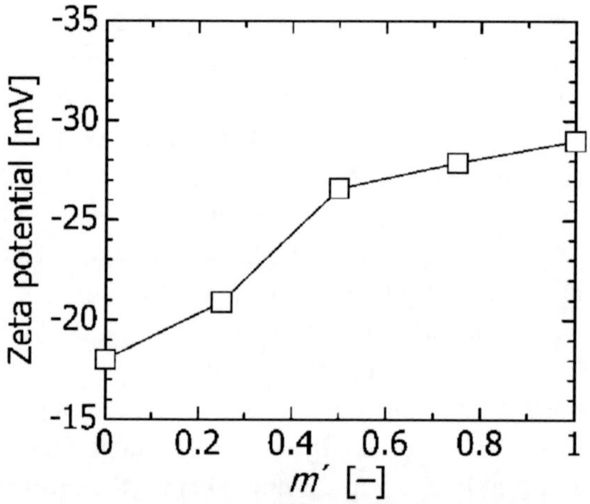

Figure 11. Change in zeta potential of Fe_3O_4 nanoparticles with concentration of lactate ion

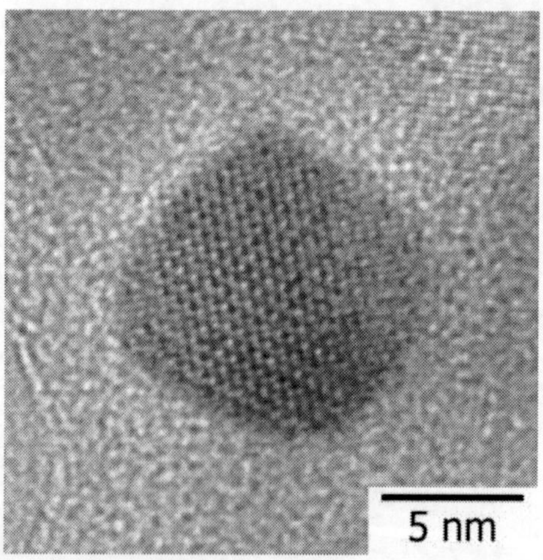

Figure 12. HR-TEM image of sample of m =1

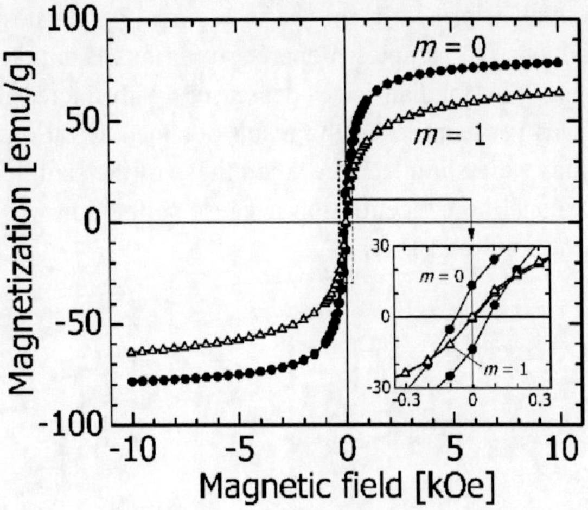

Figure 13. Magnetization-magnetic field curves of samples of $m = 0$ and 1

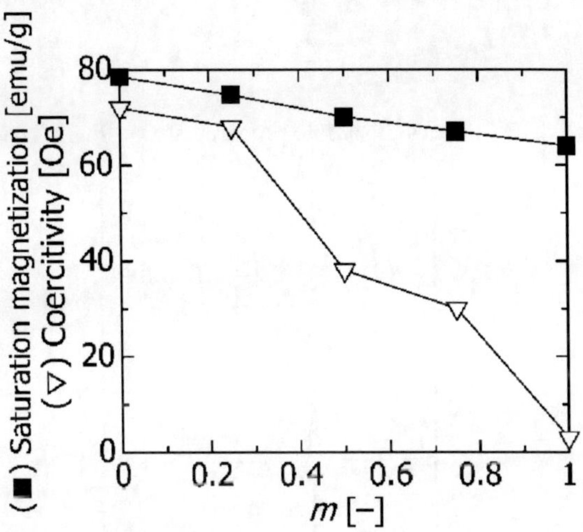

Figure 14. Effect of addition amount of $Fe(lac)_2$ on coercitivity and saturation magnetization of samples

Various molar amounts of sodium lactate (Na(lac)) and sodium sulfate (Na_2SO_4) were also added to deionized and deoxygenated water before the iron salts were dissolved. The addition molar amount of Na(lac) was varied from 0 to 2 mmol in the solution containing 1 mmol of $FeSO_4$ and 2 mmol of $FeCl_3$ (i.e., $m = 0$), in which the lactate ion concentration corresponded to that of the solution of $m = 0$ to 1. On the other hand, the amount of Na_2SO_4 was varied from 0 to 1 mmol in the solution containing 1 mmol of $Fe(lac)_2$ and 2 mmol of $FeCl_3$ (i.e., $m = 1$), in which the sulfate ion concentration corresponded to that of the solution of $m = 1$ to 0. Figure 15 shows SEM and TEM images of the sample prepared by using the starting solution containing 1 mmol of $FeSO_4$, 2 mmol of $FeCl_3$ and 2 mmol of Na(lac), in which the concentrations of sulfate ion (16.7 mM) and lactate ion (33.3 mM) corresponded to those at $m = 0$ and 1, respectively. The obtained Fe_3O_4 nanoparticles were smaller than the sample obtained at $m = 0$ (i.e., under the same sulfate ion concentration without containing lactate ion) as shown in Figure 6a. Figure 16 represents the XRD patterns of samples prepared by using the staring solution of $m = 0$ in which various amounts of Na(lac) were added. The single Fe_3O_4 phase was also formed in this case. Figure 17 indicates the average crystallite size and median particle size of samples prepared at various lactate ion concentrations. Both the crystallite size and the median size decreased with increasing in the lactate ion concentration, which was very similar to the result obtained at various m as shown in Figure 9. However, these values were slightly larger than those of the samples obtained at a given m corresponding to the lactate ion concentration because sulfate ion was contained at relatively high concentration in the solution.

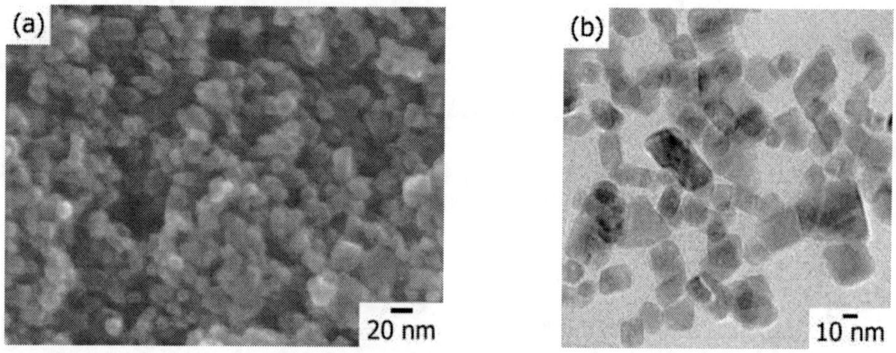

Figure 15. (a) SEM and (b) TEM images of sample prepared at $m = 0$ with Na(lac) of 33.3 mM

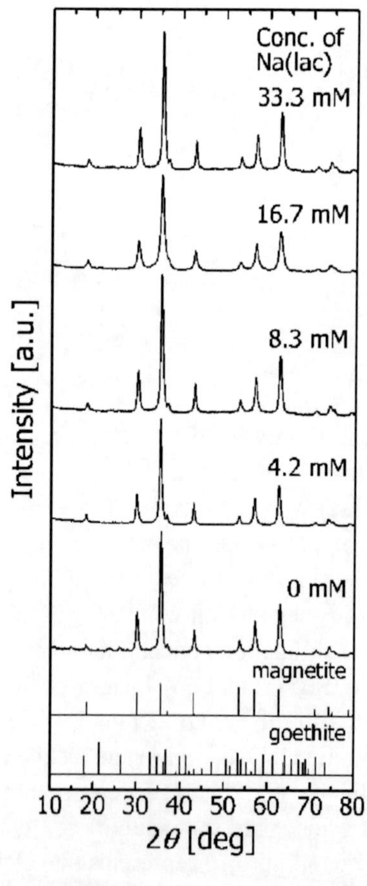

Figure 16. XRD patterns of samples prepared at various concentrations of Na(lac)

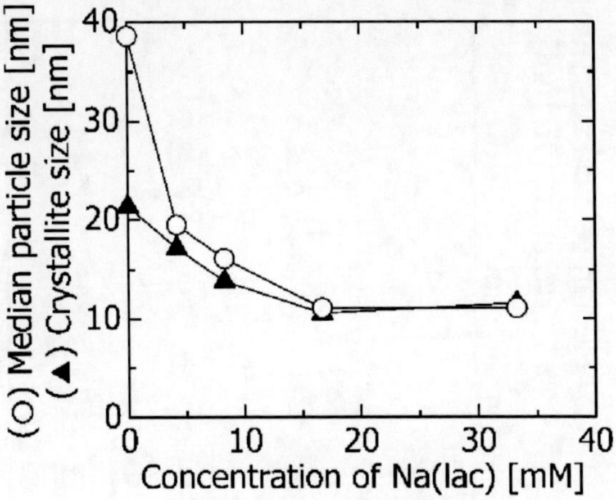

Figure 17. Effect of lactate ion concentration on average crystallite size and median particle size of samples

Figure 18 shows SEM image of the sample prepared by using the starting solution containing 1 mmol of $Fe(lac)_2$, 2 mmol of $FeCl_3$ and 1 mmol of Na_2SO_4, in which the concentrations of lactate ion and sulfate ion corresponded to those at m = 1 and 0, respectively. The obtained Fe_3O_4 nanoparticles were larger than the sample obtained at m = 1 (i.e., under the same lactate ion concentration without containing sulfate ion) as shown in Figure 6b. Figure 19 indicates the average crystallite size and median particle size of samples prepared at various sulfate ion concentrations. Both the crystallite size and the median size increased with increasing in the sulfate ion concentration and were smaller than those at a given m corresponding to the sulfate ion concentration as shown in Figure 9 since lactate ion coexisted with sulfate ion. Consequently, sulfate ion has the effect to increase the size of Fe_3O_4 nanoparticles. On the other hand, lactate ion effectively contributes to the size reduction of Fe_3O_4 nanoparticles. When sulfate ion existed in the solution, the precipitates tended to aggregate due to increasing in the surface potential, resulting in the particle growth, and the effects of lactate ion were reduced.

Figure 18. SEM image of sample prepared at m = 1 with Na_2SO_4 of 16.7 mM

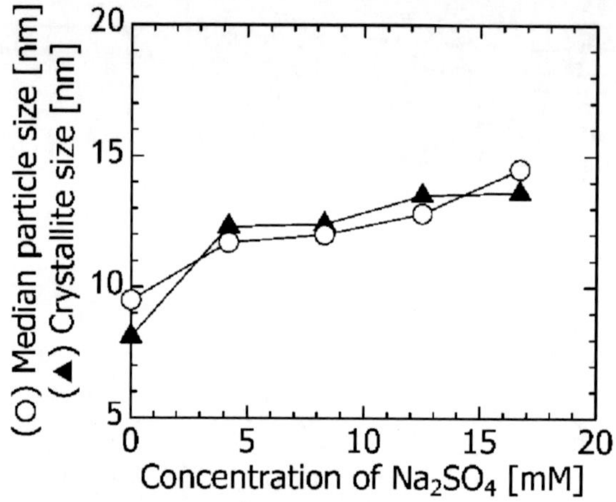

Figure 19. Effect of sulfate ion concentration on average crystallite size and median particle size of samples

In summery, the hydrothermal synthesis of Fe_3O_4 nanoparticles was carried out using the starting solution containing lactate and sulfate ions at various concentrations in order to control the particle size. In the hydrothermal synthesis via coprecipitation, the size of obtained Fe_3O_4 nanoparticles varied depending strongly on both the stability of iron complexes with the coexisting anions in the solution and the surface potential of anions-adsorbed precipitates. It was clarified that lactate ion, which forms relatively stable complex with ferric ion and tends to adsorb onto the surface of formed iron hydroxide crystals, has the effect to decrease the size of Fe_3O_4 nanoparticles. On the other hand, sulfate ion has the effect to increase the particle size because it increases the surface potential of formed Fe_3O_4 nanoparticles and promotes the aggregation (i.e., the particle growth). Under the experimental conditions in this work, the median particle size could be varied from 9.5 to 38.6 nm. Therefore, it has been demonstrated that the particle size is controlled by means of the coexistence effects of lactate and sulfate ions.

Hydrothermal Synthesis via Sequential Formation of Iron Hydroxide Precipitates

In the synthesis of iron oxides, the formation process of precursor iron hydroxides plays an important role and the particle size of obtained iron oxides depends strongly on that of iron hydroxides. For example, Žic et al. [46] synthesized hematite (α-Fe_2O_3) particles with different morphology and size by controlling the formation process of akaganeite (β-FeOOH), which is the precursor of α-Fe_2O_3, using phosphates. As mentioned above, it was confirmed that the formation mechanism of iron hydroxides precipitates changed depending on the Fe^{2+}/Fe^{3+} molar ratio in the starting solution, resulting in variation of the size of obtained Fe_3O_4 nanoparticles. Therefore, in the hydrothermal synthesis of Fe_3O_4 nanoparticles via the coprecipitation route, it is believed that the size of Fe_3O_4 nanoparticles also depends on that of the $Fe(OH)_2$-α-FeOOH composites. Thus, in order to prepare ultrafine Fe_3O_4 nanoparticles, the size of composites should be controlled to be ultrafine by controlling the formation process of $Fe(OH)_2$ and α-FeOOH. In the coprecipitation methods, however, it is difficult to precisely control the formation process because $Fe(OH)_2$ and α-FeOOH precipitate together when pH of the acidic solution containing ferric and ferrous ions is increased by adding an alkaline solution.

Therefore, a novel non-coprecipitation method for preparing ultrafine ferromagnetic Fe_3O_4 nanoparticles in water system without using any additives such as surfactant and oxidizing and reducing agents has been developed, in which ferrous and ferric hydroxides are not coprecipitated but formed separately and sequentially; one iron ion solution is added to the suspension containing another iron hydroxide formed beforehand in an alkaline solution, resulting in formation of Fe_3O_4 nanoparticles. Subsequently, the suspension is subjected to the hydrothermal treatment for improving the crystallinity. In this sequential precipitates formation (SPF) method, the supersaturation of iron hydroxides in the suspension is varied by adjusting preparation conditions of the suspension, and the size of obtained Fe_3O_4 nanoparticles is controlled by the supersaturation. The formation mechanism of iron hydroxides particles was analyzed based on the morphology of obtained Fe_3O_4 nanoparticles.

Figure 20 illustrates the synthesis processes of Fe_3O_4 nanoparticles in the SPF method. The preparation conditions of precipitates are summarized in Table 1. The concentration of ferrous and ferric ions in the starting solution was the same as that in the synthesis experiment which Wan et al. [47] performed. In Process-1 (P1), 2.0 mmol of $FeCl_3 \cdot 6H_2O$ was dissolved in 26.7 ml of deionized and deoxygenated water in a stainless steel autoclave, and then 20 ml of 1.0 $kmol/m^3$ NaOH solution was dropped slowly at a constant addition rate of 3.0 ml/min or poured rapidly (added in one portion instantaneously) into the prepared ferric ion solution under vigorous stirring using a magnetic stirrer in an argon atmosphere, resulting in formation of the α-FeOOH particles suspension according to Eqs. (2) and (3). Next, 1.0 mmol of $FeSO_4 \cdot 7H_2O$ was dissolved in 13.3 ml of deionized and deoxygenated water in a beaker, and the ferrous ion solution thus prepared was dropped slowly at a constant addition rate of 3.0 ml/min or poured rapidly into the α-FeOOH particles suspension under vigorous stirring using a magnetic stirrer in an argon atmosphere. This resulted in formation of the $Fe(OH)_2$ particles in the α-FeOOH particles suspension according to Eq. (1). Finally, the suspension containing the α-FeOOH and $Fe(OH)_2$ particles was obtained as the starting suspension for the hydrothermal synthesis of Fe_3O_4 nanoparticles. On the other hand, in Process-2 (P2), the addition sequence of $FeCl_3$ and $FeSO_4$ was replaced in the procedure as shown in Figure 20; the ferric ion solution was added to the $Fe(OH)_2$ particles suspension. The stainless steel autoclave with the resulting suspension was sealed and maintained at 120°C for 20 h in an electric oven. Heating of the suspension caused the solid phase reaction between $Fe(OH)_2$ and α-FeOOH, resulting in formation of the Fe_3O_4 nanoparticles according to Eq. (4). After the hydrothermal treatment, the stainless steel autoclave was cooled down to room temperature. The obtained nanoparticles were washed and then dried as described earlier. In addition, in order to investigate effect of the formation process of iron hydroxides precipitates on the size of Fe_3O_4 nanoparticles, the suspension prepared by conventional coprecipitation of α-FeOOH and $Fe(OH)_2$ was also used as the resulting suspension (Process-3: P3). In all the processes, the Fe^{2+}/Fe^{3+} molar ratio in the resulting suspensions was 0.5 which was the chemical stoichiometric ratio in the formation reaction of Fe_3O_4 expressed by Eq. (4), and the pH value of resulting suspensions was more than 12.

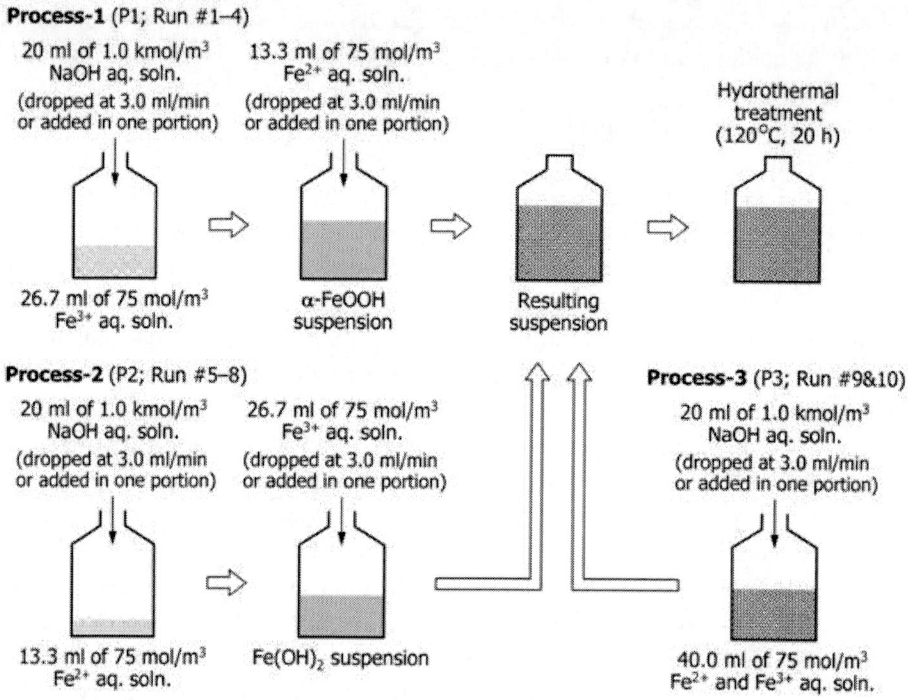

Figure 20. Schematic of formation processes of iron hydroxides precipitates

Table 1. Preparation conditions of starting suspension

Run #	Process	Addition of aqueous solution		
		NaOH	Fe^{2+}	Fe^{3+}
1	P1	dropwise	dropwise	–
2	P1	dropwise	in one portion	–
3	P1	in one portion	dropwise	–
4	P1	in one portion	in one portion	–
5	P2	dropwise	–	dropwise
6	P2	dropwise	–	in one portion
7	P2	in one portion	–	dropwise
8	P2	in one portion	–	in one portion
9	P3	dropwise	–	–
10	P3	in one portion	–	–

Figure 21 shows the SEM images of samples prepared in P1. The samples contained coarse angular-shaped particles of approximately 100 nm and had a broad (bimodal) size distribution regardless of the addition method of NaOH solution and ferrous ion solution. However, when the NaOH solution was poured rapidly into the ferric ion solution for preparing the α-FeOOH suspension, i.e. in Run #3 and #4, the amount of fine Fe_3O_4 nanoparticles tended to increase.

Figure 21. SEM images of samples prepared in Process-1: Run (a) #1, (b) #2, (c) #3 and (d) #4

Figure 22 shows the SEM images of samples prepared in P2. Their morphology differed remarkably from that of the samples prepared in P1. The amount of coarse Fe_3O_4 nanoparticles considerably decreased in all the samples of P2, and the particle size was relatively uniform. Especially, when the ferrous ion solution was poured rapidly into the α-FeOOH suspension, i.e. in Run #6 and #8, the Fe_3O_4 nanoparticles became extremely fine. Accordingly, the results obtained from P1 and P2 reveal that coarse Fe_3O_4 nanoparticles are obtained when first α-FeOOH is formed, following formation of $Fe(OH)_2$; on the contrary, when the addition sequence is reversed, the amount of coarse Fe_3O_4 nanoparticles decreases and the size distribution of obtained Fe_3O_4 nanoparticles relatively narrows.

For preparing uniform ultrafine Fe_3O_4 nanoparticles such as the samples of Run #6 and #8 in P2, it is needed that the precursor ($Fe(OH)_2$-α-FeOOH composite) is also ultrafine. Thus, it is believed that under the conditions of Run #6 and #8 ultrafine particles of $Fe(OH)_2$ and α-FeOOH were surely formed. In the formation of particles, supersaturation of the solution plays an important role and the size of particles depends strongly on the supersaturation. Under the conditions of Run #6 and #8, the ferric ion solution was poured rapidly into the $Fe(OH)_2$ particles suspension, resulting in increase of the supersaturation of $Fe(OH)_3$. This promotes the homogeneous nucleation of $Fe(OH)_3$, leading to formation of ultrafine α-FeOOH particles. However, the NaOH solution was slowly dropped and poured rapidly into the ferrous ion solution; the supersaturation of $Fe(OH)_2$ was expected to decrease in Run #6 as compared with the case of Run #8. Thus, the fact that the ultrafine Fe_3O_4 nanoparticles were obtained under both conditions suggests that ultrafine $Fe(OH)_2$ particles tend to form regardless of the addition method of NaOH solution, i.e. the supersaturation of $Fe(OH)_2$.

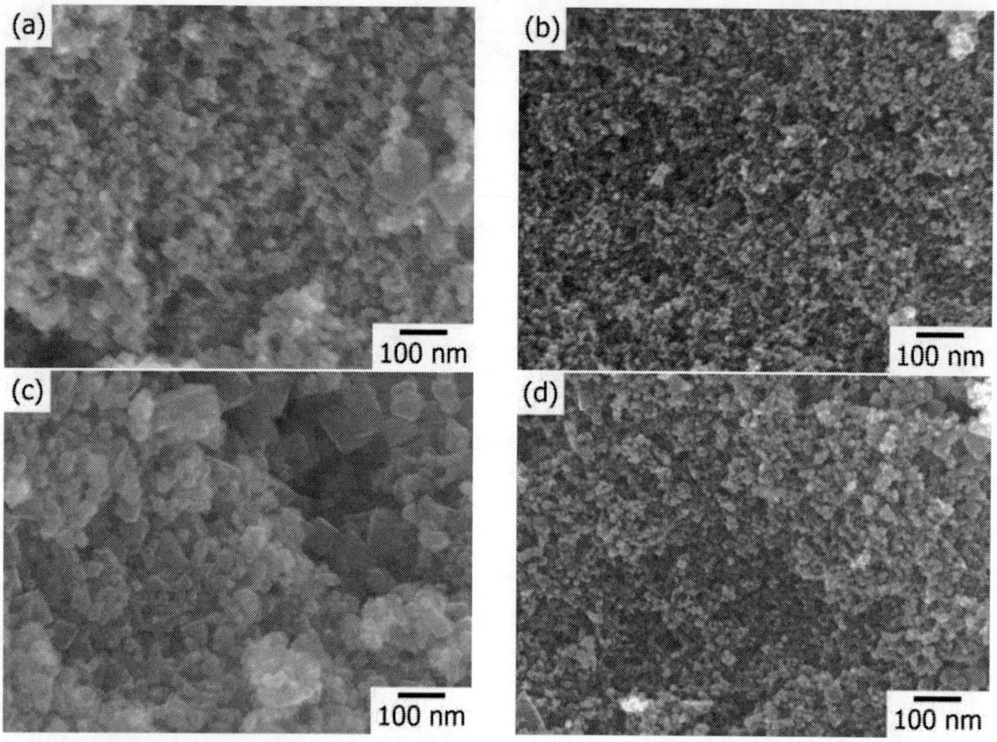

Figure 22. SEM images of samples prepared in Process-2: Run (a) #5, (b) #6, (c) #7 and (d) #8

On the other hand, in P1, coarse Fe_3O_4 nanoparticles coexisted with fine ones regardless of the addition method of NaOH solution and ferrous ion solution, although in Run #3 and #4 the NaOH solution was poured rapidly into the ferric ion solution, which increased the supersaturation of $Fe(OH)_3$ and made the α-FeOOH particles fine. Therefore, when $Fe(OH)_2$ particles form in the α-FeOOH particles suspension, a part of $Fe(OH)_2$ particles can cause the heterogeneous nucleation on the surface of α-FeOOH particles, and the $Fe(OH)_2$-α-FeOOH composite thus formed can tend to aggregate and cause the particle growth in the hydrothermal treatment.

Figure 23 shows the SEM images of samples prepared in P3. In P3, i.e. a conventional hydrothermal synthesis method, the size of obtained samples was uniform while relatively coarse particles of approximately 30 nm formed regardless of the addition method of NaOH solution as compared with those of Run #6 and #8, because the $Fe(OH)_2$-α-FeOOH composite formed via coprecipitation of $Fe(OH)_2$ and $Fe(OH)_3$.

Figure 24 indicates XRD pattern of all the samples. The XRD analysis showed that the samples had a single Fe_3O_4 phase although the crystallinity varied depending on the particle size. The lattice constant was determined from several diffraction angles showing high intensity peaks. The size distribution (number basis) was measured by the DLS particle size analyzer and the median diameter was determined. Table 2 summarizes the properties of representative samples prepared by means of the SPF method (Run #6) and the conventional method (Run #9). The lattice constants of samples of Run #6 and #9 were 8.390 and 8.372 Å, respectively, and both were close to the standard value of Fe_3O_4 (= 8.396 Å) compared to that of γ-Fe_2O_3 (= 8.345 Å). This result also supports that the samples had a single Fe_3O_4 phase.

Figure 23. SEM images of samples prepared in Process-3: Run (a) #9 and (b) #10

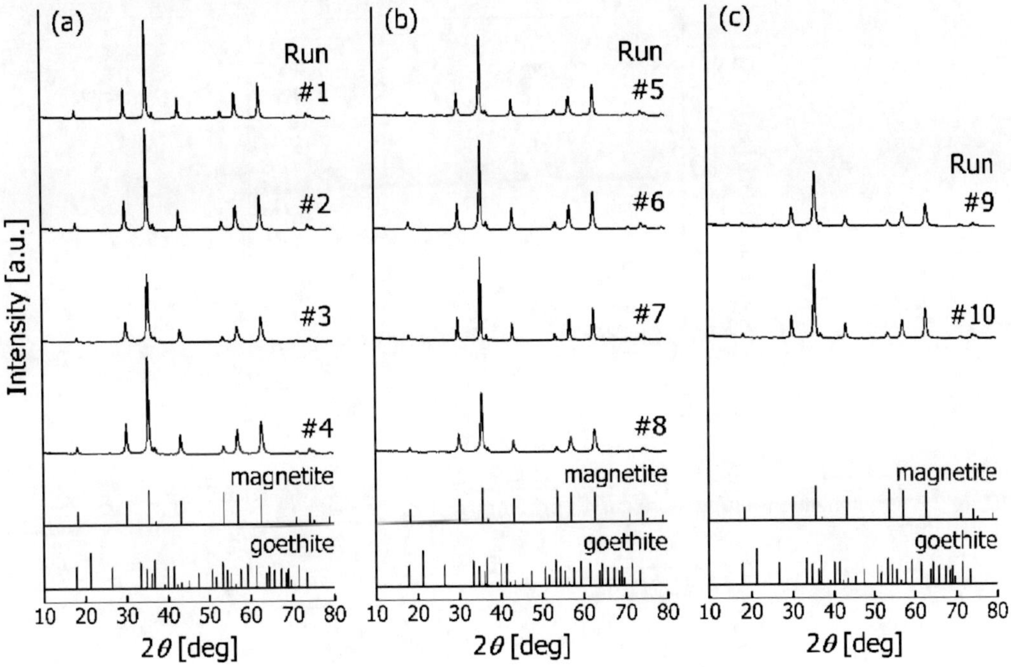

Figure 24. XRD patterns of samples prepared in (a) P1, (b) P2 and (c) P3

Table 2. Properties of samples of Run #6 and #9

Run #	Process	Median diameter [nm]	Lattice constant [Å]	Saturation magnetization [emu/g]	Coercivity [Oe]
6	P2	8.5	8.390	68.6	70
9	P3	38.6	8.372	78.5	73

Figures 25 and 26 show the magnetization-temperature (i.e. ZFC/FC) curves measured at a magnetic field of 100 Oe and the magnetization-magnetic field hysteresis measured at room temperature. The Verwey transition and hysteresis curves were observed in both samples. The samples exhibited a ferromagnetic behavior with a high saturation magnetization but the smaller particle size resulted in decrease in the value similar to the cases mentioned earlier.

Consequently, it was confirmed that the SPF method provides ferromagnetic Fe_3O_4 nanoparticles with a smaller size than the conventional method.

In summery, a novel hydrothermal synthesis method for preparing ultrafine Fe_3O_4 nanoparticles has been developed, in which the iron hydroxides particles are not coprecipitated but sequentially formed. In the sequential precipitates formation (SPF) method, both the formation sequence of $Fe(OH)_2$ and $Fe(OH)_3$ particles and the supersaturation of $Fe(OH)_3$ were essential to prepare ultrafine Fe_3O_4 nanoparticles without coexisting coarse ones; the size of obtained Fe_3O_4 nanoparticles depended strongly on the formation of $Fe(OH)_3$ particles rather than that of $Fe(OH)_2$ particles and for preventing the formation of coarse Fe_3O_4 nanoparticles it was effective to increase the supersaturation of $Fe(OH)_3$ in the $Fe(OH)_2$ particles suspension. It was confirmed that the ferromagnetic Fe_3O_4 nanoparticles of 8.5 nm could be prepared easily without using any additives by means of the SPF method.

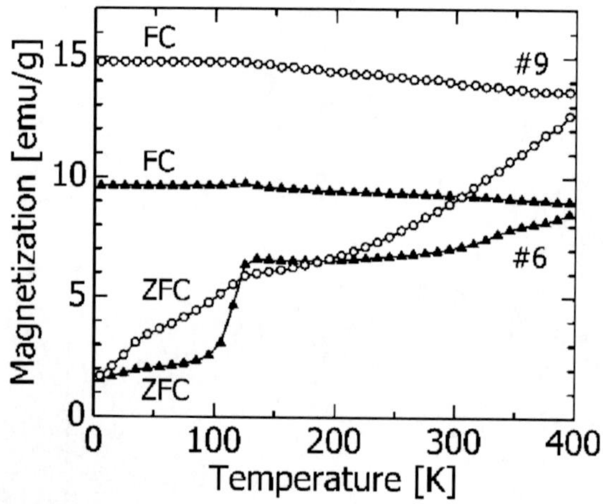

Figure 25. Magnetization-temperature curves of samples of Run #6 and #9

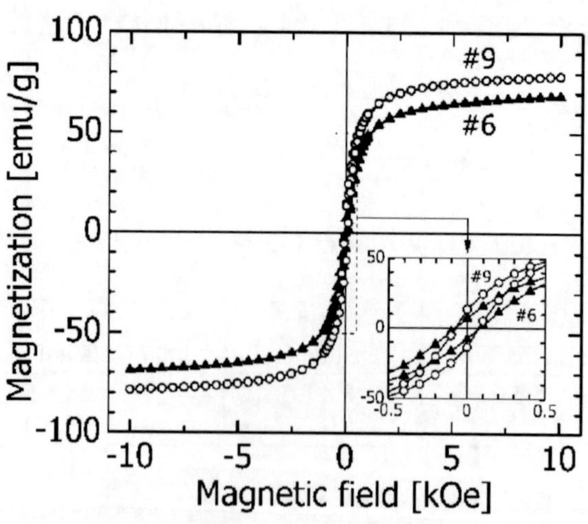

Figure 26. Magnetization-magnetic field curves of samples of Run #6 and #9

Novel Synthesis Method Using Mechanochemical Effect

Novel Mechanochemical Process for Synthesis of Magnetite Nanoparticles via Coprecipitation

As mentioned above, it is difficult to prepare simply ultrafine Fe_3O_4 nanoparticles with high crystallinity and excellent magnetic properties using conventional coprecipitation methods. In order to solve this problem, we have developed a novel coprecipitation method using a high mechanical energy field as the synthesis reaction system. In this method, a cooled tumbling ball mill is used as the reaction field for inhibiting the particle growth caused by heat energy [48]. In addition, any additives such as surfactants and oxidizing and reducing agents are not required. Both the formation reaction and the crystallization can be progressed by using the mechanical energy (i.e., the mechanochemical effect) generated by collision of ball media instead of heat energy.

The synthesis method is as follows. The concentration of ferrous and ferric ions in the starting solution was the same as that in the hydrothermal synthesis experiments using the SPF method described earlier. 1.5 mmol of $FeSO_4 \cdot 7H_2O$ and 3.0 mmol of $FeCl_3 \cdot 6H_2O$ were dissolved in 60 ml of deionized and deoxygenated water in a beaker. 30 ml of 1.0 kmol/m^3 NaOH solution was added into the solution at a constant addition rate under vigorous stirring using a magnetic stirrer in an argon atmosphere, and then the dark brown suspension containing $Fe(OH)_2$ and α-FeOOH was obtained. The pH value was higher than 12. During preparation of the starting suspension, the suspension temperature was kept below 5°C in order to inhibit progress of the solid phase reaction between $Fe(OH)_2$ and α-FeOOH, forming Fe_3O_4. The starting suspension thus prepared was poured into a milling vessel with an inner diameter of 90 mm and a capacity of 500 ml, made of stainless steel. Stainless steel balls with a diameter of 3.2 mm were used as the milling media, and the charged volume of balls containing the void formed among them was 40% of the vessel capacity, as illustrated in Figure 27. After replacement of air with argon in the milling vessel, the milling vessel was sealed. In order to promote the reaction between $Fe(OH)_2$ and α-FeOOH, the ball-milling treatment was then carried out by rotating the milling vessel at the rotational speed of 35 to 140 rpm (corresponding to 17% to 69% of the critical rotational speed (= 202 rpm) determined experimentally based on the behavior of balls containing the starting suspension) for a given time. During the ball-milling treatment, the milling vessel was cooled from its outside in a water bath. Temperature of the water bath was kept at 1.0±0.1°C, and inside temperature of the milling vessel was 1.6°C to 1.7°C regardless of the rotational speed; this means that the milling vessel was cooled thoroughly. After the ball-milling treatment, the obtained precipitate was washed and then dried at 30°C under vacuum overnight. In order to investigate the mechanochemical effect on the Fe_3O_4 formation reaction, the synthesis experiments without ball-milling were also carried out as follows: the starting suspension was sealed in a stainless steel bottle with a capacity of 200 ml, and then it was placed in the water bath kept at 1.6°C for 12 h under static condition. The sample thus obtained was compared with those from the milling treatments.

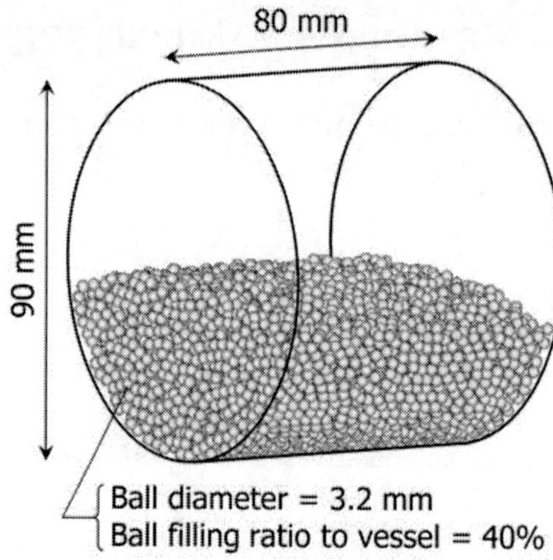

Figure 27. Schematic illustration of tumbling ball mill used in this work

Figure 28 shows the XRD pattern of samples. At the coprecipitation of $Fe(OH)_2$ and α-FeOOH, a part of them took place the solid phase reaction forming Fe_3O_4 even under cooling, as can be seen in Figure 28a. Under the static condition without ball-milling, the solid phase reaction hardly proceeded any more as shown in Figure 28b. However, when the ball-milling treatment was carried out, the solid phase reaction proceeded by the mechanical mean, i.e. the mechanochemical effect. The reaction rate depended strongly on the rotational speed of milling vessel. The reaction was almost completed for 12 h at 35 rpm, for 9 h at 70 rpm and for 6 h at 140 rpm. In the samples obtained under these conditions, the typical diffraction angles showing relatively high peaks coincided with those of Fe_3O_4.

The hydrodynamic size of samples was measured by DLS for the sample-redispersed aqueous suspension containing a small amount of sodium dodecyl sulfate, and the median diameter (number basis) was determined from the obtained size distribution. The chemical component was determined by means of an energy dispersive X-ray spectrometer (EDS; JED-2300F, JEOL) equipped with the FE-SEM. The average crystallite size was calculated from the FWHM of the Fe_3O_4 (311) diffraction peak at $2\theta \approx 35.5°$ using the Scherrer's formula. The lattice constant was also determined from several diffraction angles showing high intensity peaks. The magnetic property (magnetization-magnetic field hysteretic cycle) was analyzed using the SQUID magnetometer at room temperature. Table 3 summarizes the properties. The lattice constant was between 8.374 and 8.395 Å and close to the standard value of Fe_3O_4 (= 8.396 Å) rather than γ-Fe_2O_3 (= 8.345 Å). These results support that the samples had a single Fe_3O_4 phase regardless of the milling conditions.

Figure 29 shows SEM image of the representative samples. As can be seen in Figure 29 and Table 3, the Fe_3O_4 nanoparticles with similar properties were obtained regardless of the rotational speed. The median size agreed with the average crystallite size; this means that the crystallinity of samples was high. The samples had a low coercivity, showing superparamagnetism. The saturation magnetization was a little lower than that of the corresponding bulk (= 92 emu/g) because of the smaller size [38]. The sample contamination caused by the wear of milling vessel and balls had a tendency to increase at higher rotational speeds. However, the use of fluorocarbon resin-lined vessel and ceramic balls such as zirconia can decrease the contamination even at higher rotational speeds.

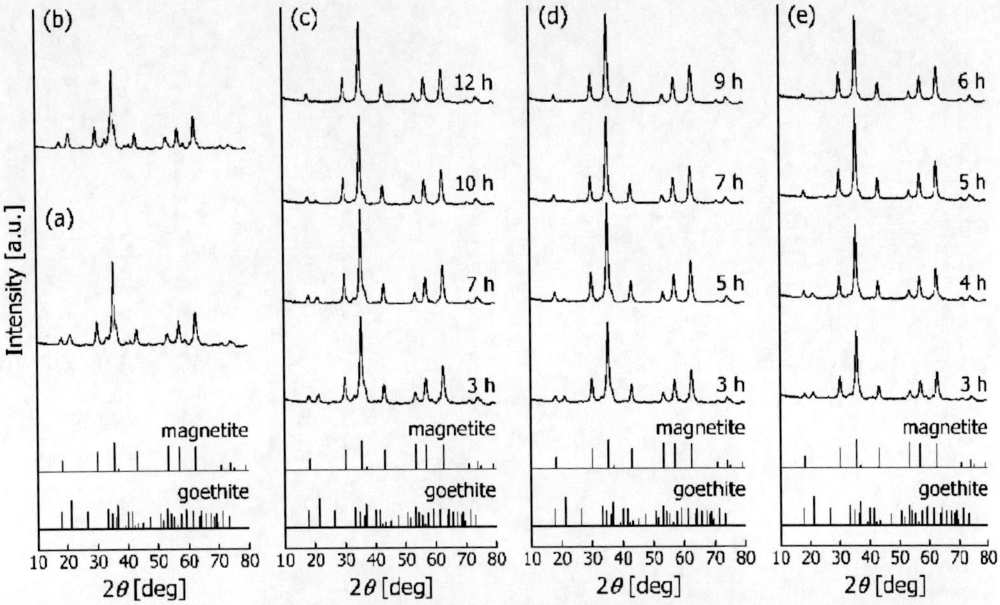

Figure 28. XRD patterns of samples (a) before ball-milling treatment and obtained (b) under static condition for 12 h at 2°C and by ball-milling at (c) 35 rpm, (d) 70 rpm and (e) 140 rpm for various milling times

Table 3. Properties of samples obtained from ball-milling treatment

Rotational speed [rpm]	Milling time [h]	Median size [nm]	Average crystallite size [nm]	Lattice constant [Å]	Saturation magneti-zation [emu/g]	Coercivity [Oe]	Contamination [wt%]	
							Cr	Ni
35	12	9.5	11.2	8.395	68.7	9	0.4	0.1
70	9	10.9	11.3	8.385	68.6	7	0.6	0.4
140	6	8.8	11.9	8.374	75.2	1	1.2	0.7

Here, the mechanical energy was numerically analyzed by simulating the behavior of balls in the milling vessel by means of the discrete element method (DEM). Based on the analysis results, contribution of the mechanical energy to progress of the Fe_3O_4 formation reaction was investigated, and the reaction mechanism in this system was analyzed.

In order to calculate the impact energy of balls, the behavior of balls in the milling vessel under wet condition was simulated using the three-dimensional DEM. This model describes the motion of each ball based on Newton's second law for individual ball, allowing for the external forces acting on the ball [49]. The interaction between ball and atmosphere gas (argon) was neglected because the contact force acting on the balls at the collision is much stronger than the drag force acting on the balls in the translational motion. The fundamental equations of translational and rotational motions of a ball are expressed as follows:

$$M\frac{\mathrm{d}^2\boldsymbol{X}}{\mathrm{d}t^2} = \boldsymbol{F} + M\boldsymbol{g} \tag{8}$$

Figure 29. SEM images of samples obtained by ball-milling at (a) 35 rpm for 12 h, (b) 70 rpm for 9 h and (c) 140 rpm for 6 h

$$I_{\mathrm{b}} \frac{\mathrm{d}\boldsymbol{\omega}}{\mathrm{d}t} = \boldsymbol{T} \tag{9}$$

where M, $\boldsymbol{X}$, I_{b} and $\boldsymbol{\omega}$ are mass, position, inertia moment and angular velocity of a ball, respectively. t, $\boldsymbol{g}$, $\boldsymbol{F}$ and $\boldsymbol{T}$ are time, gravity acceleration, contact force, and torque caused by the tangential contact force, respectively. $\boldsymbol{X}$ and $\boldsymbol{\omega}$ were calculated by integrating Eqs. (8) and (9) with respect to time between t and $t+\Delta t$.

For estimating the contact force acting on a ball, the Hertz-Mindlin contact model was used. The contact forces of normal and tangential directions, $\boldsymbol{F}_{\mathrm{n}}$ and $\boldsymbol{F}_{\mathrm{t}}$, were estimated using the following equations:

$$\boldsymbol{F}_{\mathrm{n}} = (-\kappa_{\mathrm{n}} \boldsymbol{\delta}_{\mathrm{n}}^{3/2} - \eta_{\mathrm{n}} \boldsymbol{v}_{\mathrm{n}} \cdot \boldsymbol{n})\boldsymbol{n} \tag{10}$$

$$\boldsymbol{F}_{\mathrm{t}} = -\kappa_{\mathrm{t}} \boldsymbol{\delta}_{\mathrm{t}} - \eta_{\mathrm{t}} \boldsymbol{v}_{\mathrm{t}} \ (\text{if } |\boldsymbol{F}_{\mathrm{t}}| \le \mu |\boldsymbol{F}_{\mathrm{n}}|) \tag{11}$$

$$\boldsymbol{F}_{\mathrm{t}} = -\mu |\boldsymbol{F}_{\mathrm{n}}| \frac{\boldsymbol{v}_{\mathrm{t}}}{|\boldsymbol{v}_{\mathrm{t}}|} \ (\text{if } |\boldsymbol{F}_{\mathrm{t}}| > \mu |\boldsymbol{F}_{\mathrm{n}}|) \tag{12}$$

$$\boldsymbol{v}_{\mathrm{t}} = \boldsymbol{v} - (\boldsymbol{v} \cdot \boldsymbol{n})\boldsymbol{n} + r(\boldsymbol{\omega}_i + \boldsymbol{\omega}_j) \times \boldsymbol{n} \tag{13}$$

where $\boldsymbol{\delta}$, $\boldsymbol{v}$, κ, η, μ, r and $\boldsymbol{n}$ are overlap displacement between contacting balls, relative velocity of contacting balls, stiffness, damping coefficient, sliding friction coefficient, radius of a ball and unit vector of normal direction at a contact point, respectively. The subscripts n and t mean the components of normal and tangential directions at a contact point, respectively. The subscripts i and j indicate the number of contacting balls. κ and η are determined from the following equations [50].

Stiffness for ball-to-ball collision

$$\kappa_{\rm n} = \frac{\sqrt{2rY_{\rm b}}}{3(1-\sigma_{\rm b}^{2})} \tag{14}$$

$$\kappa_{\rm t} = \frac{2\sqrt{2rY_{\rm b}}}{2(1+\sigma_{\rm b})(2-\sigma_{\rm b})}\boldsymbol{\delta}_{\rm n}^{1/2} \tag{15}$$

Stiffness for ball-to-vessel wall collision

$$\kappa_{\rm n} = \frac{(4/3)\sqrt{r}}{(1-\sigma_{\rm b}^{2})/Y_{\rm b} + (1-\sigma_{\rm w}^{2})/Y_{\rm w}} \tag{16}$$

$$\kappa_{\rm t} = \frac{8\sqrt{rY_{\rm b}}}{2(1+\sigma_{\rm b})(2-\sigma_{\rm b})}\boldsymbol{\delta}_{\rm n}^{1/2} \tag{17}$$

Damping coefficient

$$\eta_{\rm n} = \eta_{\rm t} = \alpha(M\kappa_{\rm n})^{1/2}\boldsymbol{\delta}_{\rm n}^{1/4} \tag{18}$$

where Y and σ are Young's modulus and Poisson's ratio, respectively. The subscripts b and w indicate ball and vessel wall, respectively. α is the constant depending on the restitution coefficient and was determined to be 0.20 based on the experimental value of restitution coefficient (= 0.75) according to the method proposed by Tsuji et al. [50]. The simulation parameters are summarized in Table 4. The sliding friction coefficient used in the calculation was determined based on the critical rotational speed measured experimentally under wet condition [51].

The impact energy of balls per unit time, E, was defined as the total kinetic energy of contacting balls [52].

$$E = \sum^{f} \frac{1}{2} M|\boldsymbol{v}|^{2} \tag{19}$$

$\boldsymbol{v}$ is the relative velocity of balls at the moment when the balls contact, and f is the average number of contact points per unit time. In other words, Eq.(19) means that the kinetic energy of balls just before contacting (i.e., at $|\delta| = 0$) is summed up for all the contact points within unit time but the kinetic energy during contacting (i.e., within $|\delta| > 0$) is not calculated. Therefore, the impact energy thus obtained corresponds to the maximum kinetic energy of balls at the collision. The impact energy per a single revolution of vessel, E_r, was also calculated as follows:

$$E_r = E/N \tag{20}$$

where N is the rotational speed of vessel.

Table 4. Simulation parameters

Number of ball media	7218
Ball diameter	3.2 mm
Vessel diameter	90 mm
Vessel depth	80 mm
Ball filling ratio to vessel capacity	40%
Density of ball and vessel	7930 kg/m^3
Young's modulus of ball and vessel	197 GPa
Poisson's ratio of ball and vessel	0.30
Sliding friction coefficient	0.23
Time step	1.0 μs

Table 5 gives the calculation result of the impact energy. The impact energy considerably varied depending on the rotational speed, resulting in variation of the milling time required for completing the Fe_3O_4 formation reaction. However, the properties of obtained Fe_3O_4 nanoparticles were almost the same even when the rotational speed was varied as mentioned above. This implies that the grinding of Fe_3O_4 nanoparticles did not occur in the synthesis process and that the impact energy greatly influences the formation process of Fe_3O_4 nanoparticles (especially, the reaction rate) rather than the properties of products.

Table 5. Calculation results of collision energy of ball media

Rotational speed [rpm]	Impact energy of balls	
	Energy per unit time E [J/s]	Energy per single revolution of vessel E_r [J/rev.]
35	2.78	4.76
70	7.59	6.51
140	27.08	11.61

In general, when the mechanical energy caused by shear, compression and friction actions of balls applies to particulate materials under dry condition, the surface energy of particles can increase due to the physical change such as distortion of the crystal lattice, increase of the surface area, and appearance of newly formed crystal surface. This causes the mechanochemical activation of particles. Under wet condition, however, the particles are difficult to undergo the mechanochemical activation because the increased surface energy was decreased by the solvent. Furthermore, it is very difficult to apply the mechanical energy effectively to nanoparticles. Accordingly, in this synthesis process, the mechanochemical activation of nanoparticles is difficult to occur, and the solid phase reaction between $Fe(OH)_2$ and α-FeOOH may hardly proceed by direct contribution of the mechanical energy. However, at the rotational speed of 35 rpm, i.e. in a low mechanical energy field, the Fe_3O_4 formation reaction surely proceeded while the reaction rate was relatively low. There is no doubt that the applied mechanical energy promotes the reaction; the reaction mechanism in the synthesis process is considered that the reaction proceeds by not the mechanochemical activation of precipitates but the heat energy converted from the mechanical energy.

In the conventional methods for synthesizing Fe_3O_4 nanoparticles in water system, for promoting the formation reaction and increasing the crystallinity, the starting suspension in the vessel is heated from the outside by conductive heat transfer, resulting in temperature rise of the whole system. The heating is continued to keep the reaction temperature. This causes aggregation of the precipitates, leading to the growth of Fe_3O_4 nanoparticles. On the contrary, in the synthesis process with the ball-milling treatment, the suspension is heated in the contact points between balls, which are extremely small regions, and then is cooled immediately because the vessel and the balls are thoroughly cooled. Therefore, local temperature rise of the suspension occurs instantaneously, and the heat energy is hardly stored in the system. In addition, the contact points exist discretely in the vessel; the discrete heating occurs everywhere. Consequently, the heating type in the synthesis process is the internal heating, which is regarded as a non-uniform heating from a microscopic viewpoint but a uniform heating from a macroscopic one. This inhibited aggregation of the precipitates effectively, and the nucleation frequently occurred rather than the particle growth, resulting in formation of the ultrafine Fe_3O_4 nanoparticles with high crystallinity and relatively narrow size distribution. Even at low rotational speeds, the Fe_3O_4 nanoparticles with a size of about 10 nm can be formed; this means that the aggregation-inhibition effect is confirmed in low energy fields.

In order to analyze the effect of impact energy on the reaction rate, progress of the reaction was investigated based on the XRD analysis. In the synthesis process, α-FeOOH decreased as the Fe_3O_4 formation reaction proceeded, and the typical diffraction intensity peak observed at $2\theta = 21.2°$ corresponding to the (011) plane of α-FeOOH decreased with the milling time as shown in Figure 28. Therefore, based on temporal change of the peak intensity, progress of the reaction was analyzed using the intensity ratio ξ defined by:

$$\xi = (I_m - I)/I_m \tag{21}$$

where I_m and I indicate the peak intensity at $2\theta = 21.2°$ of samples obtained before and after the ball-milling treatment under a given condition, respectively. In order to clarify the physical meaning of ξ, the correlation between ξ and the content of Fe_3O_4 in the product was investigated. Virtual products composed of Fe_3O_4, $Fe(OH)_2$ and α-FeOOH were artificially prepared by mixing proper amounts of Fe_3O_4 nanoparticles, $Fe(OH)_2$ precipitate and α-FeOOH precipitate. The mixing molar ratio of Fe_3O_4 to the mixture was determined as follows. When x mmol of Fe_3O_4 forms from 1.5 mmol of $Fe(OH)_2$ and 3.0 mmol of α-FeOOH, $(1.5–x)$ mmol of $Fe(OH)_2$ and $(3.0–2x)$ mmol of α-FeOOH remain in the product as indicated in Eq. (4). Accordingly, the content γ of Fe_3O_4 in the product is given by:

$$\gamma = x/(4.5-2x)\times 100 \tag{22}$$

XRD pattern of the virtual products with various mixing ratios was measured, and ξ was determined. Figure 30 shows the correlation between ξ and the content γ in the virtual products. ξ was proportional to γ in the range of $50\% \leq \gamma \leq 100\%$. This result indicates that change of the intensity ratio corresponds to that of the Fe_3O_4 content, i.e. the yield of Fe_3O_4, in this system. Therefore, ξ was used as an index indicating the yield in analysis of the reaction rate.

Figure 31 shows the change in ξ with the ball-milling time. ξ increased linearly with the ball-milling time, and the inclination, i.e. rate of the increase per unit time, increased with the rotational speed. The coefficient k corresponding to the reaction rate constant was determined from the inclination. Figure 32 shows the relationship between the coefficient k and the impact energy E per unit time. k increased with increasing in E. This result reveals that the reaction rate increases under high mechanical energy fields. This is analogous to increase of the reaction rate caused by temperature rise of the system, i.e., increase of the heat energy applied to the system.

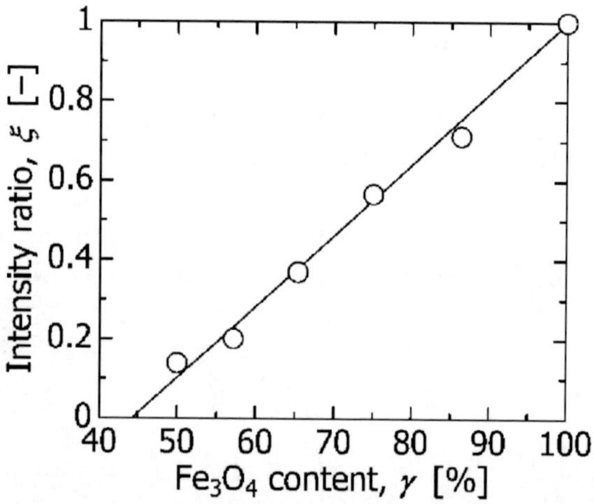

Figure 30. Relationship between diffraction intensity ratio ξ and magnetite content γ

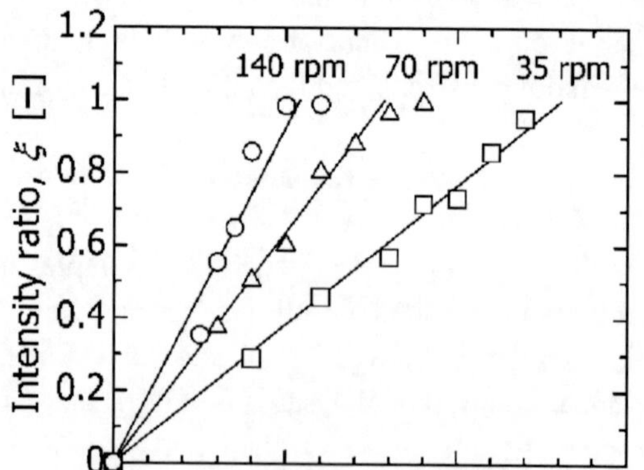

Figure 31. Change in diffraction intensity ratio ξ with ball-milling time at $N = 35$, 70 and 140 rpm

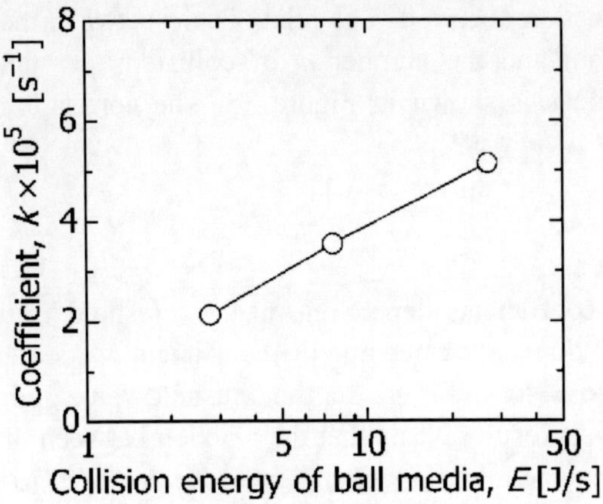

Figure 32. Relationship between coefficient k and impact energy E of balls per unit time

As can be seen in Figure 32, the value of k at the rotational speed of 140 rpm was not so large while relatively great amount of mechanical energy applied to the suspension. This suggests that the impact energy was not effectively used for progress of the reaction at high rotational speeds. Then, the intensity ratio ξ was rearranged with the number of revolutions of the milling vessel. Figure 33 shows the change in ξ with the number of revolutions. ξ increased linearly with the number of revolutions similar to that with the ball-milling time as shown in Figure 31. However, at lower rotational speeds, the increase rate of ξ per a single revolution was higher; this means that the reaction rate per a single revolution at lower rotational speeds is higher than that at higher rotational speeds. Using the inclination of the linear relation, the coefficient k' was determined, corresponding to the reaction rate per a single revolution. Figure 34 shows the relationship between the coefficient k' and the impact energy E_r per a single revolution. k' decreased with increasing in E_r. It is considered that at higher E_r the heat energy converted from the impact energy per a single revolution may be larger than that at lower E_r, resulting in higher reaction rates. However, in fact, it was found that dependence of the reaction rate on the impact energy per a single revolution was completely contrary to the prediction mentioned earlier. This result implies that relatively high impact energy did not contribute effectively to progress of the reaction. This agreed with our analysis result [53] that the contribution of mechanical energy to milling treatments depends strongly on vigorousness of the balls motion and decreases as it becomes more vigorous. Accordingly, in order to investigate contribution of the impact energy to the reaction rate, not the whole impact energy but the impact energy generated in each contact point between balls must be analyzed individually.

The formation reaction of Fe_3O_4 can occur when the impact energy exceeding a threshold value applies to unreacted $Fe(OH)_2$ and α-FeOOH precipitates at the contact points. The impact energy smaller than the threshold value cannot promote the reaction. Assuming that the region where the impact energy applies is extremely small, even though the impact energy larger than the threshold value generates at the contact point, the unreacted components far away from there may not react using the surplus energy. Thus, it is considered that progress of the reaction at the contact points depends on whether the impact energy exceeding the threshold value is given to the unreacted components or not, and that the surplus energy cannot promote the reaction. Accordingly, the reaction rate can be proportional to the number of collisions that the reaction occurred. The relationship between the coefficient k indicating the reaction rate constant and the number n_t of collisions per unit time with the energy exceeding threshold values is shown in Figure 35. The correlation between k and n_t was almost expressed by:

$$k = an_t^b \tag{23}$$

where a and b are the coefficients depending on the threshold value. Figure 36 shows the relationship between the threshold value and the coefficient b. As indicated in Eq. (23), when $b = 1$, k is proportional to n_t. From Figure 36, the threshold value giving $b = 1$ was determined to be 56.9 nJ. The analysis result reveals that the reaction between the unreacted components at a contact point occurs when the impact energy exceeding 56.9 nJ generates at the contact point regardless of the rotational speed. Consequently, this threshold value may be closely related to the activation energy in this reaction system. When both the net mechanical energy transferred from the balls to the unreacted components and the amount of them receiving the energy are found, the true activation energy of the mechanochemical reaction can be determined.

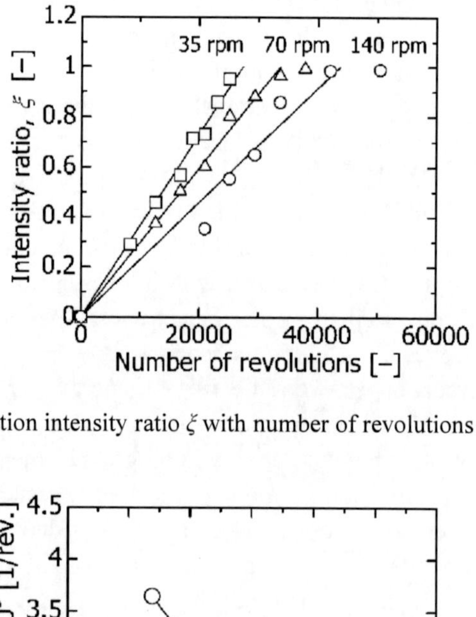

Figure 33. Change in diffraction intensity ratio ξ with number of revolutions of vessel at $N = 35$, 70 and 140 rpm

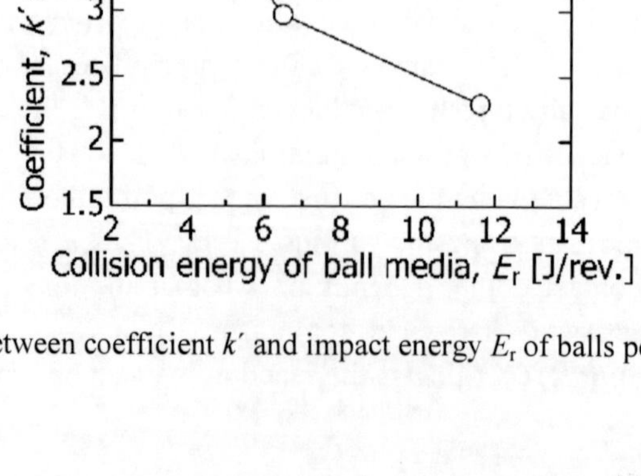

Figure 34. Relationship between coefficient k' and impact energy E_r of balls per single revolution of vessel

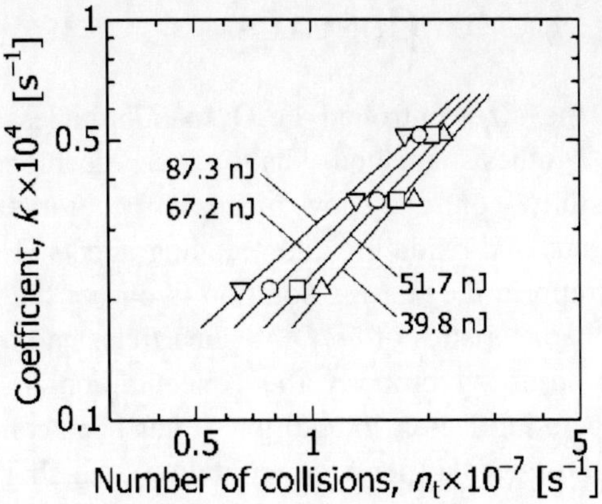

Figure 35. Relationship between coefficient k and number n_t of collisions of ball media per unit time for threshold values of collision energy of 39.8, 51.7, 67.2 and 87.3 nJ

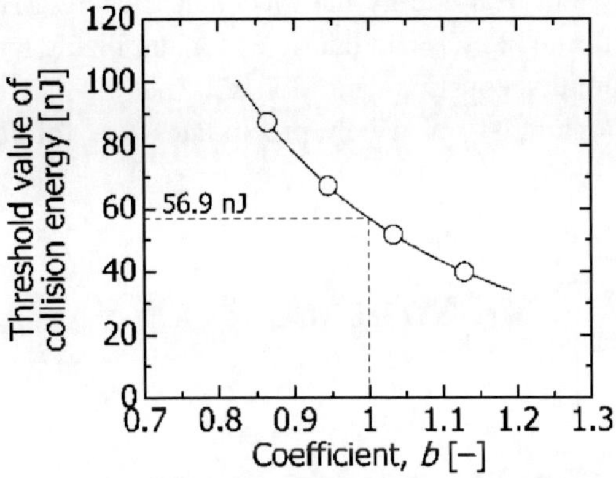

Figure 36. Relationship between threshold value of collision energy and coefficient b

In summery, a novel process for preparing the ultrafine Fe_3O_4 nanoparticles with high crystallinity in water system has been developed, in which a cooled tumbling ball mill is used as the reaction field. In the synthesis process with the ball-milling treatment, the suspension containing the precipitates of $Fe(OH)_2$ and α-FeOOH was instantaneously heated in the contact points between ball media distributing discretely in the milling vessel using the collision energy and then immediately cooled. This resulted in both progress of the Fe_3O_4 formation reaction and crystallization of the formed Fe_3O_4 nanoparticles while the growth of Fe_3O_4 nanoparticles caused by aggregation of the precipitates was inhibited. Therefore, the Fe_3O_4 nanoparticles with a size of about 10 nm could be successfully prepared without using any conventional heating techniques.

The impact energy hardly contributed directly to progress of the reaction using the mechanochemical activation of the unreacted precipitates. However, the reaction proceeded indeed by using the mechanical energy. Thus, it could be regarded as a mechanochemical reaction in a wide sense. The reaction rate depended strongly on the number of collisions with the energy exceeding a threshold value. The threshold value may be related to the activation energy of the reaction in this system.

Conclusion

In order to prepare the size-controlled Fe_3O_4 nanoparticles in water system, several modified hydrothermal synthesis methods have been developed. The size of Fe_3O_4 nanoparticles is successfully controlled by means of the methods without using any surfactants, organic solvents, and oxidizing and reducing agents. The obtained results clarify that the Fe^{2+}/Fe^{3+} molar ratio in the starting solution is one of the most important operating factors in the control of characteristics of Fe_3O_4 nanoparticles and that the size control can be carried out precisely by adjusting properly the concentration of coexisting anions in the starting solution as the operating factor. In addition, it has been confirmed that the sequential formation of $Fe(OH)_2$ and α-FeOOH (non-coprecipitation), which is a very simple method, is also effective for the size control.

Furthermore, a novel preparation process using the mechanochemical effect has been developed. In this method, formation and crystallization of Fe_3O_4 nanoparticles are promoted successfully by means of not heat energy but mechanical energy. By using this method, the surface modification of formed Fe_3O_4 nanoparticles for stabilizing the dispersion in a medium can be also carried out simultaneously. In addition, this method may be available for synthesis processes of other nanoparticles in which precipitates are formed as precursors and/or intermediate products.

Nomenclature

a	coefficient in Eq. (23) (s^{b-1})
b	coefficient in Eq. (23) (–)
E	impact energy of balls per unit time defined by Eq. (19) ($J\cdot s^{-1}$)
E_r	impact energy of balls per single revolution of vessel defined by Eq. (20) (J)
$\boldsymbol{F}_n$	contact force acting on ball in normal direction (N)
$\boldsymbol{F}_t$	contact force acting on ball in tangential direction (N)
f	average number of contact points per unit time (s^{-1})
$\boldsymbol{g}$	gravity acceleration ($m\cdot s^{-2}$)
I	peak intensity at $2\theta = 21.2°$ of sample obtained by ball-milling (–)
I_b	inertia moment of ball ($kg\cdot m^2$)
I_m	peak intensity at $2\theta = 21.2°$ of sample before ball-milling (–)
k	coefficient indicating reaction rate constant (s^{-1})
k'	coefficient indicating reaction rate constant (–)
M	mass of ball (kg)
m	initial molar ratio of ferrous lactate to total ferrous ion defined by Eq. (6) (–)
m'	addition molar amount of sodium salt defined by Eq. (7) (–)
N	rotational speed of vessel (s^{-1})
$\boldsymbol{n}$	unit vector in normal direction at contacting point (m)
n_t	number of collisions with energy exceeding threshold value (s^{-1})
r	radius of ball (m)
$\boldsymbol{T}$	torque caused by tangential contact force ($N\cdot m$)
t	milling time (s)

$\boldsymbol{v}$ relative velocity of contacting ball ($m \cdot s^{-1}$)
v_n relative velocity of contacting ball in normal direction ($m \cdot s^{-1}$)
v_t relative velocity of contacting ball in tangential direction ($m \cdot s^{-1}$)
$\boldsymbol{X}$ position of ball (m)
x amount of magnetite in product (mol)
Y_b Young's modulus of ball (Pa)
Y_w Young's modulus of vessel (Pa)

Greek Letters

α constant depending on restitution coefficient in Eq. (18) (–)
β initial molar ratio of ferrous ion to ferric ion (–)
γ content of magnetite in virtual product (%)
$\boldsymbol{\delta}$ overlap displacement between contacting balls (m)
$\boldsymbol{\delta}_n$ overlap displacement between contacting balls in normal direction (m)
$\boldsymbol{\delta}_t$ overlap displacement between contacting balls in tangential direction (m)
η_n damping coefficient in normal direction ($kg \cdot s^{-1}$)
η_t damping coefficient in tangential direction ($kg \cdot s^{-1}$)
κ_n stiffness in normal direction ($N \cdot m^{-3/2}$)
κ_t stiffness in tangential direction ($N \cdot m^{-1}$)
μ sliding friction coefficient (–)
ξ diffraction intensity ratio defined by Eq.(21) (–)
σ_b Poisson's ratio of ball (–)
σ_w Poisson's ratio of vessel (–)
$\boldsymbol{\omega}$ angular velocity of ball medium (s^{-1})

REFERENCES

[1] Buxbaum, G. & Pfaff, G. (2005). editors. *Industrial Inorganic Pigments*. Third edition. Weinheim, Wiley-Vch.

[2] Ochiai, M., Masui, M., Asanae, M., Tokunaga, M. & Iimura, T. (1994). Magnetic toner prepared by the suspension polymerization method. *J. Imaging Sci. Techn.*, *38*, 415-420.

[3] Hakata, T. (2002). A novel composite carrier for electrophotographic developers. *J. Imaging Sci. Techn.*, *46*, 591-597.

[4] Tanaka, S., Aspanut, Z., Kurita, H., Toriyabe, C., Hatuskade, Y. & Katsura, S. (2006). Bio-application of high-Tc SQUID magnetic sensor. *J. Magn. Magn. Mater.*, *300*, e315-e319.

[5] Erdem, A. Sayar, F., Karadeniz, H., Guven, G., Ozsoz, M. & Piskin, E. (2007). Development of streptavidin carrying magnetic nanoparticles and their applications in electrochemical nucleic acid sensor systems. *Electroanal.*, *19*, 798-804.

[6] Hai, N. H., Chau, N., Luong, N. H., Anh, N. T. V. & Nghia, P. T. (2008). Applications of magnetite nanoparticles for water treatment and for DNA and cell separation. *J. Korean Phys. Soc.*, *53*, 1601-1606.

[7] Chouly, C., Pouliquen, D., Lucet, I., Jeune, J. J. & Jallet, P. (1996). Development of superparamagnetic nanoparticles for MRI: effect of particle size, charge and surface nature on biodistribution. *J. Microencapsul.*, *13*, 245-255.

[8] Hong, R. Y., Feng, B., Chen, L. L., Liu, G. H., Li, H. Z., Zheng, Y. & Wei, D. G. (2008). Synthesis, characterization and MRI application of dextran-coated Fe_3O_4 magnetic nanoparticles. *Biochem. Eng. J.*, *42*, 290-300.

[9] Wan, J., Cai, W., Meng, X. & Liu, E. (2007). Monodisperse water-soluble magnetite nanoparticles prepared by polyol process for high-performance magnetic resonance imaging. *Chem. Commun.*, *47*, 5004-5006.

[10] Lin, B.-L., Shen, X.-D. & Cui, S. (2007). Application of nanosized Fe_3O_4 in anticancer drug carriers with target-orientation and sustained-release properties. *Biomed. Mater.*, *2*, 132-134.

[11] Zhang, J. & Misra, R. D. K. (2007). Magnetic drug-targeting carrier encapsulated with thermosensitive smart polymer: core-shell nanoparticle carrier and drug release response. *Acta Biomater.*, *3*, 838-850.

[12] Wang, X., Zhang, R., Wu, C., Dai, Y., Song, M., Gutmann, S., Gao, F., Lv, G., Li, J., Li, X., Guan, Z., Fu, D. & Chen, B. (2007). The application of Fe_3O_4 nanoparticles in cancer research: a new strategy to inhibit drug resistance. *J. Biomed. Mater. Res. A*, *80*, 852-860.

[13] Murakami, S., Hosono, T., Jeyadevan, B., Kamitakahara, M. & Ioku, K. (2008). Hydrothermal synthesis of magnetite/hydroxyapatite composite material for hyperthermia therapy for bone cancer. *J. Ceram. Soc. Jpn.*, *116*, 950-954.

[14] Fukumori, Y. & Ichikawa, H. (2006). Nanoparticles for cancer therapy and diagnosis. *Adv. Powder Technol.*, *17*, 1-28.

[15] Park, J., An, K., Hwang, Y., Park, J. E.-G., Noh, H.-J., Kim, J.-Y., Park, J.-H., Hwang, N.-M. & Hyeon, T. (2004). Ultra-large-scale syntheses of monodisperse nanocrystals. *Nat. Mater.*, *3*, 891-895.

[16] Paramês, M. L., Viskadourakis, Z., Rogalski, M. S., Mariano, J., Popovici, N., Giapintzakis, J. & Conde, O. (2007). Magnetic properties of Fe_3O_4 thin films grown on different substrates by laser ablation. *Appl. Surf. Sci.*, *253*, 8201-8205.

[17] Qian, Y., Xie, Y. & He, C. (1994). Hydrothermal preparation and characterization of ultrafine magnetite powders. *Mater. Bull.*, *29*, 953-957.

[18] Zhou, Z. H., Wang, J., Liu, X. & Chan, H. S. O. (2001). Synthesis of Fe_3O_4 nanoparticles from emulsions. *J. Mater. Chem.*, *11*, 1704-1709.

[19] Gotić, M., Jurkin, T. & Musić, S. (2007). Factors that may influence the micro-emulsion synthesis of nanosize magnetite particles. *Colloid Polym. Sci.*, *285*, 793-800.

[20] Visalakshi, G., Venkateswaran, G., Kulshreshtha, S. K. & Moorthy, P. N. (1993). Compositional characteristics of magnetite synthesised from aqueous solutions at temperatures upto 523K. *Mater. Res. Bull.*, *28*, 829-836.

[21] Valenzuela, R., Fuentes, M. C., Parra, C., Baeza, J., Duran, N., Sharma, S. K., Knobel, M. & Freer, J. (2009). Influence of stirring velocity on the synthesis of magnetite nanoparticles (Fe_3O_4) by the co-precipitation method. *J. Alloy. Compd.*, *488*, 227-231.

[22] Zhao, D.-L., Zeng, X.-W., Xia, Q.-S. & Tang, J.-T. (2009). Preparation and coercivity and saturation magnetization dependence of inductive heating property of Fe_3O_4 nanoparticles in an alternating current magnetic field for localized hyperthermia. *J. Alloy. Compd.*, *469*, 215-218.

[23] Iwasaki, T., Kosaka, K., Watano, S., Yanagida, T. & Kawai, T. (2010). Novel environmentally friendly synthesis of superparamagnetic magnetite nanoparticles using mechanochemical effect. *Mater. Res. Bull.*, *45*, 481-485.

[24] Iwasaki, T., Mizutani, N., Watano, S., Yanagida, T. & Kawai, T. (2010). Size control of magnetite nanoparticles by organic solvent-free chemical coprecipitation at room temperature. *J. Exp. Nanosci.*, *5*, 251-262.

[25] Wen, X., Yang, J., He, B. & Gu, Z. (2008). Preparation of monodisperse magnetite nanoparticles under mild conditions. *Curr. Appl. Phys.*, *8*, 535-541.

[26] Martínez-Mera, I., Espinosa-Pesqueira, M. E., Pérez-Hernández, R. & Arenas-Alatorre, J. (2007). Synthesis of magnetite (Fe_3O_4) nanoparticles without surfactants at room temperature. *Mater. Lett.*, *61*, 4447-4451.

[27] Wu, J.-H., Ko, S. P., Liu, H.-L., Kim, S., Ju, J.-S. & Kim, Y. K. (2007). Sub 5 nm magnetite nanoparticles: synthesis, microstructure, and magnetic properties. *Mater. Lett.*, *61*, 3124-3129.

[28] Wu, J. H., Ko, S. P., Liu, H. L., Jung, M.-H., Lee, J. H., Ju, J.-S. & Kim, Y. K. (2008). Sub 5 nm Fe_3O_4 nanocrystals via coprecipitation method. *Colloids Surf. A*, *313-314*, 268-272.

[29] Shokuhfar, A., Alibeigi, S., Vaezi, M. R. & Sadrnezhaad, S. K. Synthesis of Fe_3O_4 nanoparticles prepared by various surfactants and studying their characterizations. In: Ochsner, A. & Murch, G. E., editors. *Diffusion in Solids and Liquids III*, Defect and Diffusion Forum 273-276, Switzerland, Trans Tech Publications, 2008, 22-27.

[30] Wang, X., Zhang, C., Wang, X. & Gu, H. (2007). The study on magnetite particles coated with bilayer surfactants. *Appl. Surf. Sci.*, *253*, 7516-7521.

[31] Lian, S., Wang, E., Kang, Z., Bai, Y., Gao, L., Jiang, M., Hu, C. & Xu, L. (2004). Synthesis of magnetite nanorods and porous hematite nanorods. *Solid State Commun.*, *129*, 485-490.

[32] Fan, R., Chen, X. H., Gui, Z., Liu, L. & Chen, Z. Y. (2001). A new simple hydrothermal preparation of nanocrystalline magnetite Fe_3O_4. *Mater. Res. Bull.*, *36*, 497-502.

[33] Iida, H., Takayanagi, K., Nakanishi, T. & Osaka, T. (2007). Synthesis of Fe_3O_4 nanoparticles with various sizes and magnetic properties by controlled hydrolysis. *J. Colloid Interf. Sci.*, *314*, 274-280.

[34] Wang, J., Peng, Z., Huang, Y. & Chen, Q. (2004). Growth of magnetite nanorods along its easy-magnetization axis of [1 1 0]. *J. Cryst. Growth*, *263*, 616-619.

[35] Yu, W., Zhang, T., Zhang, J., Qiao, X., Yang, L. & Liu, Y. (2006). The synthesis of octahedral nanoparticles of magnetite. *Mater. Lett.*, *60*, 2998-3001.

[36] Schikorr, G. (1929). Über die Reaktionen zwischen Eisen, seinen Hydroxyden und Wasser. *Z. Elektrochemie*, *35*, 65-70.

[37] Shipko, F. J. & Douglas, D. L. (1956). Stability of ferrous hydroxide precipitates. *J. Phys. Chem.*, *60*, 1519-1523.

[38] Lee, J., Isobe, T. & Senna, M. (1996). Magnetic properties of ultrafine magnetite particles and their slurries prepared via in-situ precipitation. *Colloids Surf. A*, *109*, 121-127.

[39] Zaitsev, V. S., Filimonov, D. S., Presnyakov, I. A., Gambino, R. J. & Chu, B. (1999). Physical and chemical properties of magnetite and magnetite-polymer nanoparticles and their colloidal dispersions. *J. Colloid Interf. Sci.*, *212*, 49-57.

[40] Maity, D. & Agrawal, D. C. (2007). Synthesis of iron oxide nanoparticles under oxidizing environment and their stabilization in aqueous and non-aqueous media. *J. Magn. Magn. Mater.*, *308*, 46-55.

[41] Kandori, K., Fukuoka, M. & Ishikawa, T. (1991). Effects of citrate ions on the formation of ferric oxide hydroxide particles. *J. Mater. Sci.*, *26*, 3313-3319.

[42] Kandori, K., Uchida, S., Kataoka, S. & Ishikawa, T. (1992). Effects of silicate and phosphate ions on the formation of ferric oxide hydroxide particles. *J. Mater. Sci.*, *27*, 719-728.

[43] Ishikawa, T., Kataoka, S. & Kandori, K. (1993). The influence of carboxylate ions on the growth of β-FeOOH particles. *J. Mater Sci.*, *28*, 2693-2698.

[44] Sugimoto, T. & Wang, Y. (1998). Mechanism of the shape and structure control of monodispersed α-Fe_2O_3 particles by sulfate ions. *J. Colloid Interf. Sci.*, *207*, 137-149.

[45] Ali, M. A., Yasui, F. & Matsumoto, S. (2000). The lactate dependent enhancement of hydroxyl radical generation by the fenton reaction. *Free Rad. Res.*, *32*, 429-438.

[46] Žic, M., Ristić, M. & Musić, S. (2008). Effect of phosphate on the morphology and size of α-Fe_2O_3 particles crystallized from dense β-FeOOH suspensions. *J. Alloy. Compd.*, *466*, 498-506.

[47] Wan, J., Chen, X., Wang, Z., Yang, X. & Qian, Y. (2005). A soft-template-assisted hydrothermal approach to single-crystal Fe_3O_4 nanorods. *J. Cryst. Growth*, *276*, 571-576.

[48] Iwasaki, T., Kosaka, K., Mizutani, N., Watano, S., Yanagida, T., Tanaka, H. & Kawai, T. (2008). Mechanochemical preparation of magnetite nanoparticles by coprecipitation. *Mater. Lett.*, *62*, 4155-4157.

[49] Cundall, P. A. & Strack, O. D. L. (1979). A discrete numerical model for granular assemblies. *Geotechnique*, *29*, 47-65.

[50] Tsuji, Y., Tanaka, T. & Ishida, T. (1992). Lagrangian numerical simulation of plug flow of cohesionless particles in a horizontal pipe. *Powder Technol.*, *71*, 239-250.

[51] Gudin, D., Kano, J. & Saito, F. (2007). Effect of the friction coefficient in the discrete element method simulation on media motion in a wet bead mill. *Adv. Powder Technol.*, *18*, 555-565.

[52] Kano, J., Mio, H. & Saito, F. (2000). Correlation of grinding rate of gibbsite with impact energy of balls. *AIChE J.*, *46*, 1694-1697.

[53] Iwasaki, T., Kim, J. H. & Satoh, M. (2006). Characterization of media mills based on mechanical energy applied to particles. *Chem. Eng. Sci.*, *61*, 1065-1073.

In: Powder Engineering, Technology and Applications ISBN: 978-1-61761-216-9
Editor: Jason M. Barker, pp.81-124

Chapter 3

NANOSTRUCTURED FE-BASED MIXTURES PREPARED BY MECHANICAL ALLOYING

Safia Alleg [*] ***and Rachid Bensalem***
Laboratoire de Magnétisme et Spectroscopie des Solides, Physics Department, Badji-Mokhtar Annaba University, B.P. 12, 23000 Annaba, Algeria.

ABSTRACT

Nanostructured binary, ternary, and quaternary Fe-based mixtures have been prepared by mechanical alloying from pure elemental powders by the structural disintegration of coarser-grained structure through the severe plastic deformation. This later can force atoms into positions where they may not be at equilibrium and leads, therefore, to the formation of nanocrystalline solid solutions, intermetallic compounds, amorphous phases, etc. Microstructural, structural, morphological, thermal stability and magnetic properties changes during the milling process were investigated by X-ray diffraction, scanning electron microscopy, Mössbauer spectrometry, differential scanning calorimetry, and magnetic measurements. Depending on the milling conditions, the kinetics of elemental powders mixing can be described by one or two stages which are characterized by small values of Avrami parameter. *Keywords*: Nanostructured materials; Mechanical alloying; XRD; Mössbauer spectrometry; Magnetic measurements; Thermal stability.

1. INTRODUCTION

Nanocrystalline materials are single-phase or multi-phase polycrystalline solids with a crystal size of a few nanometers (typically less than 100 nm) in at least one dimension. Accordingly, they can be classified into (i) nanoparticles (zero-dimensional, 0-D), (ii) layered or lamellar structures (one-dimensional, 1-D), (iii) filamentary structures (two-dimensional, 2-D), and (iv) bulk nanostructured materials (three-dimensional, 3-D) [1]. Nanocrystalline materials have attracted extensive attention in the field of new materials design because of

[*] Corresponding author: E-mail address: safia_alleg@yahoo.fr (S. Alleg) Tel/Fax: 00 213 38 87 53 99

their remarkable magnetic, physical and mechanical properties with respect to the conventional coarse-grained polycrystals and amorphous solids [2, 3]. Such properties stem from both the crystallite size reduction to the nanometer scale and the enhanced density of defects where up to 50 % of the atoms are located in the grain boundaries and their junctions, pores and lattice defects. Consequently, nanocrystalline materials present an attractive potential for technological applications and provide an excellent opportunity to study the nature of solid interfaces and to extend knowledge of the structure-property relationship in solid materials down to the nanometer regime. Nanocrystalline materials can be produced by various methods starting from a vapor phase (e.g., inert gas condensation), a liquid phase (e.g., electrodeposition, rapid solidification) or a solid state (e.g., mechanical alloying).

The nanostructured materials consist of two components: nanocrystalline grains with different crystallographic orientations, identical arrangements and a network of boundaries the structure of which is characterized by: (i) a reduced average atomic density relative to the crystal density because of the misfit between adjacent crystallites; and (ii) the boundaries exhibit a broad spectrum of interatomic spacing ranging from closely packed to widely spaced atoms.

2. Mechanical Alloying

Mechanical alloying is a versatile process characterized by repeated fracturing, cold welding, and rewelding of powder particles in a dry high energy ball mill. Mixtures of elemental or prealloyed powders are subjected to high energy collision from the balls. The alloying process can be carried out using different apparatus such as planetary mills, attrition mills, vibratory mills, shaker mills, etc. [4]. As a result of the induced heavy plastic deformation into the powder particles during the milling process, nanostructured materials are produced by the structural decomposition of coarser-grained structure. This leads to a continuous refinement of the internal structure of the powder particles to nanometer scales.

Mechanical alloying is a non-equilibrium process resulting in solid state alloying beyond the equilibrium solubility limit. Consequently, a broad range of alloys, solid solutions, ceramics, and composites has been prepared in the nanocrystalline, quasicrystalline or amorphous state [5-12]. A significant increase in solubility has been reported in many mechanically alloyed systems [13, 14]. Several studies of the alloy formation process during mechanical alloying have lead to conflicting conclusions such as the interdiffusion of elements, the interactions on interface boundaries and/or the diffusion of solute atoms in the Fe matrix. Indeed, the alloying process is complex and hence involves optimization of a number of parameters to achieve the desired product such as type mill, raw material, milling intensity (I) or milling speed (ω), milling container, milling atmosphere, milling time, temperature of milling, ball-to-powder weight ratio (BPR), process control agent, etc.

3. Solid State Transformation

The formation of stable and/or metastable crystalline phases usually competes with the formation of the amorphous phase. For alloys with a negative heat of mixing, the phase formation has been explained by an interdiffusion reaction of the components occurring during the milling process [15]. The Fe-Co system which has a small negative heat of mixing (ΔH_{mix}= −1kJ/mol) is very interesting with respect to the phase formation during milling, so that mechanical alloying is expected to result rather in the formation of solid solutions than in amorphization. Because of their attractive soft magnetic properties, bulk Fe-Co alloys have found wide applications in a variety of areas. Since the magnetic properties are microstructure and crystal structure dependent [16-19], it has been shown that these properties are superior in the nanostructured than bulk alloys. Metastable phase formation is only observed if the formation of stable phases is suppressed under the specific milling conditions. In particular, intermetallic compounds can be energetically destabilized due to the chemical disorder introduced by the deformation during milling [20]. This often results in the formation of extended solid solutions or amorphous phases. It has been shown that the end product is not governed entirely by the starting composition, but it is also affected by numerous milling parameters in the evolution of the reaction products such as the ball-to-powder weight ratio and the milling intensity [21]. Indeed, increased milling energy which is achieved by an appropriate choice of those parameters is normally expected to introduce more strain and to increase the defect concentration in the powder and thus leads to easier amorphization. However, higher milling energies also produce more heat (higher temperature) and this can result in the crystallization of the amorphous phase. Therefore, a balance between these two effects will determine the nature of the final product phase. For example, the maximum amorphization range is observed at intermediate values of milling intensity in the Ni-Zr system [22]. A fully amorphous phase is formed at high energy of milling while, a mixture of amorphous and crystalline phases is formed at low and intermediate milling intensities. Generally, a lower milling temperature accelerates the amorphization process. Therefore, since a nanostructured material can be easily produced at lower milling temperatures, the increased grain boundary area drives the crystal-to-amorphous transformation.

The peculiarity of the Fe-Cr system is the existence of a wide miscibility gap in the phase diagram where both Fe and Cr are insoluble at room temperature. The chemical arrangement of the atoms in the alloy can either be random or exhibit clustering among similar atoms [23] according to the preparation method and further heating treatment. Beyond that, the Fe-Cr system is characterized by rich non-equilibrium behavior. The Fe-Cr system is of a great interest due to its magnetic and structural properties. Mechanical alloying of Fe-Cr system has been widely investigated and much attention has been paid to the study of the structural characteristics and of the physical properties of final products [24-26]. The Fe-C system has attracted particular attention due to its importance in the steel industry especially for the low carbon region of the phase diagram (≤ 10 at. %C). Mechanical alloying has been used to prepare high-carbon-content carbides such as Fe_3C and Fe_7C_3 [27, 28] since the formation of such phases in a metastable form is of great interest as these high-temperature equilibrium carbides are otherwise difficult to prepare by conventional methods. It has been shown that the $Fe_{100-x}P_x$ alloys have a mixture of amorphous and disordered solution of P into the bcc α-Fe lattice for $x < 15$ [29]. Ternary Fe-P-C alloys which have been investigated in a wide range of composition [30-32], exhibit a combination of remarkable properties such as corrosion resistance, hardness, soft magnetic properties, and excellent mechanical properties. Amorphous (Fe,Cr)-metalloid alloys are currently produced in ribbon form by melt spinning and planar flow [33-35] and studied for their good magnetic and mechanical properties which make them potentially useful in many industrial applications [36]. Fe-Cr, Fe-Co, Fe-Mo, Fe-

P, Fe-P-C, Fe-Cr-P-C, and Fe-Co-Nb-B powder mixtures were prepared by mechanical alloying, under argon atmosphere, in planetary ball mills Fritsch P7 and Retsch PM400. Different milling conditions such as composition, PBR and milling intensity (I) were used.

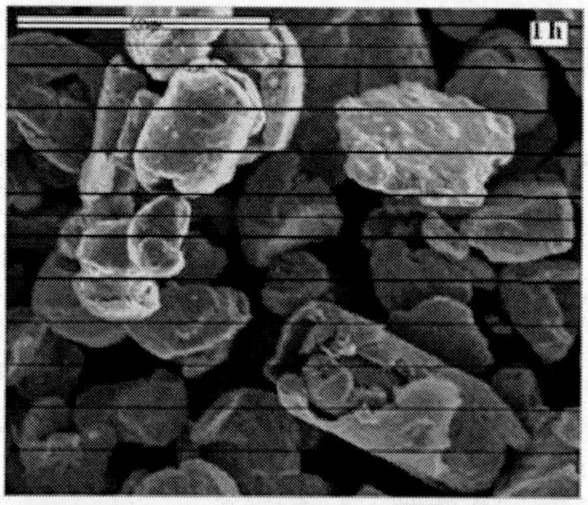

Fig. 1a

Fig. 1b

Fig. 1c

Figure 1. SEM micrographs of Fe (a), $Fe_{62}Nb_8B_{30}$ (b) and $Fe_{80}Cr_{20}$ powders (c)

4. Particle Size and Morphology

Ball-milling causes drastic changes in morphology of the powders that undergo severe plastic deformation. Due to the repeated fracturing, cold welding, agglomeration, and deagglomeration, the powder particles are plastically deformed and cold welded. Work hardening takes place because of the medium impact force of the grinding and eventually the work-hardened fracturing of the powder particles. The fresh and atomically clean surfaces created during fracture allow the particles to weld together, and this leads to an increase in particle size. At the initial stage of milling, the ductile powders can easily be plastically deformed under compressive forces and consequently, they become flattened into flakes through micro-forging (Figure 1a). Also, since the powders are soft they tend to agglomerate by cold welding and form bigger particles (Figure 1b). Simultaneously, a very large distribution in particle size can be observed. The formation of layered structures which are characteristic of ductile–ductile powders can be observed in the early stage of milling (Figure 1c). Micro-forging leads to changes in the shape of individual particles, which are impacted repeatedly by the milling balls with high kinetic energy. For intermediate milling times, important changes occur especially in particles morphology in comparison with those in the initial stage and cold welding process becomes prevailing (Figure 2). For the final stage of milling, a considerable refinement and reduction in particle size can be achieved. The particles shape and size are more homogeneous than those in the initial and intermediate stages (Figure 3a) corresponding to a steady state of the milling process. Homogeneity of the particles is an outcome of equilibrium between fracturing and cold welding. For example, true $Fe_{50}Co_{50}$ alloy is formed at this stage (Figure 3b) as confirmed by the elemental analysis data where the Fe/Co ratio is very close to the starting nominal composition (Fe/Co $\cong$ 1). In the case of the Fe-6P-1.7C powders, further milling time leads to roughly spherical agglomerate composite particles having nearly the same shape but vary in size from about 1.3 to 12.6 μm in diameter (Figure 4).

Figure 2. SEM micrograph of $Fe_{50}Co_{50}$ powders milled for 6 h

Fig. 3a

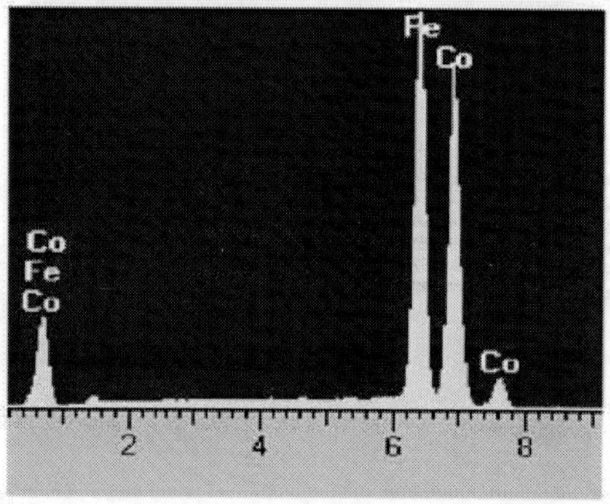

Fig. 3b

Figure 3. SEM micrograph and quantitative analysis of $Fe_{50}Co_{50}$ powders milled for 24 h

Figure 4. SEM micrograph of Fe-6P-1.7C powders milled for 104 h

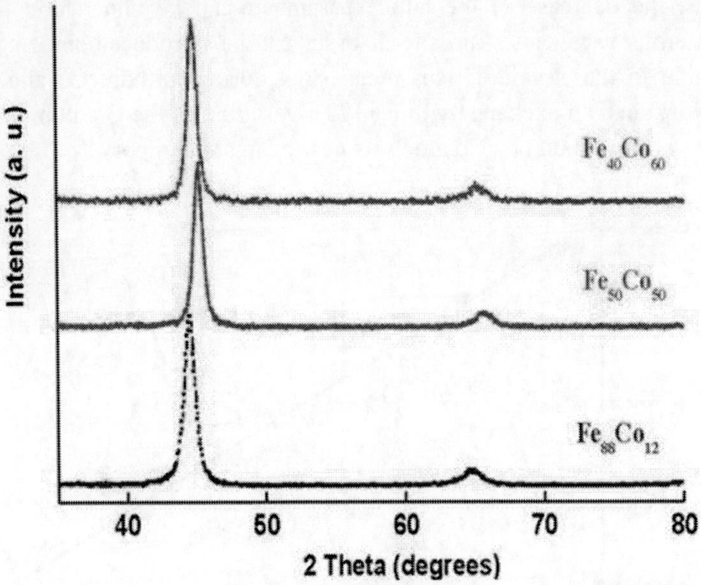

Figure 5. XRD patterns of the ball-milled Fe-Co powders for 24 h

5. Microstructural and Structural Properties

5.1. Fe-Co Alloys

Figure 5 displays the XRD patterns of the $Fe_{100-x}Co_x$ (x=12, 50 and 60 wt. %) powder mixtures milled for 24 h. The broadening of the diffraction peaks and the decrease of their intensities during the milling process can be due to the reduction of the effective crystallite size and an increase of the atomic level strain. Co is found to undergo an allotropic phase transformation from fcc to hcp form in the early stage of milling [9, 15]. This is evidenced by an increase of the hcp Co line intensities and the decrease of the fcc ones before their complete disappearance. The fcc Co phase is metastable at room temperature, becomes unstable when an external mechanical or thermal energy is introduced [37]. Hence, the fcc Co transforms to the hcp Co form owing to the accumulation of structure defects and the increase of the energy stored in the materials during the milling process. The repeated fracturing and cold welding of powder particles increase the area of contact between the reactant powder particles due to a reduction in particle size and allows fresh surfaces to come into contact repeatedly. This permits the reaction to proceed without the necessity for diffusion through the produced layer. As a consequence, the reactions that normally require high temperature do occur at lower temperature during milling without any external heating treatment. Also, the high defect densities induced during the milling accelerate the diffusion process. The disappearance of the hcp Co diffraction peaks and therefore, the interdiffusion of Co and Fe lead to the formation of nanostructured $Fe_{100-x}Co_x$ alloys with a crystallite size of about 12-17 nm. The decrease of the lattice parameter of the nanostructured $Fe_{100-x}Co_x$ alloys with increasing Co content (Figure 6) might be due to the so-called triple-defect disorder [38]. A triple defect consists of one anti-site Co atom combined with two vacancies on the Co sublattice. Thus, the decrease of the lattice parameter can be understood from relaxation around the numerous vacancies. Anti-site disorder can be introduced since no vacant lattice sites are available in the structure. This means that during milling, Co and Fe atoms will exchange position, such an exchange will lead to a variation of the Co atom neighboring, and hence a Co atom is surrounded by Co atoms as near nearest neighbors.

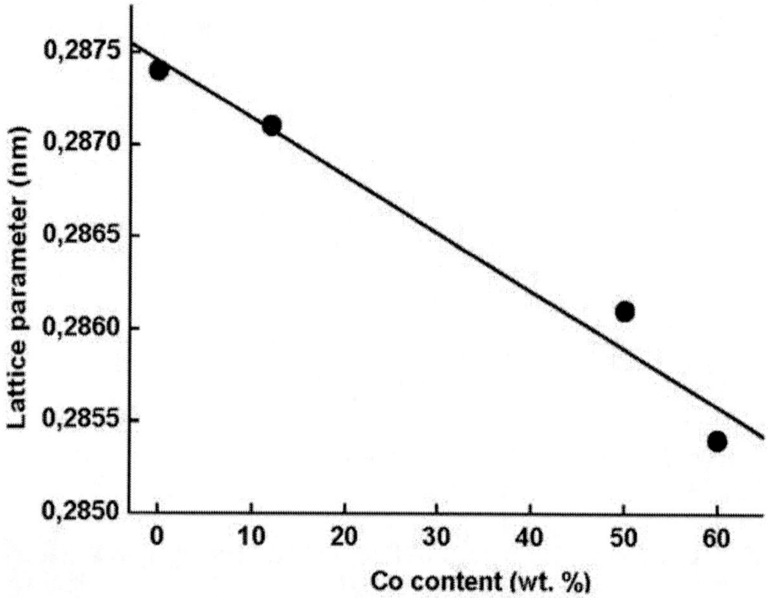

Figure 6. Lattice parameter of $Fe_{100-x}Co_x$ alloys vs. Co content for 24 h

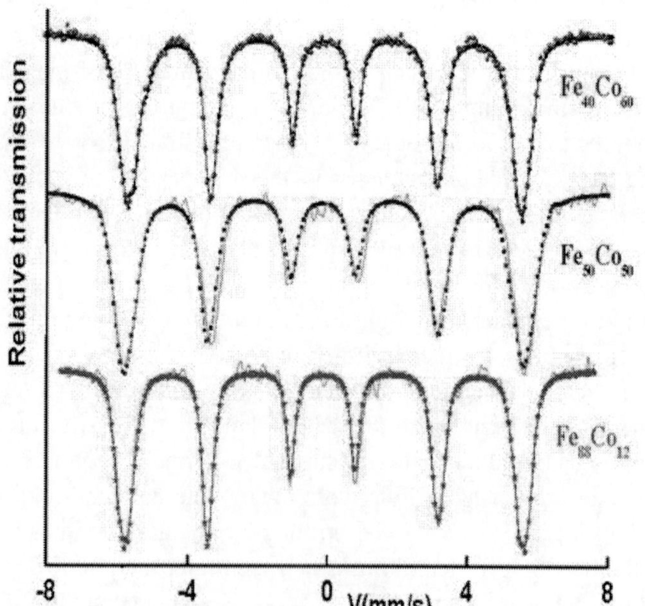

Figure 7. Mössbauer spectra, taken at 300 K, for $Fe_{100-x}Co_x$ powders milled for 24 h.

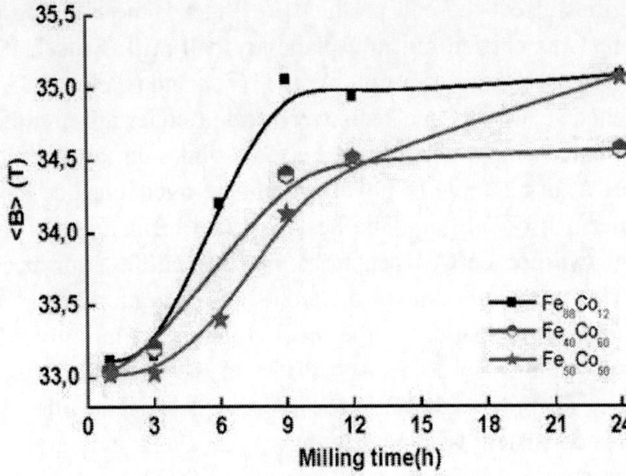

Figure 8. Variation of the average hyperfine magnetic field of the $Fe_{100-x}Co_x$ powders during the milling process

Table 1. Lattice parameter, a, average crystallite size, <L>, microstrains, $<\sigma^2>^{1/2}$, and phase proportion deduced from the Rietveld refinement of sample B

Phase	a (nm) ±0.001	<L> ±2 (nm)	$<\sigma^2>^{1/2}$ ± 10^3	Proportion (%)
bcc-Fe_{iso}	0.2873	38	0.94	53
bcc-Fe_{aniso}	0.2879	8 [100] 13 [200]	1.55 [100] 2.67 [200]	47

Mössbauer spectrometry allows a quantitative evaluation of the local structure and composition changes. So, it is used as a detailed probe of the Fe neighborhood in order to check the alloying process progress during the milling. Figure 7 shows the Mössbauer spectra of the milled $Fe_{100-x}Co_x$ mixtures for 24 h. The broadening of the Mössbauer lines and their shift towards higher hyperfine fields are attributed to the mixing of Fe and Co at the atomic level since the presence of one Co atom in the vicinity of Fe one increases the hyperfine magnetic field by about 0.8 T and the isomer shift by 0.1 mm/s when it is first or second nearest neighbor [39]. The average hyperfine magnetic field variations (Figure 8) which describe the evolution of the interdiffusion reaction between Fe and Co, can be divided into three steps: (i) a constant value of 33 T, up to 3 h of milling, related to the unreacted α-Fe; (ii) a rapid increase, between 3 and 12 h of milling, linked to the mixing of the Fe and Co powders; and (iii) a stationary state, above 12 h of milling, for 12%Co and 60%Co alloys and a slow increase for the 50%Co alloy. The obtained values of <B> = 35 T for both 12%Co and 50%Co alloys, and <B> = 34.28 T for 60%Co alloy are similar to those reported for the disordered and the ordered bulk Fe-Co alloys, respectively [40].

The milling energies affect the end product [9]. Indeed, the alloying process of the Fe-12Co powder mixtures was carried out in a planetary ball mill (Fritsch P7) at: (i) a milling intensity $I_1 = 6$ and $(BPR)_1 = 35/2$ for sample A; (ii) $I_2 = 8$ and $(BPR)_2 = 35/1$ for sample B. In both cases, the hardened steel vials and balls were sealed under argon atmosphere in a glove box to prevent from oxidation. A disordered Fe(Co) solid solution is reached after 24 h of milling for the sample A. For sample B, however, an heterogeneous bcc Fe(Co) solid solution is formed rapidly after 3 h of milling. The best Rietveld refinement of the XRD pattern is obtained by means of two-bcc Fe(Co) structures, named bcc-Fe_{iso} and bcc-Fe_{aniso} (Figure 9). These two-bcc Fe(Co) structures have different microstructural parameters (Table 1). Mössbauer spectrometry investigation of the local structure variation confirms the effect of the milling conditions (rotating speed, ω, and PBR) on the formation mechanism as well as the structural disorder. Thereby, it is observed for sample A that, after 3 h of milling, the Mössbauer spectrum is consistent with a sextet having an average magnetic hyperfine field of 33 T corresponding to pure α-Fe (Figure 10). This means that there is no reaction between Fe and Co at this stage of milling. The broadening of the Mössbauer lines and their shift towards higher hyperfine magnetic fields, as the milling process progresses, are due to the mixing of the elemental Fe and Co powders at the atomic level and consequently, to the presence of Co atoms in the vicinity of Fe ones. For sample B, nonetheless, the effect of the milling conditions is much more pronounced as shown by asymmetrical and broad lines of the Mössbauer spectra taken at 77 and 300 K (Figure 11). The components of the hyperfine magnetic field are well resolved, indicating distinct differences in the magnetic environments at the grain boundaries and in the crystallites. The range of magnetic influence of grain boundaries is expected to be rather short for transition metal alloys, since the electronic interactions involving 3d electrons are strongest over first and second nearest neighbor distances. Also, the interfacial component have a reduced atomic coordination and a wide distribution of interatomic spacing compared to the crystals and consequently, the average atomic density is considerably less than the crystal one. The fitting of the Mössbauer spectrum, taken at 300 K, was carried out with two magnetic components, each one involving a discrete distribution of hyperfine fields (Figure 11) : one component (zone1) with average hyperfine parameter values of $\langle B\rangle_1 = 34.8$ T, $\langle IS\rangle_1 = 0.014$ mm/s and a relative area of ~ 63%, assigned to the nanocrystalline grains. A second component (zone 2) with $\langle B\rangle_2 = 28.2$ T, $\langle IS\rangle_2 = 0.004$ mm/s and a relative area of ~ 37% might be related to the grain boundaries. Indeed, the distorted local environment around the ^{57}Fe atoms located at the grain boundaries and at the interface between grain boundaries and crystalline grains, contributed to a distinct signature to the Mössbauer spectrum of nanophase materials with the smallest grain size. The magnetic hyperfine field value $\langle B\rangle_2 = 28.2$ is comparable to that obtained for the milled pure Fe (B = 28.31 T) during 50 h [41]. This low hyperfine magnetic field was attributed to a lattice defect structure DS(1,1) associated with a Fe atom in the bcc Fe lattice with one nearest and one next nearest neighbor defect sites. A defect site was defined as an atom other than a Fe atom, an atomic vacancy or an interstitial atom in the bcc lattice void site. The defect structure was limited only to grain boundaries and the increased grain's lattice cell dimensions were due to grain boundaries disorder stress [41].

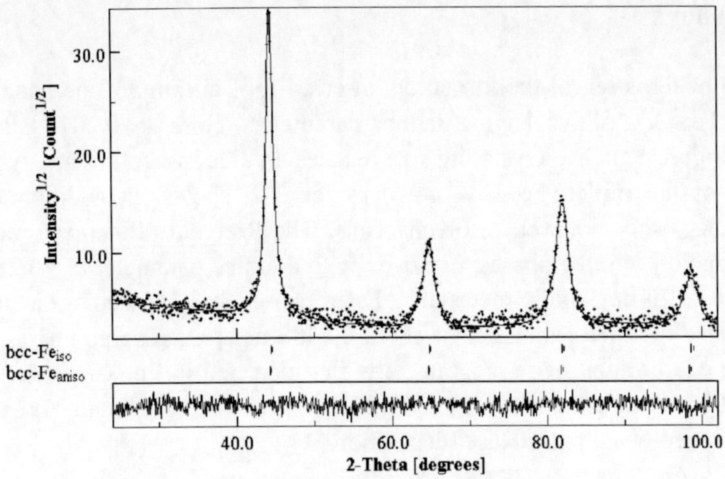

Figure 9. Rietveld refinement of the XRD pattern of sample B milled for 3h with two bcc Fe(Co) phases

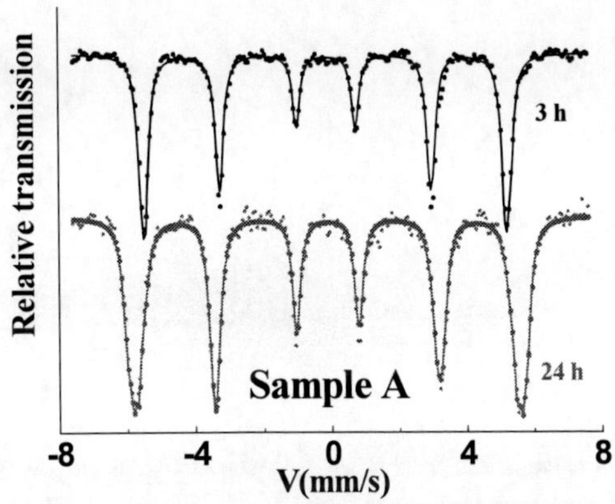

Figure 10. Mössbauer spectra, taken at 300 K, of sample A milled for 3 and 24 h

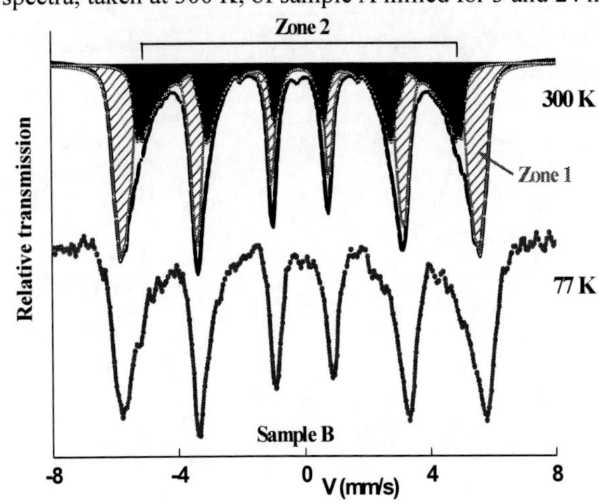

Figure 11. Mössbauer spectra, taken at 77 and 300 K, of sample B after 3 h of milling

5.2. Fe-Cr Alloys

The XRD patterns reveal the formation, after 24 h of milling, of the nanostructured bcc $Fe_{85}Cr_{15}$ and $Fe_{80}Cr_{20}$ alloys having lattice parameters close to 0.2874 and 0.2880 nm, respectively (Figure 12). The crystallite size reaches a value of about 9 nm. Nonetheless, the XRD pattern of the milled $Fe_{69}Cr_{31}$ powders for 3 h shows, in addition to the bcc-Fe diffraction peaks, other not well resolved peaks. The Rietveld refinement was obtained by using a combination of three phases: bcc α-Fe with a lattice parameter a = 0.2866 nm, fcc γ-Fe with a = 0.3591 nm and a metastable FeCr phase characterized by a primitive cubic structure (pc) and a lattice parameter of about 0.4588 nm [42]. The pc phase, which is non-existent in the equilibrium phase diagram, was first detected in fine Cr particles obtained by evaporation of Cr in high vacuum [43] and in $Fe_{1-x}Cr_x$ (x = 43.5 – 48 at. %) nanometer films prepared by crossed-beam pulsed laser deposition [44].

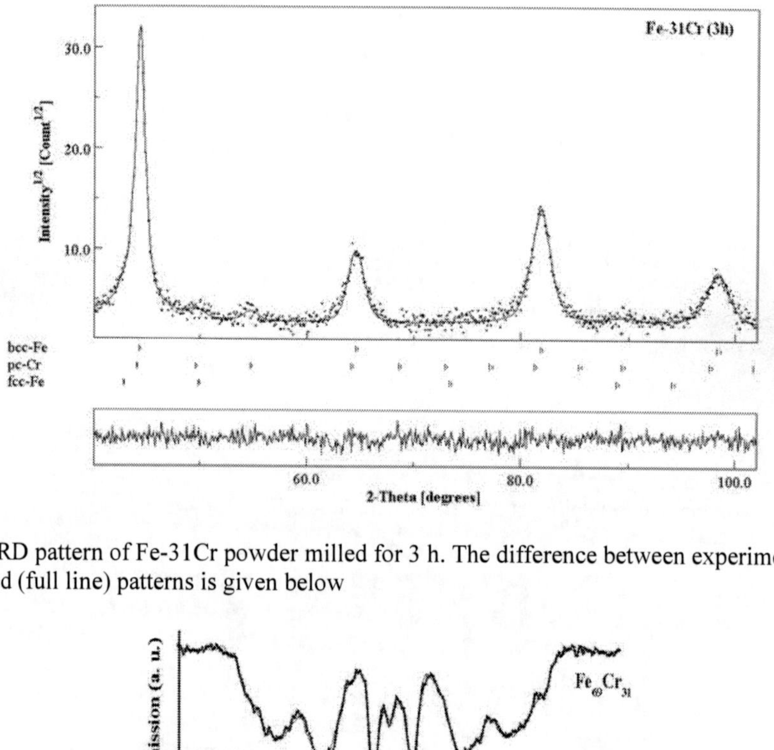

Figure 12. XRD pattern of Fe-31Cr powder milled for 3 h. The difference between experimental (dots) and calculated (full line) patterns is given below

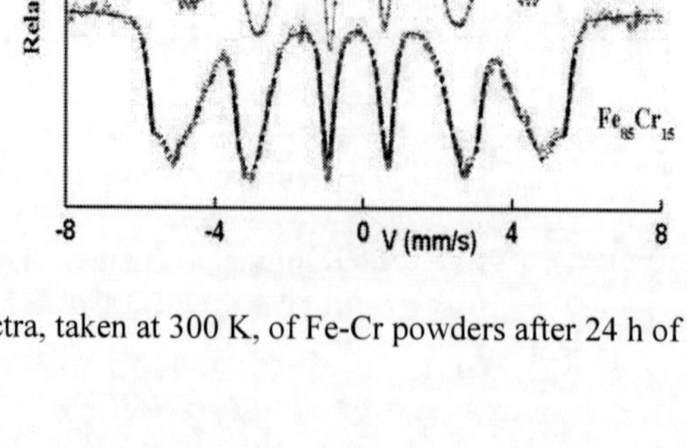

Figure 13. Mössbauer spectra, taken at 300 K, of Fe-Cr powders after 24 h of milling

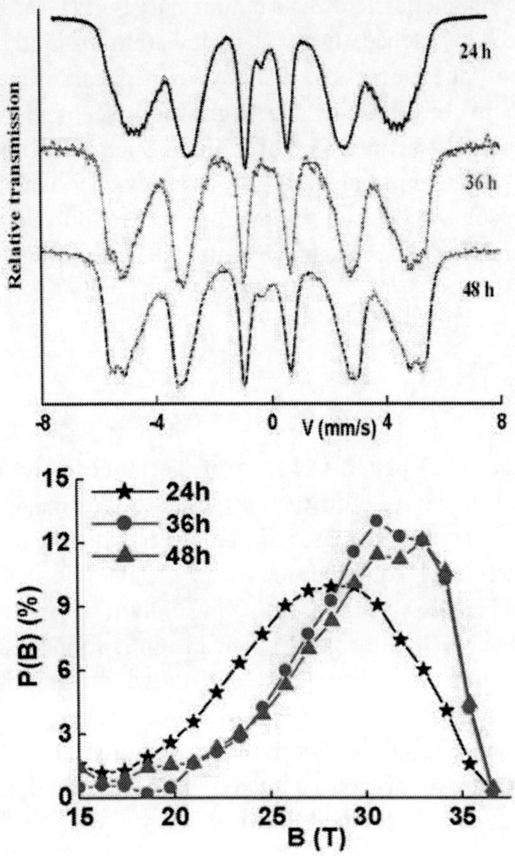

Figure 14. Mössbauer spectra, taken at 300 K, and the corresponding P(B) curves of the $Fe_{80}Cr_{20}$ powders after 24, 36 and 48 h of milling

The Mössbauer effect is a suitable technique to study the alloying process in the Fe-Cr system, because the presence of the Cr atoms in the vicinity of Fe ones is characterized by the appearance of satellites due to the reduction of the hyperfine field at Fe atom nuclei by 3 T and 2 T if the Cr atom is first or second nearest neighbor, respectively [45]. The Mössbauer spectra of the $Fe_{100-x}Cr_x$ (x = 15, 20 and 31 wt.%) alloys are consistent with broad magnetic sextets pointing out a wide variation of ^{57}Fe environments (Figure 13), such a shape is characteristic of a disordered structure where the Fe atoms are largely and diversely substituted by Cr atoms in the nearest neighborhood. For the $Fe_{80}Cr_{20}$ and $Fe_{69}Cr_{31}$ nanostructured alloys, one observes, in addition to the broad sextet lines, the presence of a paramagnetic contribution at the centre. The isomer shift values of the paramagnetic components which are about -0.097 mm/s and -0.106 mm/s, respectively, for the $Fe_{69}Cr_{31}$ and $Fe_{80}Cr_{20}$ alloys are comparable to that reported for Cr-rich α' phase in $Fe_{1-x}Cr_x$ alloys containing more than 70 at.%Cr [46, 47] and for fcc phase in Fe-Cr steels [48, 49]. Consequently, this paramagnetic component which is about four times larger in the case of the $Fe_{69}Cr_{31}$ alloy (5%) in comparison to that of the $Fe_{80}Cr_{20}$ alloy (1.3%) can be due to both fcc and pc Cr-rich phases since they are paramagnetic at room temperature. This result well supports the XRD analysis. Further milling (up to 48 h) leads to the demixtion of the nanostructured $Fe_{80}Cr_{20}$ alloy as evidenced by the shift of both the Mössbauer lines and their corresponding hyperfine magnetic field distribution curves, P(B), towards higher magnetic hyperfine fields (Figure 14). Hence, there is a deviation from the binomial distribution through the clustering of like-atoms and thereby the appearance of Fe-rich and Cr-rich environments. This later is confirmed by the slight increase of the relative fraction of the paramagnetic phase to about 2.6% after 48 h of milling. Such behavior might be explained by the high stored energy in the mechanically alloyed powders which drives the structural changes. The obtained result which is different from that reported on the same composition [50], where a stationary mixing state is achieved after 21 h of milling, can be attributed to the milling conditions.

5.3. Fe-Mo Alloys

Figure 15 exhibits the XRD patterns change of the Fe-6Mo (wt.%) powders during the milling process. The shift of the α-Fe diffraction peaks to the smaller angles, as the milling process progresses, suggests the increase of the α-Fe lattice parameter because of the structural distortions which arise from both the heavy plastic deformation and the dissolution of Mo into the Fe matrix. Hence, a bcc Fe(Mo) solid solution with a lattice parameter close to 0.2880 nm, and an average crystallite size of about 11 nm is formed, after 24 h of milling, in addition to the nanocrystalline Mo (10 nm) with a distorted lattice (a=0.3151 nm). According to the partial mixing of elemental Mo and Fe powders which is due to the Mo properties such as its high fusion temperature (2893 K) and its low solubility into α-Fe at low temperatures [51], one expects the presence of both unmixed nanocrystalline α-Fe together with the Fe(Mo) solid solution and nanocrystalline Mo [8].

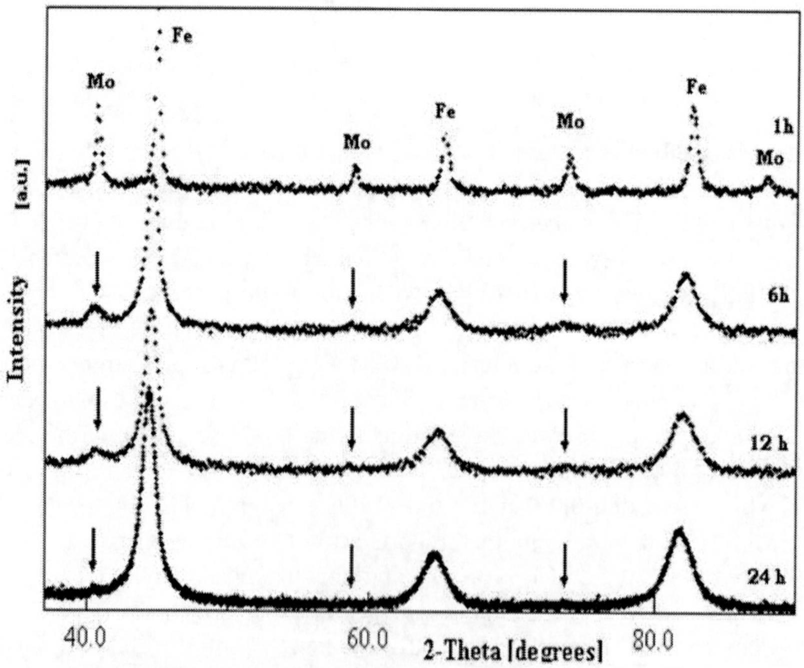

Figure 15. XRD patterns of ball-milled Fe-6Mo powder mixtures as a function of milling time

5.4. Fe-P Based Alloys

The measurement of crystalline size and lattice strain in the mechanically alloyed powders is very important since the phase constitution and transformation characteristics appear to be strongly dependent on such parameters. Compared to the binary Fe-P system, the solid state reaction is much complicated in the ternary Fe-P-C and quaternary Fe-Cr-P-C systems. This is due to the different chemical and physical properties, mixing enthalpies, affinities between elements and solubility. For example, Cr is completely soluble in the α-Fe at the solid state, so it is accommodated in the solid solution of the crystalline products obtained where Fe atoms are partly substituted by Cr ones. Also, Cr is of particular interest because its addition has always been found to be most effective in improving the corrosion resistance. Furthermore, P is known to be a stabilizer of the amorphous phase. Red phosphorus is stable at room temperature and its combustion occurs during an impact or a friction at about 250°C. The addition of a third element such as P or B to the mixture of Fe and C increases the tendency to amorphization [52, 53]. In fact, P is an interesting alloying element since the eutectic temperature is low (1050°C) compared to the melting point of Fe (1534°C).

5.4.1. Fe-P alloys

Rietveld refinement of the XRD pattern of the Fe-8P powders milled for 3 h (Figure 16), in a planetary ball-mill (P7) with a BPR of about 6/35 and I= 9, reveals the formation of hexagonal iron phosphides Fe_xP ($1<x<2$) and Fe_2P (~10%) [54]. After 21 h of milling, a mixture of Fe_3P nanophase and disordered two-bcc structures of the α-Fe(P) solid solution: designated as α-Fe1 and α-Fe2, in order to distinguish them, is obtained (Figure 17). These structures have the same lattice parameter (a=0.2861 nm) but different average crystallite sizes (Figure 18), and different microstrains. The existence of two-bcc α-Fe(P) structures which is linked to the non-equilibrium mechanical alloying process, can be assigned to the non homogeneity in concentration. Indeed, according to the very low solubility limit of the phosphorus in the iron (<0.026 at.% P at room temperature), and because the Fe-8P mixture contains more of P than the bcc α-Fe host phase can dissolve, then the surplus must separate out to give regions with high concentration of P as well as chemical compounds such as Fe_3P phosphide. The two phases are stuck together by coherent interphase boundaries with low energy and has some strain [55].

According to the thermodynamic data of the Fe-P system, the ratio of the reaction enthalpy to the heat capacity ($-\Delta H/Cp$) which represents a rough estimation of the adiabatic temperature is about 1740 and 1910 K for the Fe_2P and Fe_3P phosphides, respectively. It is thus obvious that Fe_2P adiabatic temperature is lower than that of the Fe_3P phase. The formation of the Fe_xP ($1<x<2$) and Fe_2P phosphides, in the early stage of milling, can be explained by the increase of the contact area between the reactant powder particles due to a reduction in particle size. The reaction proceeds thus without atomic thermally activated diffusion through the produced layer. Also, because ball milling route induces frequently self-sustaining thermal reaction (combustion) in systems where the adiabatic temperature is so high [56, 57] on the one hand, and since the milling process is intense (I=9) on the other hand; the ignition of a self-sustaining process is expected. Therefore, the heat generated involve the transfer of a kinetic energy of a highly energetic ball to the powder, favors the formation of the Fe_2P and Fe_xP ($1<x<2$) phases. The formation of such phosphides, in a short milling time, has been observed in the mechanically alloyed transition metal phosphides TM_xP_y (TM = Co, Mn, Fe, Ni) [58].

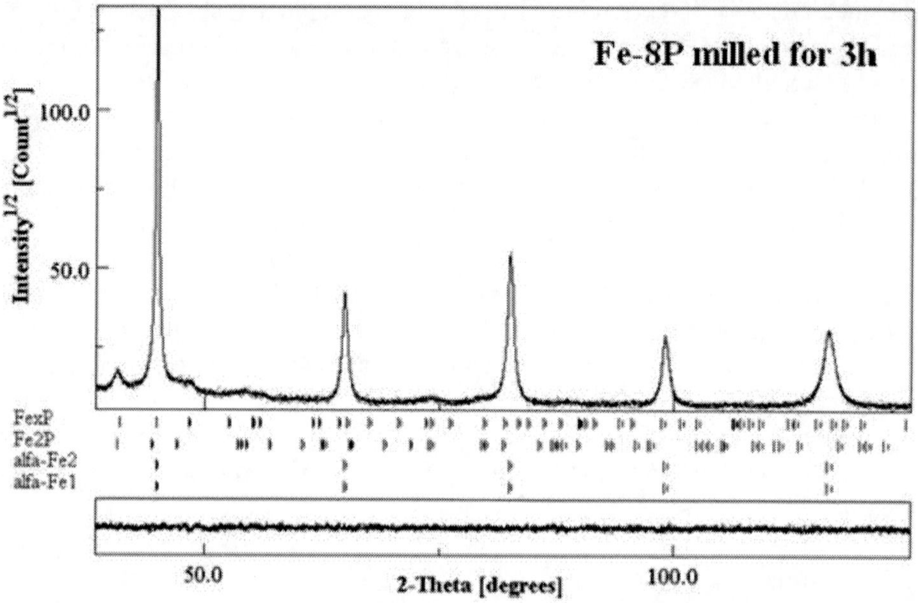

Figure 16. Rietveld refinement of the XRD pattern of the ball-milled Fe-8P powders for 3 h

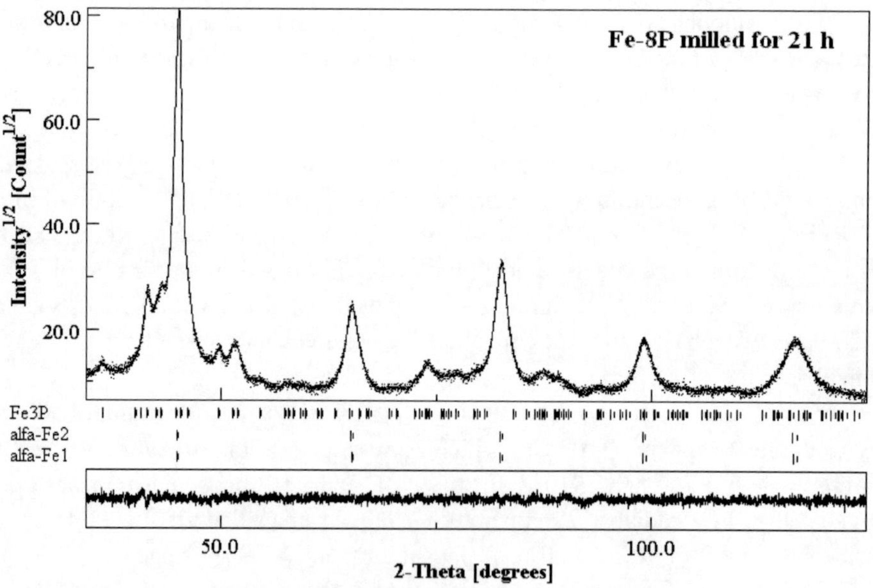

Figure 17. Rietveld refinement of the XRD pattern of the ball-milled Fe-8P powders for 21 h

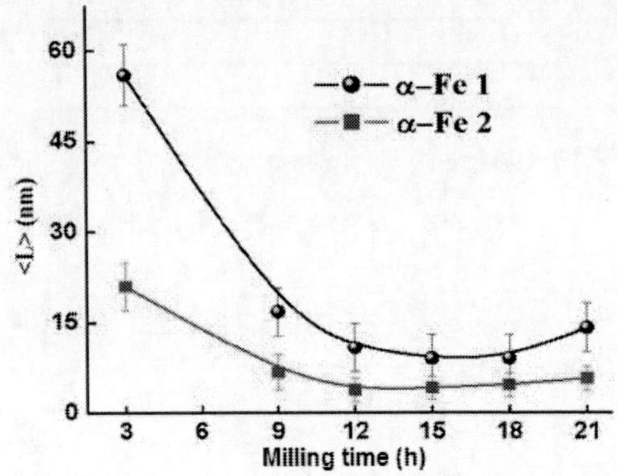

Figure 18. Milling time dependence of the mean crystallite size for α-Fel and α-Fe2 structures in the Fe-8P powder mixture

Table 2. Hyperfine parameters deduced from the fitting of the Mössbauer spectrum of the Fe-8P powder mixture milled for 3 h: IS, QS, 2ε, B and Γ correspond to isomer shift, quadrupolar splitting, quadrupolar shift, hyperfine field and line width, respectively

Phases	Sites	B (T) ±0.5	IS (mm/s) ±0.02	QS or 2ε (mm/s) ±0.02	Γ (mm/s) ±0.02	Relative area (%) ±2
Fe_xP (1<x<2)	1	0	0.23	0.75	0.30	6
Fe_2P	2	0	0.60	0.86	0.30	4
α-Fe	3	34.4	0.00	0.02	0.26	63
	4	35.3	0.00	0.02	0.25	27

The Mössbauer spectrum of the Fe-8P powder milled for 3 h reveals the existence of two paramagnetic doublets related to a non-stoichiometric Fe_xP (1<x<2) and Fe_2P phosphides according to their hyperfine parameters (Table 2). The higher hyperfine field values of the magnetic components, B_1=35.3 T and B_2=34.4 T, compared to that of pure α-Fe (B=33 T) might originate from structural distortions. Increasing hyperfine magnetic field is related to that of the magnetic moment per atom resulting from the increasing near-neighbor distance [59, 60]. After 21 h of milling (Figure 19), the paramagnetic doublet can be attributed to the FeP phosphide and the magnetic sites (1 to 4) to the three non-equivalent positions of Fe atoms in the Fe_3P tetragonal structure (Table 3). The hyperfine parameters B_1=30.0 T and IS_1=0.03 mm/s; B_2=33.3 T and IS_2= 0.01 mm/s indicate the presence of various types of atomic Fe neighboring. Since the presence of one P atom in the vicinity of Fe atom reduces the hyperfine magnetic field and increases the isomer shift by about 2 T and 0.01 mm/s, respectively [61], therefore, the site with B_2 = 33.3 T and IS_2 = 0.01 mm/s might be attributed to the Fe atom surrounded by only Fe atoms on the two first coordination shells. This corroborates with the fact that P atoms prefer surface positions, when distributed in the bcc Fe crystal, in order to minimize the free energy of the system [62]. Nevertheless, the site with B_1=30.0 T and IS_1= 0.03 mm/s can be related to the presence of at least one P atom as near-neighbor to the Fe atom.

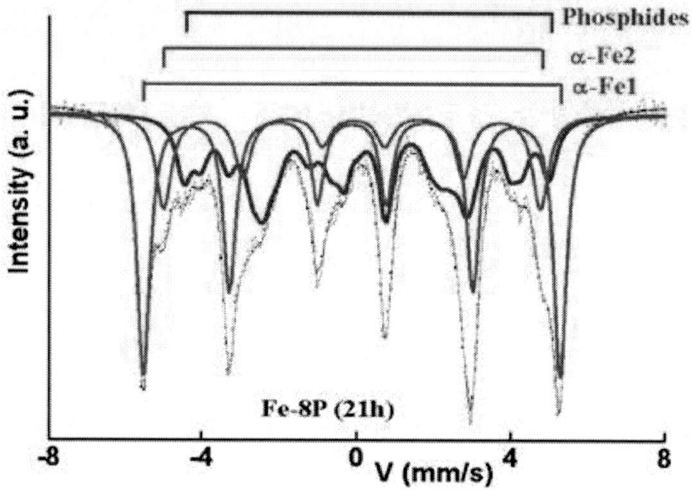

Figure 19. Mössbauer spectrum of the ball-milled Fe-8P powder for 21 h

Table 3. Hyperfine parameters deduced from the fitting of the Mössbauer spectrum of the Fe-8P powder mixture milled for 21 h: IS, QS, 2ε, B and Γ correspond to isomer shift, quadrupolar splitting, quadrupolar shift, hyperfine field and line width, respectively

Phases	Sites	B (T) ±0.5	IS (mm/s) ±0.02	QS or 2ε (mm/s) ±0.02	Γ (mm/s) ±0.02	Relative area (%) ±2
Phosphides	Doublet	0	0.33	1.00	0.30	2
α-Fe(P)	Site 1 (FeII)	16.5	0.43	~0.08	0.60	12
	Site 2 (FeIII)	22.4	0.33	0.26	0.46	8
	Site 3 (FeIII)	25.5	0.22	0.07	0.49	8
	Site 4 (FeI)	29.1	0.29	0.20	0.46	10
	Site 5	30.0	0.03	~0.03	0.49	18
	Site 6	33.3	0.01	0.01	0.40	41

5.4.2. Fe-P-C alloys

For the Fe-6P-1.7C powders, in a planetary ball-mill Retch PM400 with a BPR of about 5/37 and a rotation speed of 350 rpm, no reaction between the elemental Fe, P and C powders was observed after 3 h [63]. The dissolution of both P and C into the α-Fe matrix, after 6 h of milling, is evidenced by the formation of Fe_2P phosphide, α-Fe(C) and α-Fe(P) solid solutions having different lattice parameters a_1=0.2910 nm and a_2=0.2872 nm and, different crystallite sizes L_1=49 nm and L_2=74 nm, respectively. A mixture of Fe_3C carbide (L = 5 nm), Fe_3P phosphide (L=7 nm) and bcc α-Fe(C, P) solid solution (L=8 nm) with relative proportions of about 31, 19 and 50%, respectively, is obtained after 104 h of milling (Figure 20). The same phases have been obtained during the rapid transformation of the Fe-6P-1.7C (wt. %) amorphous alloy when heated at rates exceeding 100°C/min [64]. The Mössbauer spectrum of the Fe-6P-1.7C powder milled for 104 h reveals the existence of several components (Figure 21). According to the hyperfine parameters (Table 4), the subspectra SS1 to SS3 which represent ~ 47.5% of the total area are related to the α-Fe(C, P) solid solution with various types of Fe neighboring. While, the sites SS4 to SS7 (~43 %) can be assigned to the three non-equivalent positions of Fe atoms in the Fe_3P tetragonal structure and finally, the SS8 component corresponds to iron site in the Fe_3C carbide (~9.5%). The alloying process can be monitored by the variation of both the average hyperfine magnetic field, <B>, and the average isomer shift, <IS>, as shown in Figure 22. The average hyperfine magnetic field decreases rapidly up to 12 h and then remains nearly milling time independent. Simultaneously, the average isomer shift increases with increasing milling time, reaches a maximum value of about 0.157 mm/s after 50 h, then decreases to about 0.102 mm/s after 104 h of milling. The decrease of <B> and the increase of <IS> are related to the diffusion of C and P atoms into the α-Fe matrix since both P and C atoms reduce the hyperfine magnetic field and increase the isomer shift [61, 65].

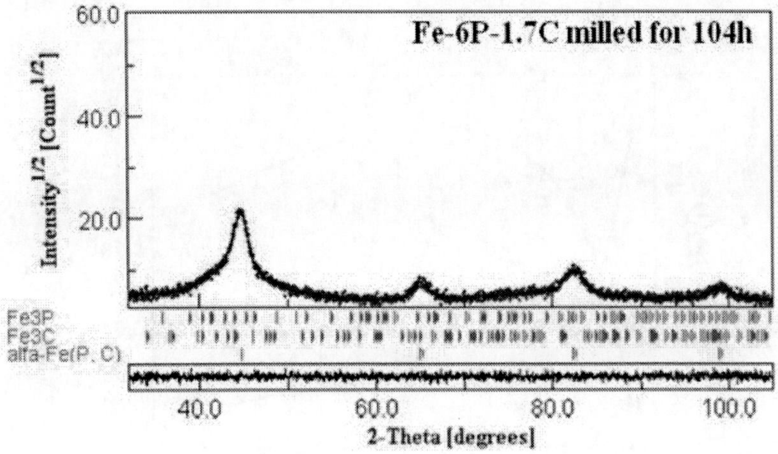

Figure 20. Rietveld refinement of the XRD pattern of the Fe-6P-1.7C powders after 104 of milling

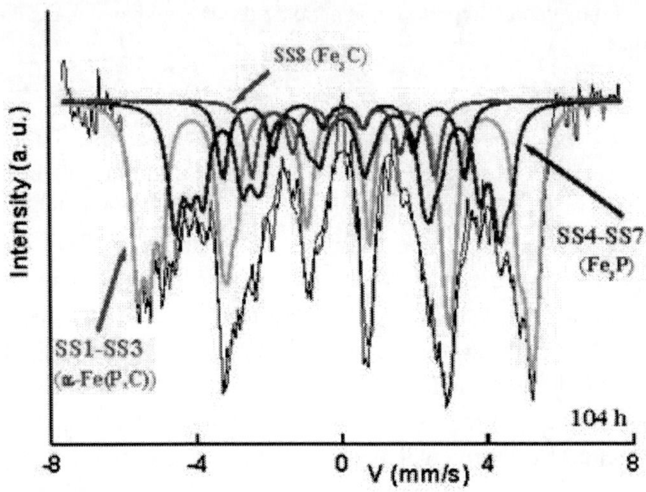

Figure 21. Fitting of the Mössbauer spectrum of the milled powders for 104 h

Table 4. Hyperfine parameters values (B, IS, 2ε, Γ and the relative area) deduced from the fitting of the Mössbauer spectrum of the Fe-6P-1.7C powders milled for 104 h

Phases	Subspectra	B (T) ±0.5	IS (mm/s) ±0.05	2ε (mm/s) ±0.05	Γ (mm/s) ±0.02	Area (%) ±1
α-Fe(C, P)	SS1	33.8	0.014	0.185	0.48	20.0
	SS2	32.3	0.081	0.157	0.31	12.8
	SS3	30.3	0.159	0.18	0.36	14.6
Fe_3P	SS4	27.5	0.154	0.22	0.44	16.6
	SS5	27.2	0.2	0.185	0.58	7.6
	SS6	23.4	0.375	0.185	0.58	10.0
	SS7	15.5	0.35	0.185	0.58	9.0
Fe_3C	SS8	20.5	0.235	0.185	0.58	9.4

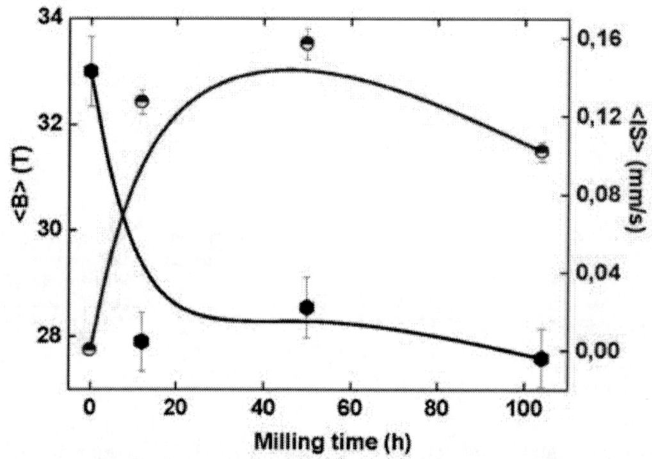

Figure 22. Variation of the average hyperfine magnetic field, <B>, and average isomer shift, <IS>, as a function of milling time

5.4.3. Fe-Cr-P-C alloys

Amorphous $Fe_{77}Cr_4P_8C_{11}$ alloys have been prepared by mechanical alloying [66, 67]. The milling process was carried out in a planetary ball mill P7, under argon atmosphere, using the following conditions: (i) Fe, P, Cr and activated carbon, I = 7 and a BPR of 35/2 for sample A-$Fe_{77}Cr_4P_8C_{11}$, and (ii) Fe, Cr, P and graphite, I = 9 and a BPR of 35/5 for sample G-$Fe_{77}Cr_4P_8C_{11}$. In order to understand how the alloying of the quaternary Fe-Cr-P-C mixture occurs during the milling process, Mössbauer spectrometry measurements have been used jointly to XRD since their patterns are not adequate and conclusive enough to determine if the phase only results of fine-grained structures with grain boundaries or contains small crystallites embedded in an amorphous matrix. For the G-Fe-Cr-P-C powders, there is only a crystallite size refinement of elemental powders with a small fraction of the paramagnetic Fe_2P phase in the first stage. After 12 h of milling, 24%Fe is still discerned by both XRD (Figure 23) and Mössbauer spectrometry (Figure 24) while the other 76%Fe are mixed with P, C and Cr resulting in the formation of α-(Fe,Cr), $(Fe,Cr)_2P$, $(Fe,Cr)_3P$ and $(Fe,Cr)_3C$ type phases. The milling process results in a more or less homogeneous crystallite size of all nanocrystalline phases ranged from 40 to 50 nm in addition to the amorphous-like phase after 24 h. The amorphization may be related to the instantaneous high-pressure effect. The new crystalline phase on the background of amorphous halo is attributed to the $(Fe,Cr)_7C_3$ type carbide according to its lattice parameters (Figure 25). The Mössbauer spectrum is consistent of a broad and symmetrical six spectral lines characteristic of a distribution of the hyperfine parameters (B, IS) where the Fe atoms are largely and diversely surrounded by C, Cr and P atoms. According to the disappearance of the paramagnetic doublet and the Mössbauer spectrum shape (Figure 24) which looks like those of the amorphous FePC alloys [52] prepared by conventional techniques, one can deduce that the alloying process at the atomic level and, consequently, the formation of an amorphous-like alloy is achieved. From the variation of individual phase's weight fractions as a function of milling time (Figure 26), it is obvious that the proportions of α-Fe and α-(Fe,Cr) decrease with increasing milling time and vanish completely above 12 h leading to the alloying of elemental Fe, Cr, P and C powders. One also observes that the fraction of the $(Fe,Cr)_3P$ phase exhibits a different behavior in comparison with those of the other phases. Thus, the former increases continuously up to 96 h, while those of $(Fe,Cr)_2P$, $(Fe,Cr)_3C$ and $(Fe,Cr)_7C_3$ decrease. Above 96 h milling, the plateau may correspond to a steady state arising from the complete diffusion process and/or the equilibrium state.

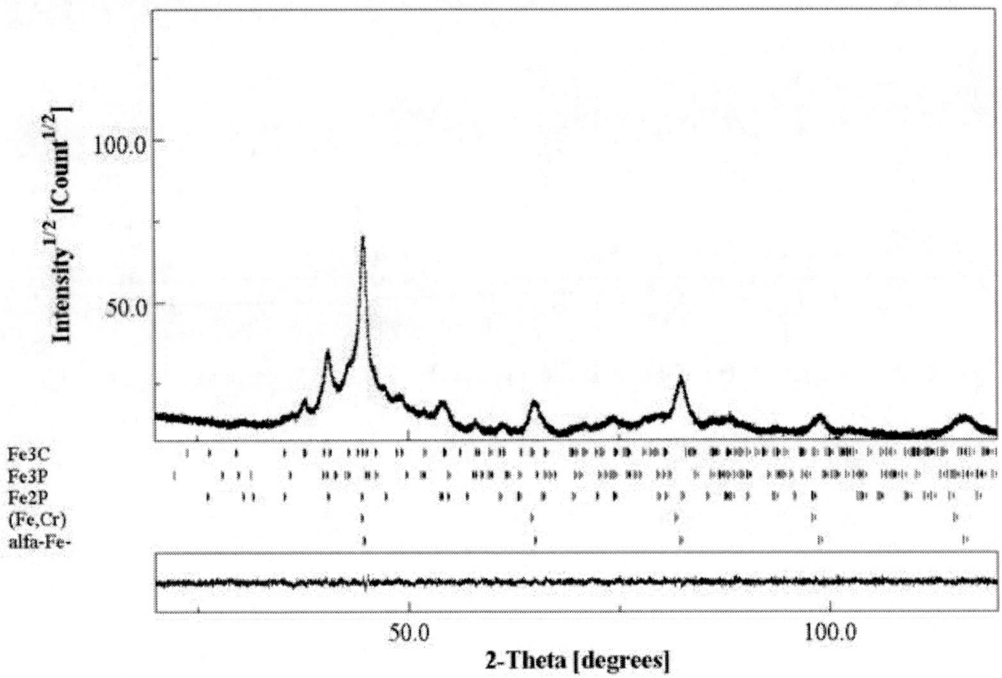

Figure 23. XRD pattern of the G-FeCrPC powder mixture milled for 12 h. Experimental (dots) and calculated (full line) patterns are shown

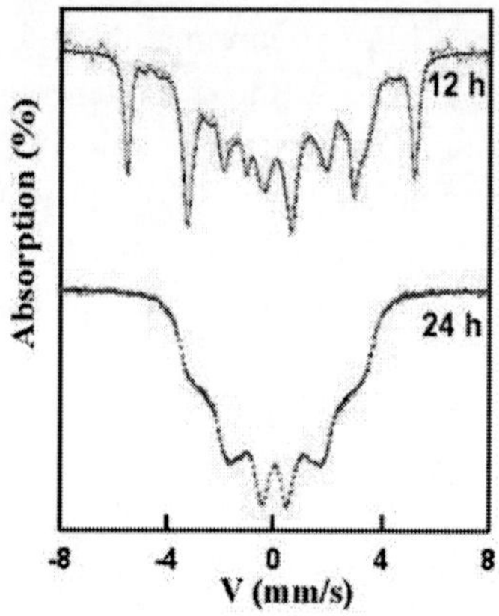

Figure 24. Mössbauer spectra, taken at 300 K, of the G-FeCrPC powder mixture milled for 12 and 24 h

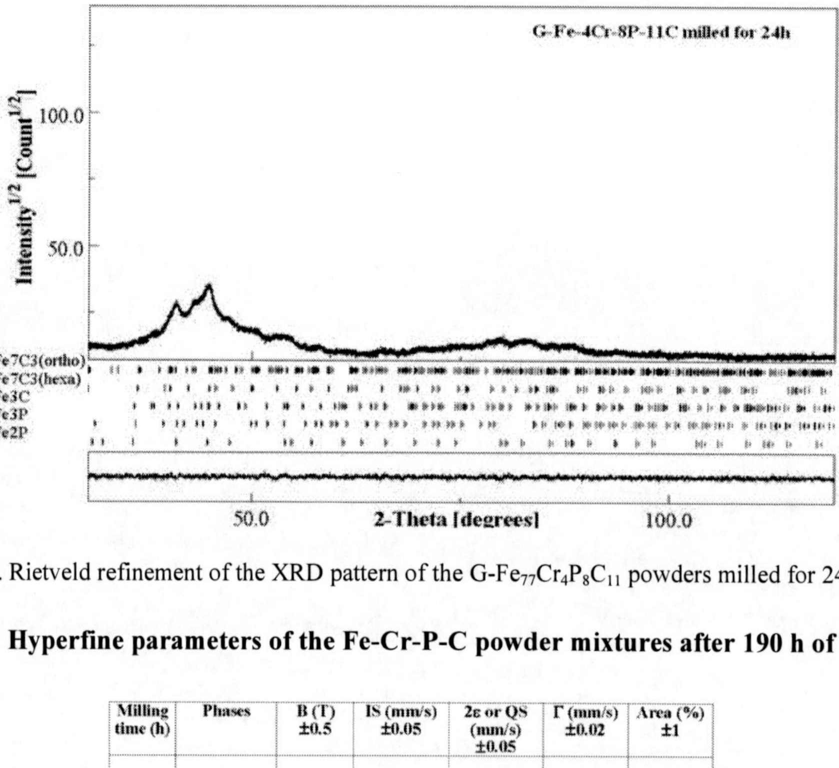

Figure 25. Rietveld refinement of the XRD pattern of the G-$Fe_{77}Cr_4P_8C_{11}$ powders milled for 24 h

Table 5. Hyperfine parameters of the Fe-Cr-P-C powder mixtures after 190 h of milling

Milling time (h)	Phases	B (T) ±0.5	IS (mm/s) ±0.05	2ε or QS (mm/s) ±0.05	Γ (mm/s) ±0.02	Area (%) ±1
190	$(Fe,Cr)_{2.2}C$	0	0.25	0.91	0.66	12
	$(Fe,Cr)_3P$	17.1	0.34	-0.77	0.85	14
		23.1	0.37	-0.51	0.47	8
	$(Fe,Cr)_3C$	21.1	0.30	-0.07	0.64	17
	$(Fe,Cr)_7C_3$	16.4	0.21	0.29	0.66	34
	Amorphous like phase	10.3	0.48	-0.26	0.58	15

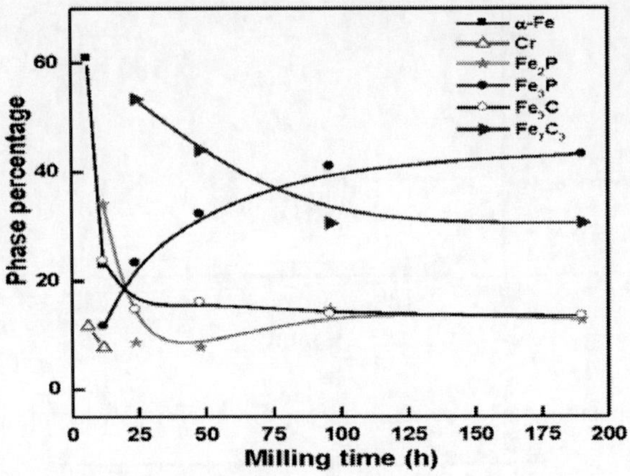

Figure 26. Variation of individual phase's weight fractions as a function of the milling time

As mechanical alloying is a high energy process involving the transfer of kinetic energy of highly energetic balls to the powder, further milling time (up to 190 h) will generate heat resulting in a mechanical recrystallization of the amorphous phase. This is evidenced by the increase in both the XRD peak intensities (Figure 27) and the crystallite size of $(Fe,Cr)_7C_3$ (L = 176 nm) and $(Fe,Cr)_3C$ carbides (L = 144 nm) as well as the reduction of the amorphous like phase amount. The significant change in the shape of the Mössbauer spectrum (Figure 28) results in the emergence of a paramagnetic doublet and other magnetic sextets attributed to the phosphides and carbides (Table 5). The hyperfine parameters of the paramagnetic doublet are close to those of the superparamagnetic ε'-$Fe_{2.2}C$ hexagonal carbide [68], in agreement with that obtained during milling of the $Fe_{75}C_{25}$ powders mixture using a spex mill [69]. In fact, it has been reported that amongst the carbides formed during milling: the cementite, Fe_3C, and the disordered hexagonal carbide (either magnetic or superparamagnetic) with approximate composition Fe_2C.

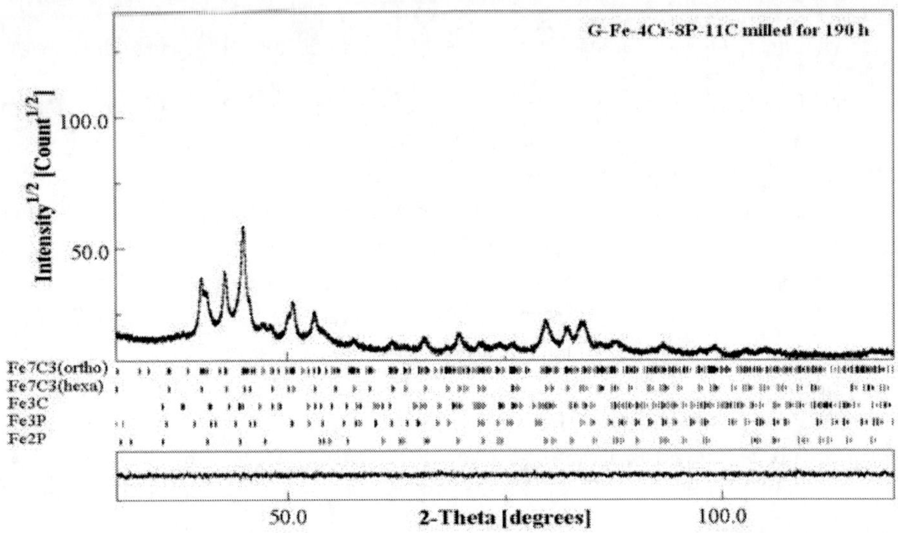

Figure 27. Rietveld refinement of the XRD pattern of the G-$Fe_{77}Cr_4P_8C_{11}$ powders milled for 190 h

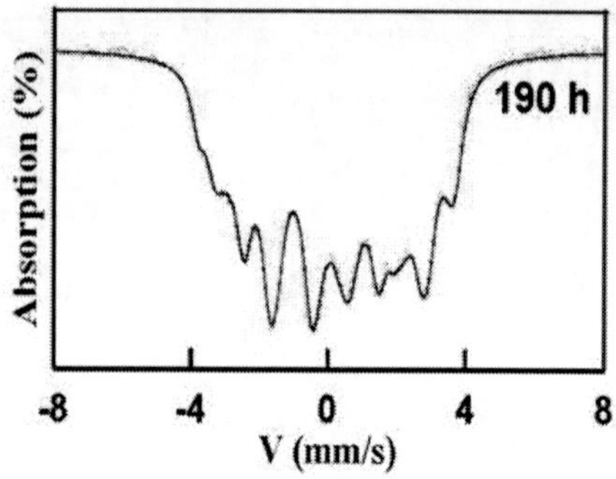

Figure 28. Mössbauer spectrum, taken at 300 K, of the G-FeCrPC powder mixture milled for 190 h

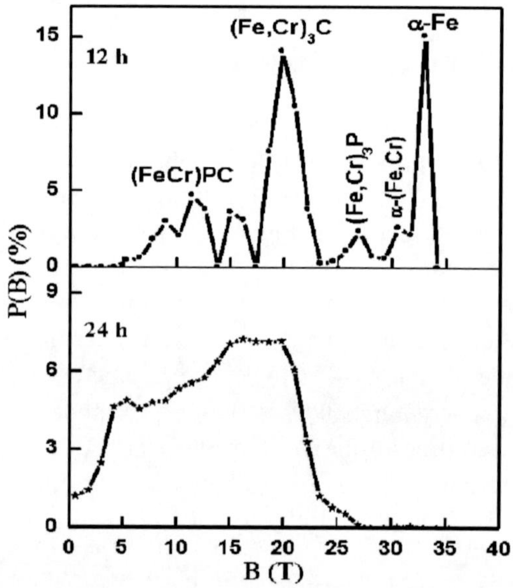

Figure 29. Hyperfine field distributions of the powder mixtures milled for 12 h and 24 h

The hyperfine field distributions, P(B), of the powders milled for 12 and 24 h provide evidence of the Fe environments change versus milling time and thus the structural transformation (Figure 29). Accordingly, after 12 h, besides the peak with 33 T attributed to α-Fe and having a relative proportion of about 24%, there clearly appear several and distinct peaks ranging from 5 to 31 T corresponding to different Fe atoms surroundings and/or phases, such as α-(Fe,Cr), $(Fe,Cr)_3P$, $(Fe,Cr)_3C$ and may be amorphous (Fe,Cr)PC having relative fractions of about 3, 7, 42 and 21%, respectively. Upon milling for 24 h, a significant change is observed in the hyperfine field distribution. This later consists of a broad and weak peak enlarged upon the domain 0-25 T. The absence of the main peaks might be related to more or less homogeneous distribution of the Cr, P and C atoms into the Fe lattice and to the amorphization process. The broadening of the P(B) curve after 24 h of milling has to be associated to the high lattice strains, the density of defects and of grain boundaries increasing.

The effect of the starting powders, especially the activated carbon, is evidenced by the complete mixing of all elemental powders of the A-Fe-Cr-P-C mixture after 12 h of milling. Separate alloying of P, Cr and C with Fe leads to the formation of α-Fe(Cr), Fe_xP ($1< x <2$), Fe_3P and Fe_3C phases according to heir mixing enthalpy. Indeed, it seems that the attraction

Cr-P is rather moderate [70], whereas the Cr forms several types of carbides, as usual. It is well established that the reaction between Fe and Cr is always observed in the early stage of milling because of the negative mixing enthalpy ($\Delta H_{mix}= -1kJ/mol$) and their bcc structures and close lattice parameters. The presence of Cr in steels favors the P segregation at the grain boundaries, while its concentration increases when the global C content decreases [70].

6. MAGNETIC PROPERTIES

Microstructural factors such as grain size, defects, precipitates and structural order parameter, significantly change both magnetic and mechanical properties in FeCo-based alloys, especially in samples with small Nb additions [71]. It has been found experimentally that grain boundaries, precipitates, and disordered phases act as pinning sites for the magnetic domain wall movement, which increase the coercivity, whereas the precipitates and order parameters alter the saturation magnetization, Ms. The changes in magnetic properties might be related to changes in local order. These properties include Curie temperature variation, Tc, and rate of saturation magnetization decrease with solute concentration.

6.1. Pure Fe

The Curie temperature, T_c, of the ball-milled Fe powders is deduced from the differential scanning calorimetry (DSC) scans (first endothermic peak) and plotted versus milling time in Figure 30. One observes that the Curie temperature decreases rapidly between 3 and 12 h and slightly upon further milling time. The obtained value of about 761°C, after 40 h milling, is slightly lower than that of the bulk iron (T_c = 770°C). The decrease of Curie temperature might be due to the grain size decreasing. In fact, if the grain sizes are small enough, the structural distortions associated with surfaces and/or interfaces can lower the Curie temperature. The structural distortions can be confirmed by the Fe lattice parameter increasing with the milling time. It has been reported on far-from-equilibrium nanostructured metals that interfaces present a reduced atomic coordination and a wide distribution of interatomic spacing compared to the crystals and consequently, the atomic arrangement at the grain boundary may be considered close to the amorphous configuration and should therefore alter the Curie temperature [19]. For most observations in the literature, the Curie temperature does not deviate strongly from that of the bulk materials. Host et al. have reported a Tc value of 1093°C for carbon arc produced Co[C] nanoparticles, in good agreement with the 1115°C value for bulk Co [72]. The Curie temperature of 10 nm Gd is decreased about 10 K from that of coarse-grained Gd while the magnetic transition is broader [73].

Figure 31 exhibits the dependence of coercivity H_c with grain size for the ball milled Fe. The increase of coercivity during the first 3 h of milling (up to 35 Oe), can be related to the plastic deformation effect and consequently to the introduction of internal strain and different types of defects, which serve as pinning sites for the magnetic domain wall movement. It has been observed at this stage of milling, that the Fe powder particles are flattened (Figure 1a) by compressive forces [74, 75]. Above 6 h of milling, H_c decreases with decreasing grain size up to 24 Oe then levels of. The reduction of the coercivity which can be attributed to the deviation of interatomic spacings in the interfacial regions as compared with the crystalline component is consistent with the predictions of the random anisotropy model. For grain sizes comparable with the magnetic exchange length, L= $(A/K)^{1/2}$ where A is the exchange stiffness and K is the magnetocrystalline anisotropy, there is a plateau on the H_c versus d curve; in addition, for smaller grain sizes a sharp decrease ($Hc \sim d^6$) is expected because of the drastic decrease of the effective K determined by averaging over the grains above which the domains are extended [76]. The observed drop on the curve of H_c versus grain size is the first experimental verification of the model for ball-milled elemental Fe powders. The soft magnetic behavior of the as-milled pure iron powder has been reported [77]. It is observed that the coercivity value of about 20 Oe for crystalline grain size of around 11 nm does not change with the milling time up to 48 h.

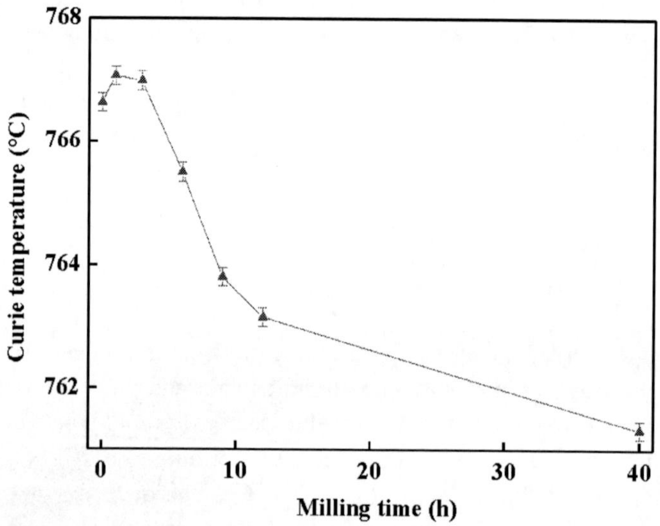

Figure 30. Curie temperature variation of milled pure Fe as a function of milling time

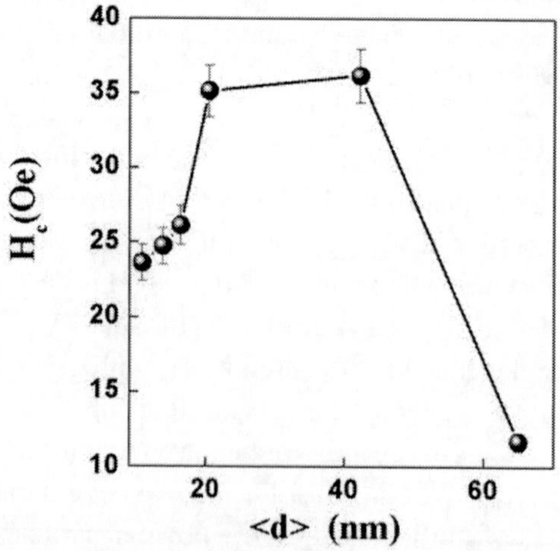

Figure 31. Dependence of the coercivity, Hc, with grain size in ball milled pure Fe

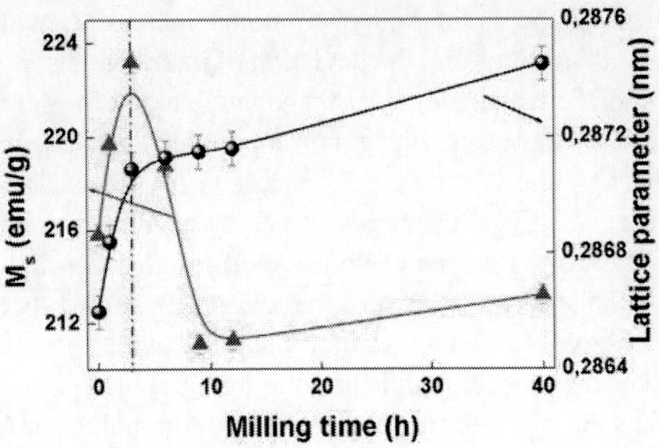

Figure 32. Saturation magnetization, Ms, and lattice parameter changes of pure Fe as a function of milling time

The increase of the saturation magnetization of the ball-milled Fe indicates the introduction of atomic disorder during the first 3 h of milling (Figure 32). The heavy plastic deformation of the ductile Fe powder particles which is followed by repeated fracturing and cold welding leads to the particle size refinement and the introduction of high density of defects, particularly dislocations. Such defects cause a 0.5% volume expansion and consequently changes in geometry of the atomic arrangement and interatomic distances which have an effect on the exchange interaction and hence the mode of spin alignment. Since the magnetization is created by domain wall movement and spin rotation, it can be thus affected by the induced structural defects. The increase of the Fe lattice parameter provides additional evidence for the occurrence of atomic disorder during milling (Figure 32). It has been reported on a good correlation between the saturation magnetization and the lattice parameter (increase of both the lattice parameters and the saturation magnetization) in the Fe-Al alloys, which suggests that the observed magnetic transition may be partially related to the changes in the density of states at the Fermi level [78]. Magnetization decreasing with grain size refinement, above 3 h of milling, may be due to the influence of grain boundary atoms whose arrangement is quite different from that within the grains and results in lower magnetization. The obtained Ms value after 40 h of milling (213 emu/g) is comparable to that found by Brüning et al. [79]. The authors have found that the M_s of the milled Fe powders in a planetary ball mill P5 decreased up to 209 emu/g after 212 h milling [79]. However, the Ms of 6 nm Fe was reported to be 130 emu/g compared to 220 emu/g for normal polycristal [80]. This reduction was attributed to the deviation of interatomic spacing in the interfacial regions as compared with the crystalline component.

6.2. Fe-P Alloys

The temperature dependence of the magnetic susceptibility, χ, of the milled powders for various times is shown in Figure 33. The magnetic susceptibility remains nearly constant up to 500 K then decreases monotonously. A two-magnetic phase state is revealed with different Curie temperatures Tc_1 and Tc_2. The first, Tc_1, is nearly constant during the milling process while the second, Tc_2, is milling time dependent (Figure 34). The lower Curie temperature values might be related to the grain boundary atoms increasing with progressing grain refinement, the structural distortions and the deviation of interatomic spacing in the interfacial regions. The depression of Tc which are lower than that obtained for the milled pure Fe (Tc = 1033 K) [75], indicates that the nearest-neighbor coordinates are essentially changed in the magnetic nanocrystallites by the P additions. The reason for the existence of two-magnetic phase state and therefore, two Curie temperatures can be attributed to inhomogeneities since the Curie temperature is sensitive to the chemical short range order and subsequently to the local Fe environment. This result confirms the existence of two-bcc α-Fe1 and α-Fe2 structures as discussed above. In nanocrystalline materials with two phase microstructures, the Curie temperature of the nanocrystalline and the intergranular phases are both important parameters in describing the magnetic response. The intergranular phase which is typically richer in non-magnetic species can have a suppressed magnetic order parameter and lower Tc of α-Fe [81]. It has been reported that Curie temperature varies with P, B and C content in the Fe-P, Fe-B and Fe-C systems [82]. Curie temperature decreases with increasing Fe content suggesting that the structure is behaving like close-packed rather than the packing of bcc α-Fe. Tc_1 can be attributed to an equilibrium structure with unchanged chemical short range order, while the variation of Tc_2 can be related to the diffusion of P into the Fe matrix.

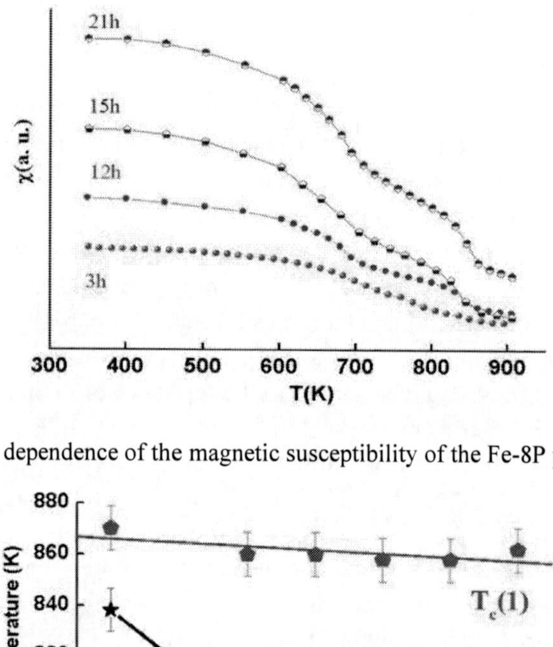

Figure 33. Temperature dependence of the magnetic susceptibility of the Fe-8P powder mixture

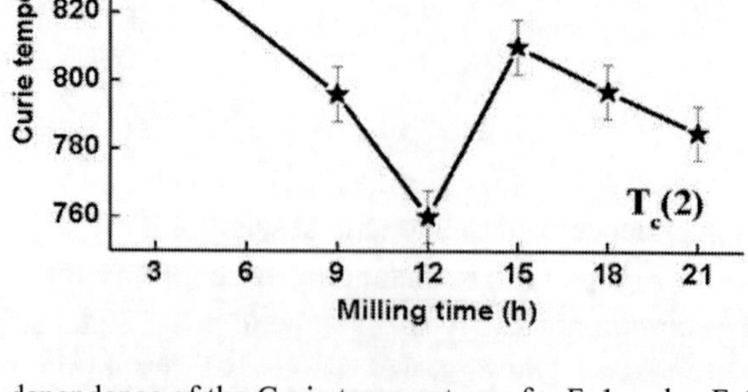

Figure 34. Milling time dependence of the Curie temperature of α-Fe1 and α-Fe2 structures

6.3. FeCo-Based Alloys

From the magnetic susceptibility variation as a function of temperature (Figure 35), it is noticeable that the magnetic susceptibility value, at 300 K, of sample B milled for 3 h (1.3 emu/mole) is about two times larger than that of sample A milled for 24 h (0.64 emu/mole). The magnetic susceptibility remains constant up to 500 K and then decreases monotonously. The higher magnetization of sample B might be due to the highly atomic disorder and/or to the heterogeneity of the Fe(Co) structure. A prominent Curie tail is observed above 800 K indicating paramagnetic response. The obtained Curie temperature, Tc, of about 845 and 851 K for samples A and B, respectively, are milling time independent for each sample. The lower Tc values might be related to the grain boundaries atoms increasing with progressing grain refinement, the structural distortions and the deviation of interatomic spacing in the interfacial regions. The depression of Tc in the Fe-12Co (wt.%) powder mixtures which are some 183-189 K lower than that obtained for the milled pure Fe (1034 K) [75], indicate that the nearest-neighbor coordinates are essentially changed in the magnetic nanocrystallites. This reflects, to some extent, that there are more open disordered spaces or the nearest-neighbour coordination distance in the nanometer-sized crystallites is increased caused by lattice distortion. Also, the Tc values confirm the fact that the Fe-rich alloys typically have lower Curie temperatures which is reached before the $\alpha \rightarrow \gamma$ transition for Co concentration less than 17%. However, for Co content more than 17%, the phase transformation to the nonmagnetic γ phase occurs while the γ phase is still ferromagnetic, leading to "virtual" Tc values for the α phase which are greater than the $\alpha \rightarrow \gamma$ transition temperature. Hence, for the nanostructured $Fe_{50}Co_{50}$ and $Fe_{40}Co_{60}$ alloys, Tc cannot be unambiguously observed because there is a phase transformation from the bcc to the fcc form at 985°C according to the Fe-Co phase diagram.

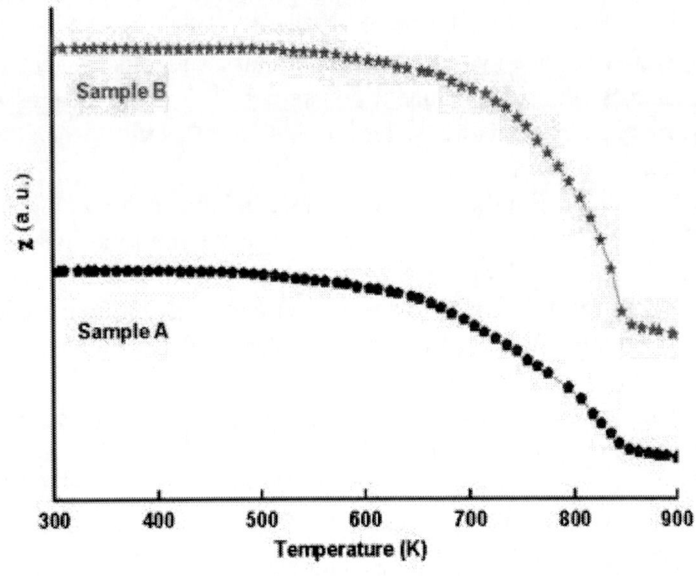

Figure 35. Magnetic susceptibility vs. temperature for samples A and B milled for 24 and 3 h, respectively

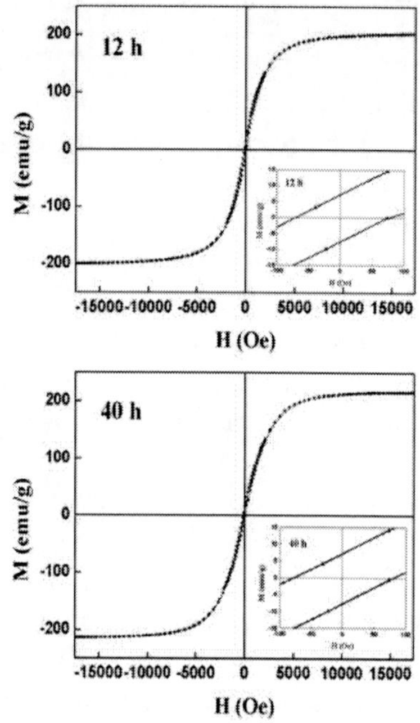

Figure 36. Typical hysteresis loops dependence on milling time, at T = 100 K, of $Fe_{50}Co_{50}$ powders mixture. The insert is an enlargement of low field region

The changes in the magnetic properties of nanoscale ferromagnetic materials usually originate from the microstructure and are attributed to the large fraction of atoms associated with the grain boundaries/interfaces, linking single domain crystalline grains. Figure 36 illustrates the milling time dependence of the hysteresis loops, at 100 K, of the ball-milled $Fe_{50}Co_{50}$ powders mixture. The hysteresis cycles exhibit a sigmoïdal shape, which is usual in nanostructured samples with small magnetic domains. This is due to the presence of structural distortions inside grains. The small hysteresis losses are properties generally desired in soft magnetic materials.

The coercivity of the ball-milled $Fe_{50}Co_{50}$ powder mixture attains a maximum value after 1 h of milling (Figure 37) [75]. H_c decreases with decreasing grain size, up to 9 h, then increases slightly for longer milling times while the grain size levels off. The H_c enhancement for short milling times can be a consequence of the transformation from fcc to hcp Co, since, in general, hcp structures are known to have larger magnetocrystalline anisotropy than fcc ones [83]. The grain size reduction resulted in the decrease of the coercivity in good agreement with the theoretical predictions of the random-anisotropy model (RAM) [76, 84]. This result differs from that obtained by Kuhrt et al. [85]. The authors have reported that the coercivity of the Fe-Co powders is milling time independent but pronounced composition dependence was observed similar to that of the Slater-Pauling curve established for bulk alloys.

The variation of the coercivity, Hc, of the $(Fe_{50}Co_{50})_{62}Nb_8B_{30}$ powder mixtures during the milling process can be divided into three stages (Figure 38) [11]: (i) a rapid decrease up to 5 h; (ii) an increase between 5 and 50 h, and (iii) a continuous reduction to a minimum value of about 160 Oe on further milling time. The increase of coercivity between 5 and 50 h of milling can be attributed to the plastic deformation effect and consequently, to the introduction of internal strain and of different types of defects as well as to the Fe-borides formation which can act as additional obstacles to the magnetic domain wall movement [86]. The decomposition process inside the ferromagnetic grains might produce ferromagnetic regions with small size where the Fe-rich ferromagnetic grains are separated by Fe-boride phases with weaker ferromagnetic properties. Thus, the coupling between the ferromagnetic grains of small nanocrystals with enough intergranular material becomes less effective, giving rise to the observed increase of Hc. The decrease of Hc, above 50 h of milling, can be ascribed to the deviation of interatomic spacing in the interfacial regions as compared with the crystalline component.

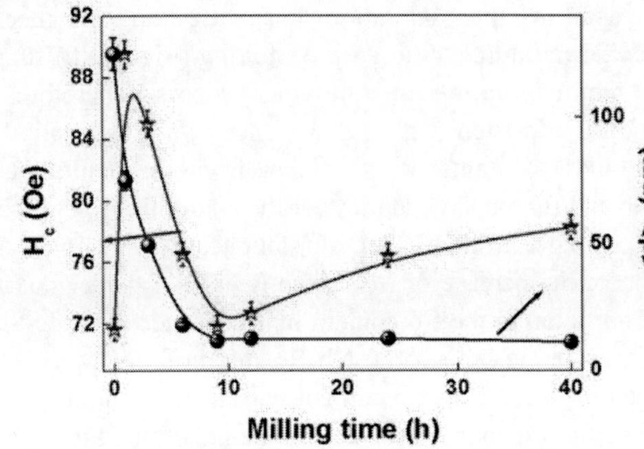

Figure 37. The coercivity, Hc, and the grain size, <d>, variations of the ball milled $Fe_{50}Co_{50}$ powder mixture as a function of milling time

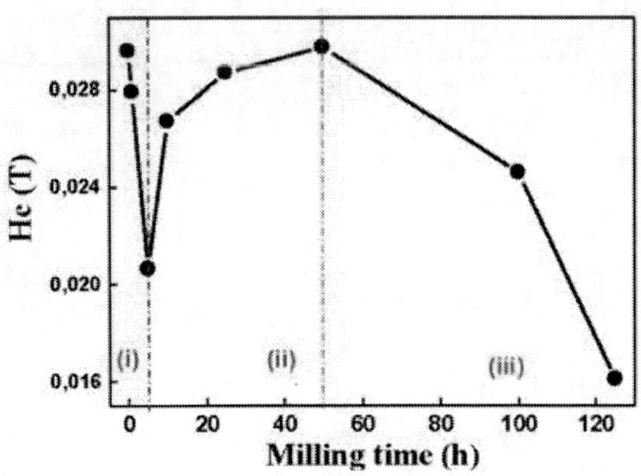

Figure 38. Variation of the $(Fe_{50}Co_{50})_{62}Nb_8B_{30}$ coercivity as a function of milling time

The progress of alloying reaction between elemental Fe and Co powders during the milling process can be monitored by means of saturation magnetization changes (Figure 39). The increase of the magnetization is found to be related to the grain size reduction to about 12 nm. Each grain may be considered as being a single magnetic domain avoiding the influence from magnetic walls [76]. For much smaller grains, the ferromagnetic exchange interaction predominates and prevents the magnetic moments from alignment parallel to the easy axis of each individual grain. This results in a considerable reduction of the effective anisotropy by averaging over several grains presuming that the randomly oriented grains are perfectly coupled via the exchange interaction (RAM). The continuous increase of the magnetization during the milling process is accompanied by a decrease of the lattice parameter (Figure 39). Thus, clustering is favored with more and more Co atoms surrounding a Co atom. Such a Co-rich cluster will bear a larger magnetic moment because of an increasing probability of Co-Co exchange interactions. The number of Co-rich clusters will gradually increases with increasing milling time. This will lead to an increase of the average magnetization which clearly shows a significant change of magnetic moment after ball milling due to the change of the nearest neighbor configuration of the magnetic elements Fe and Co.

The rapid decrease in the saturation magnetization of mechanically alloyed $(Fe_{50}Co_{50})_{62}Nb_8B_{30}$ powder mixtures up to 25 h of milling (Figure 40) suggests a remarkable change of magnetic moment during the alloying process. Therefore, the reduction of saturation magnetization is consistent with the existence of non magnetic Nb and B elements in the vicinity of Fe and Co atoms. It is evident from the amorphous volume fraction increasing that both Nb and B atoms, which favour the amorphization process, are responsible for the Ms decrease. For longer milling times, Ms saturates at about 15.24 Am^2/kg.

The observed monotonic increase of Ms of the $Fe_{50}Co_{50}$ powder mixtures, as a function of milling time, is proportional to the Co content in accordance with the Slater-Pauling curve of Fe-Co alloys. However, in the $(Fe_{50}Co_{50})_{62}Nb_8B_{30}$ mixture**,** Co atoms would prefer to go to boron-rich sites. Therefore, increasing Co content will drive Fe atoms away from boron-rich sites, causing the distribution of number of B near neighbors of Fe to narrow down, which will cause the magnetic moment to decrease. This explains the reduction of the magnetization saturation.

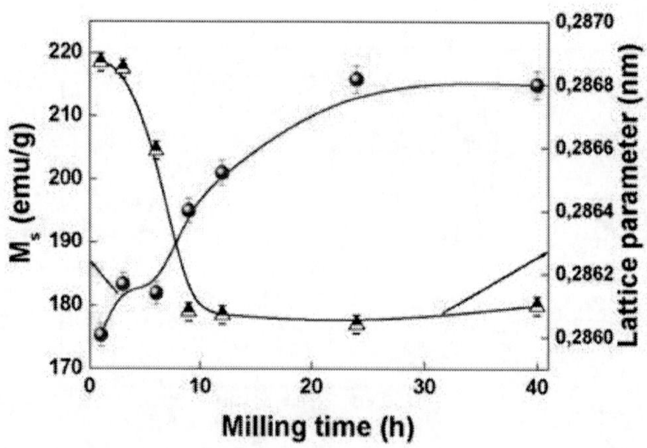

Figure 39. Saturation magnetization, Ms, and lattice parameter changes of $Fe_{50}Co_{50}$ powder mixture as a function of milling time

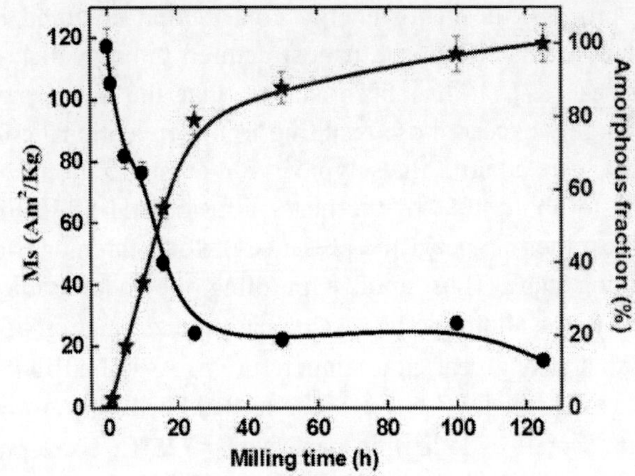

Figure 40. Saturation magnetization and amorphous volume fraction changes of the $(Fe_{50}Co_{50})_{62}Nb_8B_{30}$ powders versus milling time

7. THERMAL STABILITY

Thermal analysis is widely used in the study of reaction of the mechanically alloyed powders. The obtained disordered structures by mechanical alloying are metastable and therefore, they will experience an ordering transition during heating resulting in exothermic reactions. The thermal stability of nanostructured Fe powders is followed by DSC scans as shown in Figure 41. The broad exothermic reaction which consists of two or three overlapping exothermic peaks occurs for all samples and spreads over the temperature range 100-600°C. This behavior which is very similar to that observed for other crystalline metals and nanostructured materials [16-18] originates from recovery, strain relaxation and grain growth. Since the transition heats evolving in the atomic reordering and in the grain growth are comparable in value [16], both atomic disorder and nanocrystallite boundaries are thus major sources of mechanical energy storage. The decrease of the exothermic peak temperature with increasing milling time indicates a decrease of crystallite size.

The evolution of the bcc α-Fe to fcc γ-Fe phase transformation temperature (second endothermic peak), $T_{\alpha\rightarrow\gamma}$, is presented as a function of milling time (Figure 42). Noticeable is the monotonically increase of the $\alpha\rightarrow\gamma$ transformation temperature above the first 3 h of milling, which reaches a value of about 970°C after 40 h. According to the variations of both the Curie temperature and the $\alpha\rightarrow\gamma$ Fe phase transformation temperature ($T_{\alpha\rightarrow\gamma}$), it is obvious that the paramagnetic nanostructured bcc α-Fe domain is extended by about 50°C at the expense of both magnetic bcc α-Fe and nonmagnetic fcc γ-Fe as compared to coarse-grained bcc α-Fe.

For the mechanically alloyed $Fe_{50}Co_{50}$ powders, typical DSC scans are given in Figure 43 as a function of milling time. The transition shape, the transition heat and the peak temperature are sensitive to the milling time. The non-equilibrium state is revealed by the broad exothermic reaction for all the samples, in the temperature range 100-600°C, which is consistent with the energy release during heating due to recovery, grain growth, and relaxation processes within the grain boundaries. As a result of the cold work during the mechanical alloying process, the major energy contribution is stored in the form of grain boundaries and related strains within the nanostructured grains which are induced through grain boundary stresses [87]. It has been shown that the final energies stored during mechanical alloying largely exceed those resulting from conventional cold working of metals and alloys, since they can attain values typical for crystallization enthalpies of metallic glasses corresponding to about 40% of the heat of fusion ΔH_f [87]. Figure 43 exhibits, in addition to the broad exothermic peak, the presence of four, three or two endothermic peaks depending on the milling stage. Thus, up to 1 h milling, the endothermic peaks related to the Curie temperature and to the allotropic bcc→fcc phase transformation of the elemental Fe are observed together with a strong peak at a temperature of 980°C attributed to the α→γ FeCo phase transition. Thereafter, the DSC curve of the milled $Fe_{50}Co_{50}$ powders during 3 h reveals the existence of a new small endothermic peak at T_1=725°C, corresponding to the α→α' disorder-order phase transformation of the FeCo solid solution and indicating, thus, the beginning of the atomic mixing between elemental Fe and Co powders. Thereby, beyond 6 h milling the DSC traces show only two endothermic peaks related to the α→α' disorder-order and to α→γ FeCo, T_2, phase transformations.

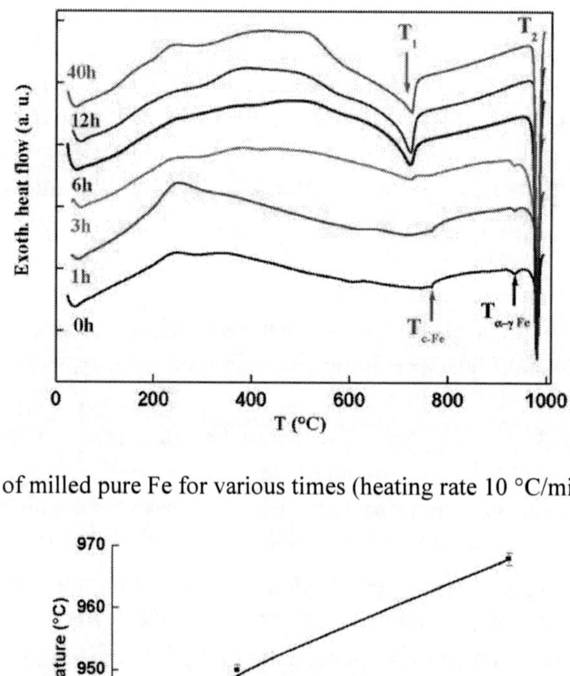

Figure 41. DSC traces of milled pure Fe for various times (heating rate 10 °C/min)

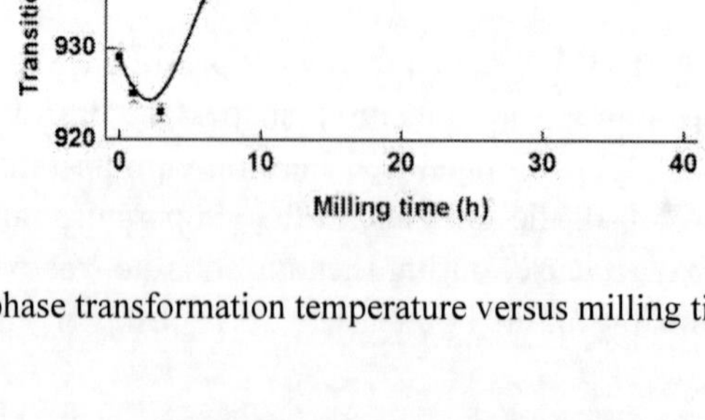

Figure 42. The α →γ–Fe phase transformation temperature versus milling time

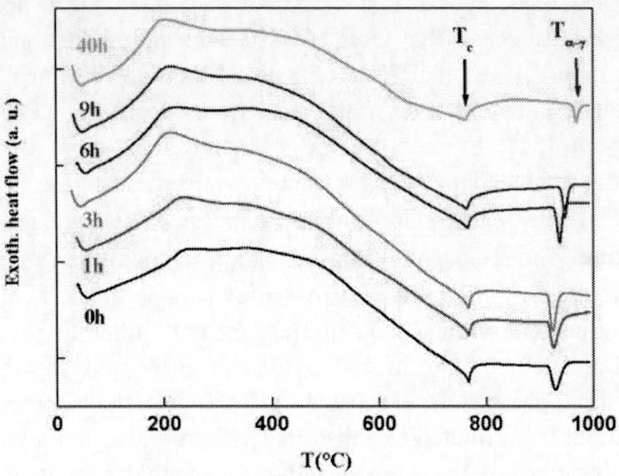

Figure 43. DSC traces of mechanically alloyed $Fe_{50}Co_{50}$ (heating rate 10 °C/min)

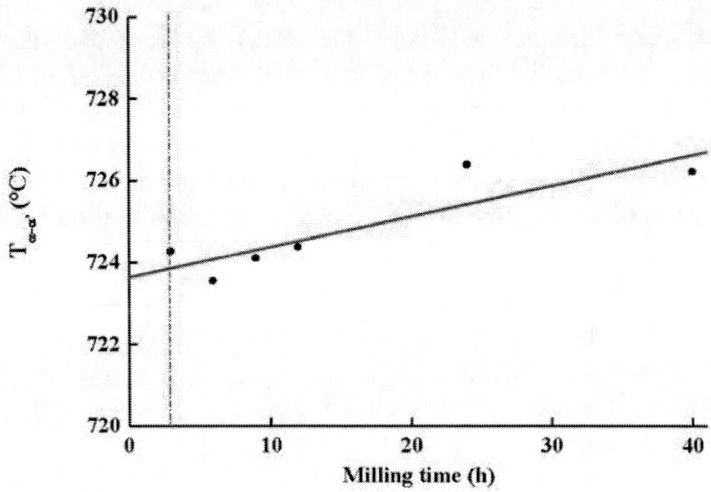

Figure 44. The $\alpha \rightarrow \alpha'$ disorder-order phase transformation temperature of the $Fe_{50}Co_{50}$ powder mixture versus milling time

The $\alpha \rightarrow \alpha'$ disorder-order phase transformation temperature of the Fe(Co) solid solution which is nearly constant (~724°C) along of the milling process (Figure 44), is comparable to that of bulk Fe-Co alloys. It is commonly accepted that Fe-Co undergoes an ordering transition at around 730°C, where the bcc structure takes the ordered α'-CsCl(B2)-type structure [51]. This result confirms the disordered bcc structure of the nanostructured FeCo powders. Similar behavior has been observed in the DTA data of the carbon-coated $Fe_{0.5}Co_{0.5}$ alloy nanoparticles produced using an RF plasma torch [88]. The ordering effect in the FeCo nanocrystals, at about ~730°C, has been also revealed by the changes in the magnetization upon heating and the temperature variation of the coercivity on heating and cooling [88].

The variation of the $\alpha \rightarrow \gamma$ phase transformation temperature in the nanostructured $Fe_{50}Co_{50}$ powders is plotted against milling time as shown in Figure 45. Noticeable is the nearly constant temperature value of 982°C which is milling time independent. The lower resistivity of $Fe_{50}Co_{50}$ compared to that of pure Fe at 300 K [89] and the higher Curie temperature of $Fe_{50}Co_{50}$ suggests that there is less scattering of the conduction electrons by the magnetic excitations. Thus, the Curie temperature cannot be unambiguously observed because there is a phase transformation from the bcc to the fcc form at 985°C according to the Fe-Co phase diagram [51]. In fact, for Co concentrations more than 17%, the phase transformation to the nonmagnetic γ phase occurs while the α phase is still ferromagnetic, leading to "virtual" Curie temperatures for the α phase which are greater than the $\alpha\rightarrow\gamma$ transition temperature. The $\alpha\rightarrow\gamma$ phase transformation temperature (982°C) is rather higher than that obtained in the carbon-coated $Fe_{0.5}Co_{0.5}$ alloy nanoparticles, $T_{\alpha\rightarrow\gamma}$ ~950°C [88].

The Rietveld refinement of the XRD pattern of the ball milled $Fe_{92}P_8$ powder mixture for 21 h and annealed at a temperature of 210°C for half an hour is shown in Figure 46. The refined microstructural parameters are listed in Table 6. The appearance of small new diffraction peaks might be attributed to a distorted orthorhombic FeP phosphide structure. In fact, the lattice parameters of the nanocrystalline phosphide exhibit a deviation from the equilibrium values of the perfect lattice. The relative deviations attain as much as $\Delta a = 0.07$ %, $\Delta b = -0.13\%$ and $\Delta c = 0.28$ % with a grain size of about 50 nm. The existence of only one α-Fe(P) solid solution after annealing might confirm the above cited non homogeneity concentration of the milled powders. Also, the relative proportion of the α-Fe(P) solid solution increases after annealing from 58 to 71%, while that of the Fe_3P one decreases from 42 to 23.5%. On annealing, no significant grain growth occurs for both the α-Fe(P) solid solution and the Fe_3P phase. This can be explained by the lattice distortion of the nanometer-sized crystallites and the decreasing interfacial energy with the grain refinement which will reduce the driving force for a grain growth process and then stabilize the grain size.

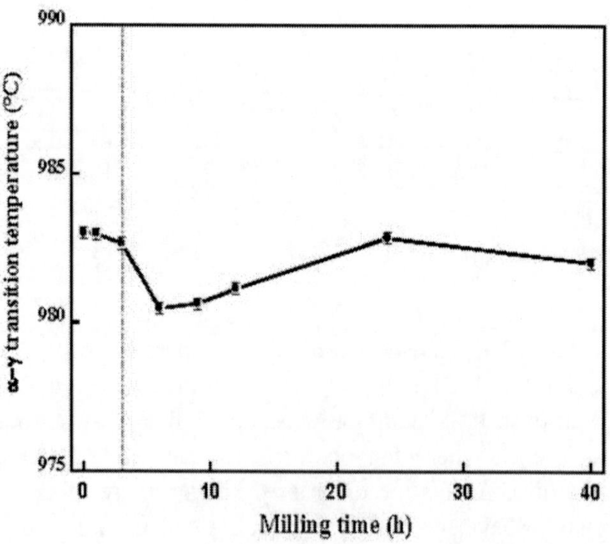

Figure 45. The $\alpha\rightarrow\gamma$ FeCo phase transformation temperature against milling time

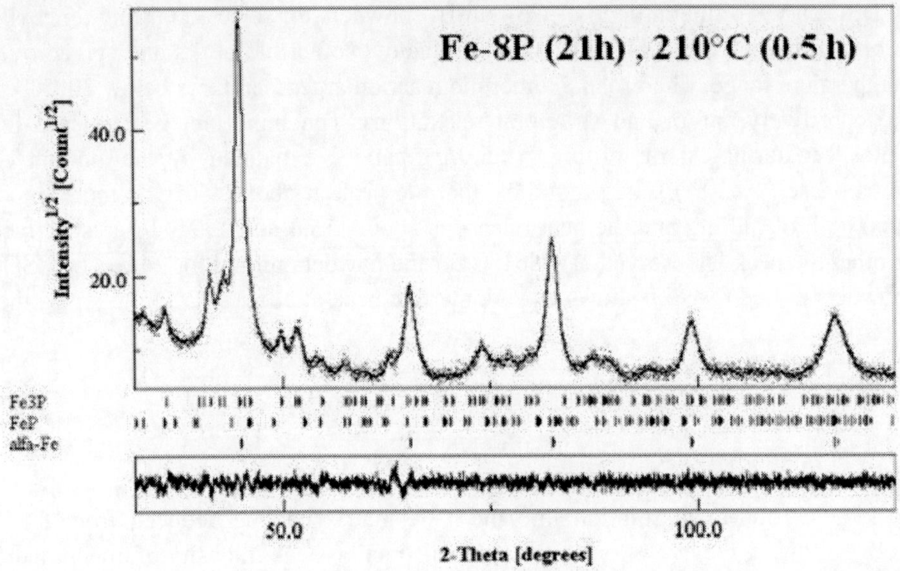

Figure 46. Rietveld refinement of the XRD pattern of the $Fe_{92}P_8$ powders milled for 21 h and heat treated at 210°C for ½ h

Table 6. Lattice parameters (a, b, c), microstructural parameters (<L> and $<\sigma^2>^{1/2}$, average crystalline grain size and microstrains, respectively) and phase percentage of the annealed ball milled $Fe_{92}P_8$ powder mixture for 21 h, at a temperature of 210°C for half an hour

Phases	a (nm) $\pm 10^{-3}$	b (nm) $\pm 10^{-3}$	c (nm) $\pm 10^{-3}$	<L> (nm) $\pm 10^{-1}$	$<\sigma^2>^{1/2}$ $\pm 10^{-3}$	Percentage (%) ±1
α-Fe(P)	0.2869	0.5052	----	8.5	0.62	71
FeP	0.5585	----	0.3958	50	6.24	5.5
Fe_3P	0.9098	----	0.4488	8	1.13	23.5

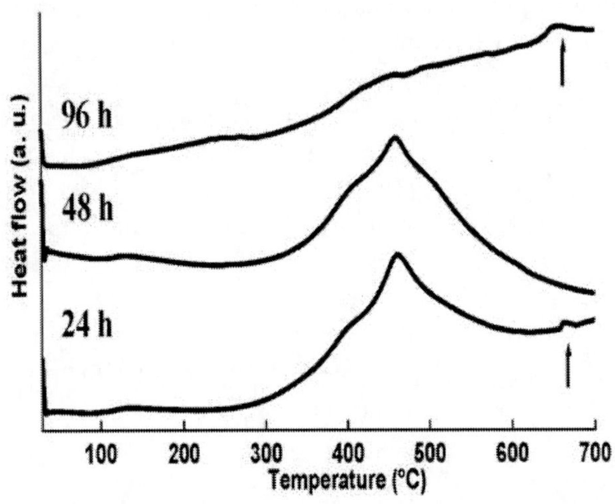

Figure 47. DSC scans of the $Fe_{57}Co_{21}Nb_7B_{15}$ powders milled for selected milling times

The DSC curves of the milled $Fe_{57}Co_{21}Nb_7B_{15}$ powders for selected milling times (Figure 47) exhibit exothermic effects characterized by many exothermic peaks and spread over the entire temperature range. The broad exothermic reaction at temperatures below 300°C can be ascribed to recovery, strain and structural relaxations. The important heat release for the milled powders during 96 of milling (10.6 J/g), can be explained by the amount of the amorphous phase (~ 82 %). The second exothermic peak at about 460°C is more important for 24 and 48 h of milling and the heat release is about 155.4 and 169.5 J/g, respectively. A small exothermic peak is observed at ~ 661°C for the powders milled for 24 h. The DSC scan of the powders milled for 96 h shows an endothermic peak at about 572°C and an exothermic one at 653°C.

8. KINETICS

The kinetics of Mo dissolution into the α-Fe matrix can be deduced from the XRD analysis by following the evolution of the (110) diffraction peak intensity of the unmixed Mo as a function of milling time (Figure 15). Since the milling process occurs at room temperature, one can suppose that the temperature is constant. In addition, the milling time can be considered as the necessary time for phase transformation. Consequently, the mixed fraction of Mo which is considered as the fraction transformed, X, (Figure 48) can be described by the Johnson-Mehl-Avrami formalism [90-92] in which the fraction transformed exhibits a time dependence of the form:

$$X = 1 - \exp[-(kt)^n]$$

where n is the order of reaction or Avrami parameter; X, the volume of transformed fraction; t the milling time, and k the rate constant. The double logarithmic plot ln(-ln(1-X)) versus lnt leads to the order of reaction or Avrami parameter n, and the rate constant k. Two stages can be distinguished according to the kinetics parameter values: (i) a first stage with $n_1 = 0.83$ and $k_1 = 0.34$; and (ii) a second stage with $n_2 = 0.33$ and $k_2 = 0.73$. The former proves that the Mo dissolution is very slow even non-existent up to 6 h of milling, while the later can be linked to the increased diffusivity by decreasing crystallite size and increasing the grain boundaries area.

The amorphization kinetics of the mechanically alloyed $Fe_{27.9}Nb_{2.2}B_{69.9}$ (at.%) powders can be deduced from the Mössbauer spectrometry results through the transformed fraction of α-Fe as a function of milling time [93]. The amorphization process can be described by one stage with an Avrami parameter of about n~1 (Figure 49). This value is comparable to those obtained for transformations controlled by the diffusion at the interface and dislocations segregation with $0.45 < n < 1.1$. This might be correlated to the existence of a high density of dislocations and various types of defects as well as to the crystallite size refinement. Comparable values of the Avrami parameter were obtained for the primary crystallization of the amorphous FeCoNbB alloy prepared by melt spinning [94].

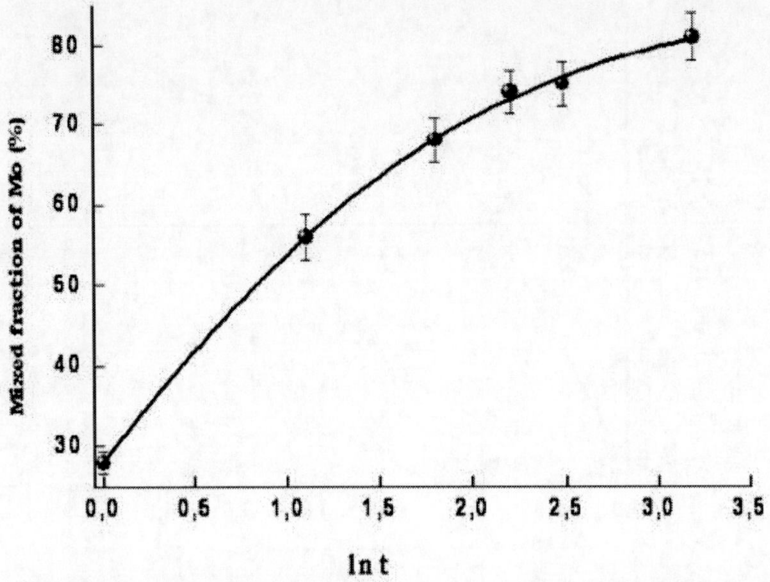

Figure 48. Mixed fraction of Mo vs. milling time

The mixing kinetics of the $Fe_{57}Co_{21}Nb_7B_{15}$ powders mixture can be described by two stages with different Avrami parameters n = 1.08 and n' = 0.34 (Figure 50). The lower values of the Avrami parameters can be attributed to the presence of both Nb and B which favor the grain size refinement and the formation of a highly disordered state [95]. For the $(Fe_{50}Co_{50})_{62}Nb_8B_{30}$ mixture, two stages have been obtained with different Avrami parameter values $n_1 = 1.41$ and $n_2 = 0.34$ [96]. The former value is comparable to those obtained for the

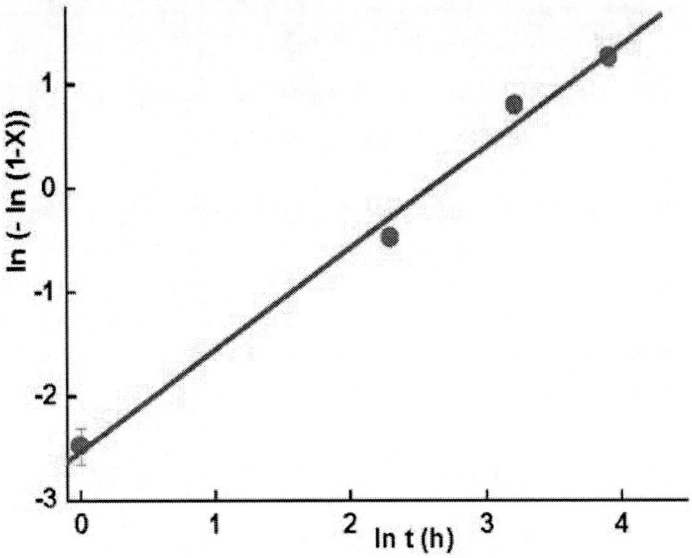

Figure 49. JMA plot of the ball-milled $Fe_{27.9}Nb_{2.2}B_{69.9}$ versus milling time

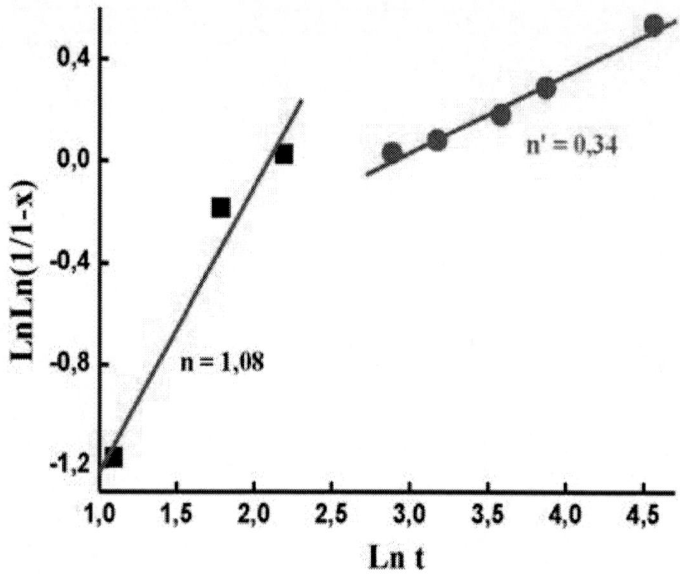

Figure 50. JMA plot of the ball-milled $Fe_{57}Co_{21}Nb_7B_{15}$ versus milling time

Finemet and Nanoperm [81]. However, it is higher than that obtained during the crystallization of the amorphous FeCoNbB alloy where α-(Fe,Co) nanocrystals with grain size of 15 nm are distributed in the amorphous matrix [94]. Bigot et al. have obtained a value of $n = 1.5$ for the nanocrystallization of the Finemet [97]. Comparable kinetics parameters have been obtained in the Ni-15Fe-5Mo ($n = 1.049$ and $k = 0.57$) [98]. The important fraction of structural defects which is introduced during the milling process favors the phase formation through the diffusion at the surface which is dominant, at lower temperatures, in comparison to the diffusion by the grain boundaries and the lattice parameter (vacancy's diffusion).

9. Conclusion

A review on some Fe-based nanostructured alloys prepared by mechanical alloying has been presented in this chapter. The phase transformation, morphological, structural, microstructural and magnetic properties of the mechanically alloyed mixtures are milling conditions dependent. The kinetics of the elemental powders mixing is described by lower values of the Avrami parameter comparable to those obtained for transformations controlled by the diffusion at the interface and dislocations segregation, in agreement with the mechanical alloying process.

Acknowledgment

The authors are very grateful to all researchers who have participated to the obtained results.

REFERENCES

[1] Siegel, RW. *Nanostruct. Mater*., 1994, 4, 121.

[2] Gleiter, H. *Nanostruct. Mater*., 1995, 6, 3.

[3] Birringer, R; Gleiter, H; Klein, HP; Marquardt, P. *Phys. Lett.,* 1984, 102A, 365.

[4] Kuhn, WE; Friedman, IL; Summers, W; Szegvari, A. *ASM Metals Handbook*, Vol. 7, Powder Metallurgy, Metals Park, (OH), 1985, 56.

[5] Gleiter, H. *Prog. Mater. Sci*., 1990, 33, 223.

[6] Suryanarayana, C. *Prog. Mater. Sci*., 2001, 46, 1.

[7] Sherif El-Eskandarany, M. *Mechanical Alloying for Fabrication of Advanced Engineering Materials,* Noyes Publications/William Andrew Publishing, Norwich New York 2001.

[8] Moumeni, H; Alleg, S; Greneche, JM. *J. alloys Compd*., 2006, 419, 140.

[9] Alleg, S; Bentayeb, FZ; Bensalem, R; Djebbari, C; Bessais, L; Greneche, JM. *Phys. Stat. Sol*. (a), 2008, 205(7), 1641.

[10] Suñol, JJ; González, A; Saurina, J; Escoda, L; Fernández-Barquin, I. *J. Non-Cryst. Solids*, 2007, 353, 865.

[11] Alleg, S; Azzaza, S; Bensalem, R; Suñol, JJ; Khene, S; Fillion, G. *J. Alloys Compd*., 2009, 482, 86.

[12] Djekoun, A; Bouzabata, B; Alleg, S; Greneche, JM; Otmani, A. *Ann. Chim. Sci. Mat*., 1998, 23 557.

[13] Iwanabe, H; Lu, B; McHenry, ME; Laughlin, DE. *J. Appl. Phys*., 1999, 85B, 4424.

[14] Turgut, Z; Scott, JH; Huang, MQ; Majetich, SA; McHenry, ME. *J. Appl. Phys*., 83, 1998, 6468.

[15] Moumeni, H; Alleg, S; Greneche, JM. *J. Alloys Compd*., 2005, 386, 12.

[16] Zhou, GF; Bakker, H. *J. Phys.: Condens. Matter,* 1994, 6, 4043.

[17] Yavari, AR. *Acta Metall. Mater*., 1993, 41, 1391.

[18] Zhou, GF; Bakker, H. *Scripta Mater*., 1996, 34, 29.

[19] Lu, K; Sun, NX. *Philos. Mag. Lett*., 1997, 75, 389.

[20] Pochet, P; Tominez, E; Chaffron, L; Martin, G. *Phys. Rev*., 1995, B52, 4006.

[21] Campbell, SJ; Wang, GM; Calka, A; Kaczmarek, WA. *Mater. Sci. Eng*., 1997, A226, 75.

[22] Calka, A; Nikolov, JI; Ninham, BW. In: J.J. DeBarbadillo et al. (eds.) Mechanical alloying for structural applications. Materials Park, OH: *ASM International*, 1993, 189.

[23] Dubiel, S; Sauer, C; Zinn, W. *Phys. Rev*., 1984, B 30, 6285.

[24] Otmani, A; Bouzabata, B; Djekoun, A; Alleg, S. *Ann. Chim. Sci. Mat*., 1997, 22, 201.

[25] Lemoine, C; Fnidiki, A; Lemarchand, D; Teillet, J; *Phys, J. Condens. Mater*., 1999, 11, 8341.

[26] De Nys, T; Gielen, PM. *Met. Trans*., 1971, 2, 1423.

[27] Wang, GM; Campbell, SJ; Calka, A; Kaczmarek, WA. *Nanostruct. Mater*., 1995, 6, 389.

[28] Campbell, SJ; Wang, GM; Calka, A; Kaczmarek, WA. *Mater Sci. Eng*., A226-228, 1997, 75.

[29] Garcia-Arribas, AM; Fdez-Gubieda, L; Orue, I; Barandiaran, JM. *Phys. Rev*. B, 1995, 52(17), 12805.

[30] Tsuei, CC; Lilienthal, H. *Phys. Rev., B*, 1976, 13(11), 4899.
[31] Kato, Y; Yamada, S; Okamoto, T. *Rapidly Quenched Metals*, S. Steeb and H. Warlimont (eds.), 1985, 699.
[32] Stewart, JW; Charles, JA; Wallach, ER. *Mater. Sci. Technol.*, 2000, 16, 283.
[33] Bouzabata, B; Alleg, S. *J. Alloys Compd.*, 1992, 178, 117.
[34] Kadiri, H; Djega-Mariadassou, C; Rougier, P; Dorman, JL; Berrada, A; Renaudin, P. *J. Phys.*, 1988, C8 49, 1371.
[35] Tan, J; Gao, YF; Liu, GD; Zhang, CI; Liu, T; Ma, RZ. *J. Magn. Magn. Mater.*, 1996, 164, 211.
[36] Egami, T; Vitek, V. & Srolovitz, D. in T. Masumoto and K. Suzuki (eds.), Proc. 4_{th} Int. Conf. On Rapidly Quenched Metals, *Jap. Inst. Met., Sendai*, 1982, 517.
[37] Huang, JY; Wu, YK; Ye, HQ; Lu, K. *Nanostruct. Mater.*, 1995, 6, 723.
[38] Di, LM; Bakker, H; De Boer, FR. *Physica*, 1992, B182, 91.
[39] Vincze, I; Campbell, IA; Meyer, A. *Sol. State Comm.*, 1974, 15, 1495.
[40] Hamdeh, H; Fults, B; Pearson, DH. *Phys.Rev.*, 1989, B39 (16), 11233.
[41] Passamani, EC; Lartica, C; Nascimento, VP. *J. Mater. Sci.*, 2002, 37, 819.
[42] Bentayeb, FZ; Alleg, S; Bouzabata, B; Greneche, JM. *J. Magn. Magn. Mater.*, 2005, 288, 282.
[43] Villars, O; Calvert, LD. *Pearson's Handbook of Crystallographic Data for Intermetallic Phases*, American Society for Metals Park, OH, 1991.
[44] Levin, AA; Meyer, DC; Tselev, A; Gorbunov, A; Pompe, W; Paufler, P. *J. Alloys Compd.*, 2002, 334, 159.
[45] Dubiel, SM; Zukrowski, J. *J. Magn. Magn. Mater.*, 1981, 23, 214.
[46] Kuwano, H. *Trans. Jap. Inst. Met.*, 1985, 26, 473.
[47] Dubiel, SM; Cielak, J; Sepiol, B. *Mössbauer Spectroscopy in Materials Science*, Kluwer Academic Publishers, 1999, 107.
[48] Kuwano, H; Ishikawa, Y; Yoshimura, T; Hamaguchi, Y. *Nucl. Instr. Meth. Phys*. Res. 1993, B 76, 290.
[49] Boubeker, B; Eymery, JP; Denanot, MF; Sayouty, ELH. *J. Magn. Magn. Mater.*, 1994, 133, 470.
[50] Lemoine, C; Fnidiki, A; Lemarchand, D; Teillet, J. *J. Magn. Magn. Mater.*, 1999, 203, 184.
[51] Massalski, TB. Binary Alloys Phase Diagrams, *ASM*, Metals Park, OH, USA, 1990, 1726.
[52] Tsuei, CC; Longworth, G; Lin, SCH. *Phys. Rev.*, 1968, 170(2), 603.
[53] Chen, HS. *Rep. Prog. Phys.*, 1980, 43, 353.
[54] Tebib, W; Alleg, S; Bensalem, R; Bensebaa, N; Bentayeb, FZ; Suñol, JJ; Greneche, JM. *J. Nanosci. Nanotechnol.*, 2008, 8, 2029.
[55] Ashby, MF; Jones, DRH. Engineering Materials 2: An introduction to microstructure, processing and design, 2nd ed., *Butterworth-Heinemann*, Oxford, 1998, 17.
[56] Shim, JH; Byun, JS; Chao, YW. *Scripta Mater.*, 2002, 47, 493.
[57] Byun, JS; Shim, JH; Chao, YW. *Scripta Mater.*, 2004, 50, 279.
[58] Kim, YK; Cho, YW. *J. Alloys Compds.*, 2005, 393, 211.
[59] Gonser, U. *Hyperfine Interact.*, 1991, 68, 71.
[60] Herr, U; Jing, J; Birringer, R; Gonser, U; Gleiter, H. *Appl. Phys. Lett.*, 1987, 50, 472.
[61] Maurer, M; Cadeville, MC; Sanchez, JP. *J. Phys. F: Met. Phys.*, 1979, 9, 271.

[62] Hutchinson, TS; Baird, DC. *The physics of engineering solids*, John Wiley & Sons, New York, 1963.
[63] Dekhil, L; Alleg, S; Suñol, JJ; Greneche, JM. *Adv.Powd.Technol.*, 2009, 20, 593.
[64] Duwez, P; Lin, SCH. *J. Appl. Phys.*, 1967, 38, 4096.
[65] Bauer-Grosse, E; Le Caër, G. *Philos. Mag.* B, 1987, 56, 485.
[66] Bensebaa, N; Alleg, S; Bentayeb, FZ; Bessais, L. & Greneche, JM. *J. Alloys Compd.* 2005, 388, 41.
[67] Bensebaa, N; Alleg, S; Greneche, JM. *J. Alloys Compd,* 2005, 393, 194.
[68] Yu. V; Maksimov, IP; Suzdalev, RA; Arents, SM. Loktev, *Kinet. Catal.*, 1974, 15, 1144.
[69] Basset, D; Matteazzi, P; Miani, F; Hadjipanayis, GC; Siegel, RW. (eds.), *Nanophase Materials*, 1994, 145.
[70] Guttman, M; McLean, D. Interfacial segregation, WC. Johnson, JM. Blakely (eds.), *ASM*, Metals Park (1979) 251.
[71] Di, LM; Bakker, H; De Boer, FR. *Physica*, 1992, B182, 91.
[72] Host, JJ; Teng, MH; Elliot, BR; Hwang, JH; Mason, TO; Johnson, DL. *J. Mat. Res.*, 1997, 12, (5), 1268.
[73] Krill, CE; Merzoug, F; Krauss, W; Birringer, R. *Nanostruct. Mater.*, 1997, 9, 455.
[74] Moumeni, H. *Ph.D. Thesis* University of Badji-Mokhtar Annaba Algeria, 2005.
[75] Azzaza, S; Alleg, S; Moumeni, H; Nemamcha, AR; Rehspringer, JL; Greneche, JM. *J. Phys.: Condens. Matter,* 2006, 18, 7257.
[76] Herzer, G. *Handbook of Magnetic Materials* (ed.) K. H. Buschow, 1997, Vol. 10, 415.
[77] Ślawska-Waniewska, A; Grafoute, M. & Greneche, JM. *J. Phys.: Condens. Matter,* 2006, 18, 2235.
[78] Amils, X; Nogués, J; Suriñach, S; Baró, MD. *J. Magn. Magn. Mater.*, 1999, 203, 129.
[79] Brüning, R; Samwer, K; Kuhrt , C; Schultz , L. *J. Appl. Phys.*, 1992, 72(7), 2978.
[80] Gleiter, H. *Prog. Mater. Sci.*, 1989, 33, 223.
[81] McHenry, ME; Willard, MA; Laughlin, DE. *Prog. Mat. Sci.*, 1999, 44, 291.
[82] Beekman, O; Lundgren, L. in: K.H.J. Buschow (Ed.) *Handbook of Magnetic Materials* (Elsevier Science Publishers B.V.), 1991.
[83] Weller, D; Harp, GR; Farrow, RFC; Cebollada, A. *J. Sticht, Phys. Rev. Lett.*, 1994, 72, 2097.
[84] Herzer, G. IEEE *Trans. Magn.*, 1990, 26, 1397.
[85] Kuhrt, C; Shultz, L. *J. Appl. Phys.*, 1992, 73, 6588.
[86] Birringer, R. *Mater. Sci. Eng.*, 1989, A117, 33.
[87] Fecht, HJ. *Nanophase Materials* eds. GC. Hadjipanayis and R.W. Siegel, 1994, 125.
[88] Turgut, Z; Huang, MQ; Gallagher, K; McHenry, ME. *J. Appl. Phys.*, 1997, 81(8), 4039.
[89] Persiano, AIC; Rawlings, RD. *J. Mat. Sci.*, 1991, 26, 4026.
[90] Avrami, M. *J. Chem. Phys.*, 1939, 7, 1103.
[91] Avrami, M. *J. Chem. Phys.*, 1940, 8, 212.
[92] Avrami, M. *J. Chem. Phys.*, 1941, 9, 177.
[93] Alleg, S; Hamouda, A; Azzaza, S; Bensalem, R; Suñol, JJ; Greneche, JM. *Mat. Chem. Phys.*, 2010, 122, 35.
[94] Blázquez, JS; Conde, CF; Conde, A. *J. Non-Cryst. Solids,* 2001, 287, 187.
[95] Souilah, S. *Master Thesis,* Badji-Mokhtar Annaba University Algeria, 2006.
[96] Azzaza, S. *Ph.D. Thesis*, Badji-Mokhtar Annaba University, Algeria, 2009.

[97] Bigot, J; Lecaude, N; Perron, JC; Milan, C; Ramiarinjaona, C; Rialland, JF; Magn. J. *Magn. Mater*., 1994, 133, 299.
[98] Shen, SY; Hng, HH; Oh, JT. *Mater. Letter,* 2004, 58, 2824.

In: Powder Engineering, Technology and Applications ISBN: 978-1-61761-216-9
Editor: Jason M. Barker, pp.125-147

Chapter 4

MILLING: PARTICLE ENGINEERING BEYOND SIZE REDUCTION

***Wai Kiong Ng* [1], *Jin Wang Kwek* [1] *and Reginald B.H. Tan* [1,2,*]**

[1] Institute of Chemical and Engineering Sciences, Agency for Science, Technology and Research (A*STAR), Jurong Island, Singapore
[2] The National University of Singapore, Dept of Chemical and BiomolecularEngineering, Singapore

ABSTRACT

Milling or physically dividing solids into fine fragments is the best-known method for the production of powders. It is found widely across industries from pharmaceuticals, nutraceuticals, agrochemicals, aerosols, personal care products, pigments, ceramics, cement to semi-conductors. Traditionally, milling (or also known as grinding) is regarded as a particle size reduction process, which increases the specific surface area of a solid phase for desired interaction with the neighboring phase. Advanced powder characterization techniques have revealed that milling performs more than just a size reduction role. The mechanical and thermal energies imposed onto a milled powder alter other physicochemical properties like morphology, crystallinity, polymorphism, glass transition temperature, surface energy, melting properties, as well as lead to particle size shift and changes in specific surface area during subsequent storage. These changes often have a pronounced effect on the product performance such as powder flow, aerosolization, dissolution, physical and chemical stability. The underlying scientific knowledge behind these changes has created avenues to engineer new applications as well as mitigate the milling-induced effects that are detrimental to product performance. This review presents recent developments in milling technologies such as nanomilling, co-milling and post-milling conditioning. Nanomilling using stirred media mills with suitable surfactants has produced stabilized nanosuspensions below the postulated grinding limit of half a micron, and thereby increases the specific surface area by up to 10-fold and the saturated solubility by 6-fold. Co-milling or simultaneous milling of two or more solid component has been effective in controlling the solid-state properties of

* Corresponding author: Tel: +65-6796-3863/41. Fax.: +65-6316-6183. Email addresses: ng_wai_kiong@ices.a-star.edu.sg; reginald_tan@ices.a-star.edu.sg.

milled powders. Milling-induced amorphization or the loss of long-range crystalline structure is often associated with stability problems as the amorphous phase is thermodynamically unstable and tends to re-crystallize to the stable crystalline form. This effect can be mitigated by co-milling with crystalline seeds. On the other hand, when an amorphous product is desired as it has advantages of higher solubility, dissolution rate and better compression characteristics over the crystalline form, the addition of polymeric additives during milling has led to stable amorphous solid dispersions. Co-milling with mesoporous materials also results in an amorphous form, which is entrapped inside nanosized channels. Co-grinding has been used to prepare co-crystals, which possess different physicochemical properties from the original components. To alleviate undesired milling-induced effects, new findings in post-milling conditioning as well as recent research to alternative processes to particle size reduction will be discussed.

1. Introduction: Milling Induced Disorder, Amorphization, Effects and Stability

Ten thousand years ago, during the Stone Age, human beings learned to grind corn using millstones and then classify the powders according to their different sizes (Arai, 1996). In the age of industrialization, communition is the oldest mechanical unit operation for the size reduction of solid materials (Tanaka and Kanda, 2006). In order to meet stringent particle size requirements, milling is often used to break up large agglomerates formed. Advanced particle characterization techniques such as differential scanning calorimetry and powder X-ray diffraction showed that milling actually induces other changes besides increasing the surface areas through mechanical and thermal activation (Colombo et al., 2009). In an inverse gas chromatography study, Heng and co-workers (2006) reported that milling exposes a hydrophobic surface for paracetamol crystals which becomes increasingly more dominant with decreasing particle size. In fact, the particle size achieved by milling may also shift to a coarser or finer fraction during subsequent storage (Ng et al., 2008).

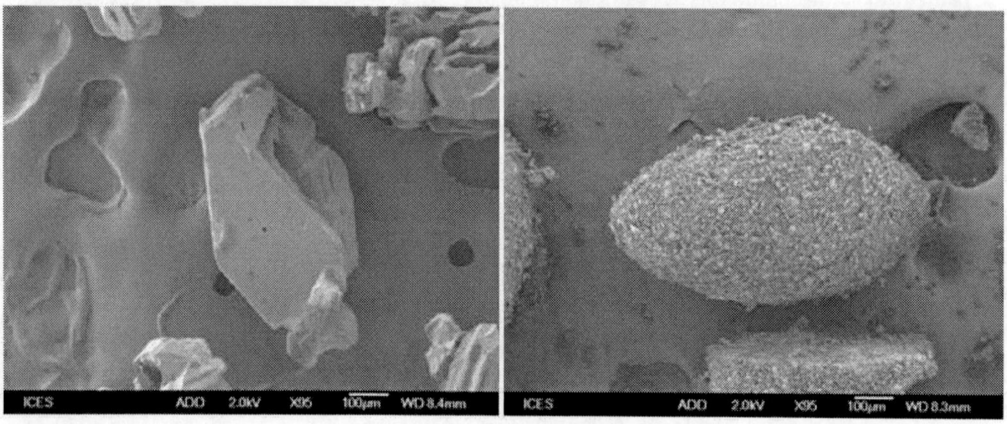

Figure 1. Scanning electron micrographs (SEM) of crystallized adipic acid (left) before milling and (right) after milling

From the perspective of energy balance, most of the input energy is lost during milling as heat, whilst the energy used directly in grinding the solids is only 0.6% of the whole. Most of the mechanical energy supplied as power is changed to heat by collision and friction (Arai, 1996). It is, therefore, not surprising that milling can cause disruption or activation of the crystal structure, leading to various degrees of disorder (Sebhatu et al., 1994; Saleki-Gerhardt et al., 1994) depending on the milling conditions (Brodka-Pfeiffer et al., 2003; Descamps et al., 2007; Begat et al., 2003) and material (Willart et al., 2001, 2004; Chamarthy and Pinal, 2008). Figure 1 shows the morphologies of crystallized adipic acid before and after being subject to ball-milling for 30 min. They illustrate that radical changes in morphology can take place beyond just size reduction.

It is demonstrated in metal alloying that solid state amorphization or disruption of the crystal order could be achieved without undergoing melting of the crystalline state (Weeber and Bakker, 1988; Bakker et al., 1995). In some pharmaceutical materials, it is also possible to obtain glassy amorphous state from the crystalline state if milling is performed below the glass transition temperature. Willart and co-workers (2001, 2004) carried out ball milling on crystalline sugars: alpha lactose and trehalose and found that a direct transformation from crystal to glass can be induced similar to the amorphous states produced by the usual thermal route such as quenching of the liquid. On the other hand, Chamarthy and Pinal (2008) concluded that the disorder induced by cryogenic milling of griseofulvin or felodipine consists of crystal defects rather than amorphous regions. However, if milling of crystalline compounds is done above the glass transition temperature, polymorphic transformations may occur instead such as in sugars sorbitol (Willart et al., 2005) or mannitol (Willart et al, 2006) and active pharmaceutical ingredient (API) sulfamerazine (Zhang et al., 2002). Polymorphism is described as the property of a chemical entity that can exist in various forms having different molecular packing arrangements in the crystal lattice (York, 2002). Nevertheless, amorphous solids have advantageous pharmaceutical properties such as enhanced dissolution rate (Saleki-Gerhardt et al., 1994; Chiou and Kyle, 1979; Burt and Mitchell, 1981) and sometimes better compression characteristics (Yu, 2001). However, such amorphous materials can be thermodynamically unstable and solid-state recrystallization or transformation to other forms may occur if the molecular mobility is high enough. Milling studies conducted on hydrates facilitates the removal of lattice water resulting in the formation of an amorphous product but it also results in the lowering of the dehydration temperatures of the ground products (Brittain, 2002). Milled API cephalexin monohydrate (Otsuka and Kaneniwa, 1983, 1984) exhibited a greater tendency to adsorb moisture and re-form the hydrate crystalline form. The amorphous material created during milling could potentially re-crystallize with time as in the case of milled API xemilofiban that subsequently recrystallized over a short period of time resulting in failed pre-lubrication acceptance criteria (Mackin et al., 2002). Amongst the crystalline disorder, milling can also cause either an increase or reduction in the enthalpy of fusion and melting temperature (Révész, 2005; Chikhalia et al., 2006; Ng et al., 2010). These changes could affect theoretical predictions of aqueous solubility of new pharmaceutical compounds (Hojjati and Rohani, 2006) as well as input parameters for the design of melt crystallizers and hot melt extruders.

In order to achieve stability of the milled powders, one of the techniques used is to introduce a conditioning step following milling to convert the unstable amorphous regions into crystalline solids under controlled relative humidity and temperature (Brodka-Pfeiffer et al., 2003). These conditions will help to reduce the glass transition temperature by adsorption of water and to promote molecular mobility through setting temperature to above glass transition values so that the re-crystallization process is accelerated. However, not all powder samples can be treated in this way as discussed earlier for the case of re-crystallized xemilofiban having different physical properties hence resulting in failed specifications (Mackin et al., 2002). Ng and co-workers (2008) reported that the storage of milled adipic acid below the critical relative humidity led to localized disintegration from the agglomerate surface and particle size reduction, which was not influenced by moisture sorption or loss. This evidence supports the “stress relaxation” mechanism behind breakage of postmilled particles. Appropriate storage conditions are important in maintaining the stability of milled powders. An alternative would be to stabilize the amorphous form or to suppress initial milling-induced amorphization by co-milling the material of interest with an additive.

2. Co-Milling

In metallurgy, mechanical alloying was developed by accident in 1968 at INCO’s Paul D. Merica Research Laboratory as a technique to produce oxide-dispersion strengthened (ODS) nickel- and iron-base superalloys known for its increased stress rupture properties in the aerospace industry (Benjamin, 1970). It involves repeated welding, fracturing, and rewelding of powder particles in a high-energy ball mill (Suryanarayana, 2001). During high energy milling, the force of impact will deform the powder particles plastically leading to work hardening and fracture. With continued deformation, an increasing number of crystal defects and grain boundaries will enhance the diffusivity of solute elements into the solid matrix. If there is a slight rise in temperature, the diffusion behavior will be aided and consequently true alloying from its constituent elements takes place.

As an analogue to mechanical alloying in intermetallics, molecular drug alloys produced by co-milling drug with excipients or carriers are potentially attractive systems that offer new materials with desired properties such as sustained crystallinity/amorphicity over time (Bahl and Bogner, 2006; Gupta et al., 2003; Descamps et al., 2007; Balani et al., 2010a), enhanced dissolution (Otsuka et al., 1998; Mura et al., 2002; Cirri et al., 2004; Allesø et al., 2009) and aerosolization (Giry et al., 2006) behavior. Co-milling has also been used to produce mesoporous structures (Le Bolay et al., 2009) and co-crystals (Chadwick et al., 2007).

2.1. Role of Excipients in Co-Milled Mixtures

The key reason to form amorphous drugs is to increase solubility and therefore the bioavailability of the poorly water soluble compounds. The bioavailability of a drug is described in terms of the extent and the rate at which the drug appears intact in the systemic circulation after the administration of a known dose (Ashford, 2002). Another potential technique would be to co-mill the drug with an excipient or carrier to prepare amorphous materials which are more prone to degradation either by heat or solvent which preclude other preparation methods (Crowley and Zografi, 2002; Watanabe et al., 2003). API Indomethacin has reportedly higher apparent equilibrium solubility when co-ground with silica or silicates (Watanabe et al., 2002; Bahl and Bogner, 2008; Bahl et al., 2008). A stable amorphous indomethacin-silica alloy was formed due to plasticization of amorphous drug leading to mechanical transfer of drug to silicate, vapor-phase mass transfer and water in facilitating particle-particle surface migration of the drug to the silicate. API naproxen could not be transformed to the amorphous state by mechanical activation but upon co-milling with API cimetidine at 4 °C for 60 min, the mixture became amorphous with enhanced dissolution rate and synchronized release of the two drugs (Allesø et al., 2009). The stabilization of the amorphous binary system is reported to be dictated by molecular level interactions rather than bulk level phenomena. Other co-milled API mixtures with improved dissolution properties include glyburide-cyclodextrin (Cirri et al., 2004) and glisentide-polyvinyl pyrrolidone (PVP) (Mura et al., 2002).

The selection of carriers or excipient, drug-excipient ratios and the processing humidity are important considerations towards the physical stability of the co-milled mixture (Bahl and Bogner, 2006). Balani and co-workers (2010a) discovered that α-lactose monohydrate, adipic acid and magnesium stearate could induce complete re-crystallization of the amorphous inhalant drug salbutamol sulphate formed by milling. Watanabe and his team (2001) suggested that one of the reasons behind the re-crystallization of the amorphous domains could be due to the presence of residual microcrystallites serving as seed crystals. Another reason might be due to localized heating in the ball milling as can be seen with API fanaserine as well (Gusseme et al., 2008).

On the other hand, co-milling salbutamol sulphate with polyvinyl pyrrolidone (PVP) could stabilize the amorphous form of salbutamol sulphate (Balani et al., 2010b). PVP is a hydrophilic polymer that is reported elsewhere to inhibit re-crystallization and stabilize the amorphous dispersions (Aso et al., 1996; Watanabe et al., 2003; Cirri et al., 2004). However, a major concern with the use of PVP is its hygroscopic nature. PVP has been reported to have an equilibrium moisture content of 27.8 wt% at 25°C/75% RH and classified as "very hygroscopic" (Umprayn and Mendes, 1987). The hygroscopic nature of PVP could be detrimental when it is used to stabilize solid dispersions. Marsac and co-workers (2008) observed that 25 wt% PVP in the presence of water increased the nucleation rate of the solvent evaporated amorphous nifedipine-PVP molecular dispersions leading to re-crystallization. Vapor sorption studies indicated that pure amorphous nifedipine with a moisture uptake of around 1-2 wt% showed a significant increase in moisture uptake up to 6 wt% when PVP content was increased from 0 to 25 wt%. In an Atomic Force Microscopy study, Mahlin and team (2006) found that the more PVP present in spray-dried lactose-PVP increased the propensity of moisture-induced re-crystallization. The glass transition temperature (Tg) of pure amorphous lactose was below 25°C at 40% RH. This RH referred to as the critical crystallization RH increased steadily with more PVP in the composite mixture. PVP when used at high concentrations impaired the stability of co-formulated amorphous phase. On the other hand, Shakhtshneider's group (2007b) recently reported that lower PVP content resulted in less stable cryogenic co-ground mixture of indomethacin and PVP. PVP concentration above 50 wt% was required to keep the amorphous indomethacin stable for 10 months.

The molecular mobility, inter and intra molecular interactions and variations in processing conditions play an important role (Bhugra and Pikal, 2008) behind the physical stability of the co-milled mixtures. With reference to drug-polymer systems, the stability of the amorphous form primarily depends on factors such as drug and polymer interaction, viscosity of polymer and glass transition temperature of the mixture (Choksi et al., 2008). In addition, the presence of polymer has been shown to increase the activation energy towards crystallization and could act as a physical barrier against crystallization. Although Tg serves as an important indicator of crystallization tendency, modes of molecular mobility not reflected by Tg should also be considered. This was clearly observed in the case of amorphous solid dispersions of PVP with drugs such as nifedipine and felodipine, where Tg alone could not explain the reduced rate of crystallization in the presence of polymer and water (Marsac et al., 2008). On the other hand, PVP- indomethacin cryoground systems (Shakhtshneider., 2007b) were found to show an elevated Tg but at higher concentration of PVP (>50 wt%). Similar results were reported in case of PVP-piroxicam cryoground systems (Shakhtshneider et al., 2007a).

Intra and intermolecular hydrogen bonding interactions could be instrumental in achieving stable amorphous dispersions. Miyazaki et al (2004) attributed the ability of polyacrylic acid (PAA) to stabilize amorphous dispersions of acetaminophen to the strength of interaction between hydroxyl group of acetaminophen and the carboxyl group in PAA in comparison to PVP. Tang et al (2002) investigated the differences in hydrogen bonding tendencies in the crystalline and amorphous states of seven dihydropyridines analogues. The authors concluded that for some compounds, hydrogen bonding was stronger in the crystalline state while for others, it was stronger in the amorphous state. A strong relation of hydrogen bonding in molecules to molecular mobility was also found in case of acetaminophen glass (Gunawan et al., 2006). This meant that the strength of the hydrogen bonding could significantly impact the structural relaxation time i.e. the molecular mobility and hence impact the tendency to crystallize. Hence, a precise knowledge of such interactions could be useful if the acceptor or donor groups are targeted by polymeric excipients in order to achieve physical stabilization. Hydrogen bonding has also been reported in case of PVP-drug co-ground systems. Hydrogen bonding with PVP has played an important role in stabilizing amorphous co-ground systems of drugs such as indomethacin, piroxicam and sulphathiazole (Boldyrev et al., 1994; Shakhtshneider et al., 2007a, b). Fourier Transformed Infrared (FTIR) analysis on PVP stabilized co-milled amorphous salbutamol sulphate mixtures (Balani et al., 2010b) supported the formation of hydrogen bonding between SS and PVP. The extent of inter-molecular interactions increased with higher PVP content suggesting that they could contribute towards a stabilized amorphous phase even when stored at higher RH (75%RH) for 7 days. Raman microscopy showed that SS and PVP were homogeneously distributed (at minimum PVP content of 25 wt%) and could not be distinguished from each other even at 5 μm resolution. This evidence suggested that PVP could be dispersed in very fine clusters or at molecular interaction level, resulting in a very stable solid dispersion. Figure 2 shows that the morphology of co-milled salbutamol sulphate-PVP remains unchanged after storage at 75%RH for 7 days.

Co-milling of a poorly water-soluble drug, flurbiprofen using a planetary ball mill with water-soluble additives (D-mannitol, lactose, and erythritol), light anhydrous silicic acid and starch was found to improve dissolution. The approach was considered to be novel, industrially applicable methods for improving the dissolution rate of poorly water-soluble drugs (Sonoda et al., 2008). In another recent development, Shen and co-workers (2010) described the use of mesoporous silica, SBA-15, to stabilize the amorphous state of API ibuprofen by drug entrapment inside nanosized pores. Although the main method reported was co-spray drying, co-grinding with SBA-15 also produced partially amorphous ibuprofen, which was likely to be confined in a similar manner.

2.2. Producing Composite Structures

Conventional methods used to produce polymeric materials include extrusion and injection moulding (Reyene, 1995). But the methods rely on heating the polymers till they can flow at a reasonable rate under stress (Zapata et al., 2004). Ultra high molecular weight polyethylene cannot be treated using the conventional methods due to its high molecular weight and melt viscosity. Similarly, new polymeric alloys cannot be made using the conventional methods as phase separation would occur due to thermal incompatibility between different polymers. To overcome the problems, mechanical alloying using co-milling or co-grinding was used to produce polymeric composite materials. Zapata and co-workers (2004) found that when a filler such as calcium carbonate is co-ground with poly(vinyl acetate) as a matrix, the mechanical properties of the composite materials were enhanced. This might be explained by an increase in the dispersion homogeneity of the filler or by a modification of the matrix/filler interface.

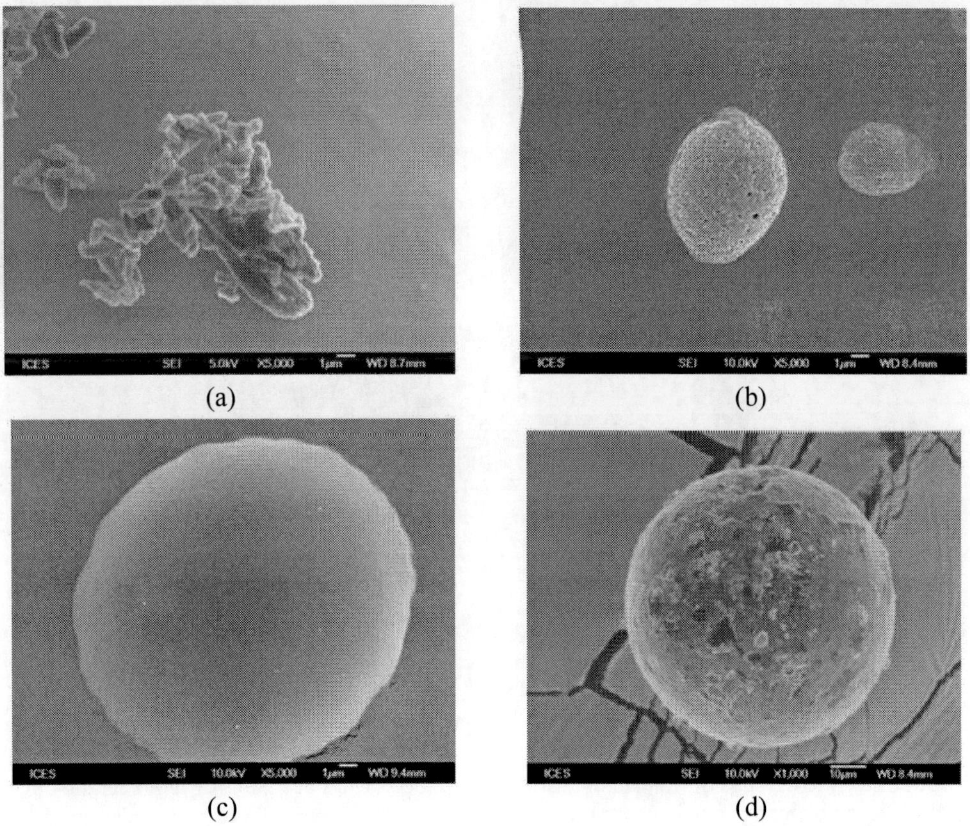

Figure 2. SEM micrographs of (a) unmilled salbutamol sulphate, (b) milled amorphous salbutamol sulphate, (c) amorphous salbutamol sulphate co-milled with PVP, and (d) amorphous salbutamol sulphate co-milled with PVP after storage at 75%RH for 7 days

Le Bolay and co-workers (2009) proposed the mechanism of alternated fragmentation and agglomeration of particles to explain the co-grinding phenomenon. Since the materials used in the co-grinding were of different hardness, they will not have the same grinding rates.

Based on the system of the mineral filler (calcium carbonate) and the polymer matrix (poly(vinyl acetate)), the fragmentation of the mineral is faster than that of the polymer. As the mineral fragments, they will adhere to the polymer particles which in turn would undergo fragmentation. Consequently, the mineral fines would agglomerate on the free faces of the polymer particles and slow down the agglomeration rate of the polymer particles. The agglomeration of the mineral fines on the polymer particles depends very much on the electronic nature of the filler surfaces in relation to that of the polymer surfaces (Zapata-Massot and Le Bolay, 2007). Successive fragmentation and agglomeration of the minerals continue till a progressive coating of the polymer particles by the mineral particles results. The following figures show the individual SEM images of calcium carbonate, poly(vinyl acetate) and coated polymer particle surface.

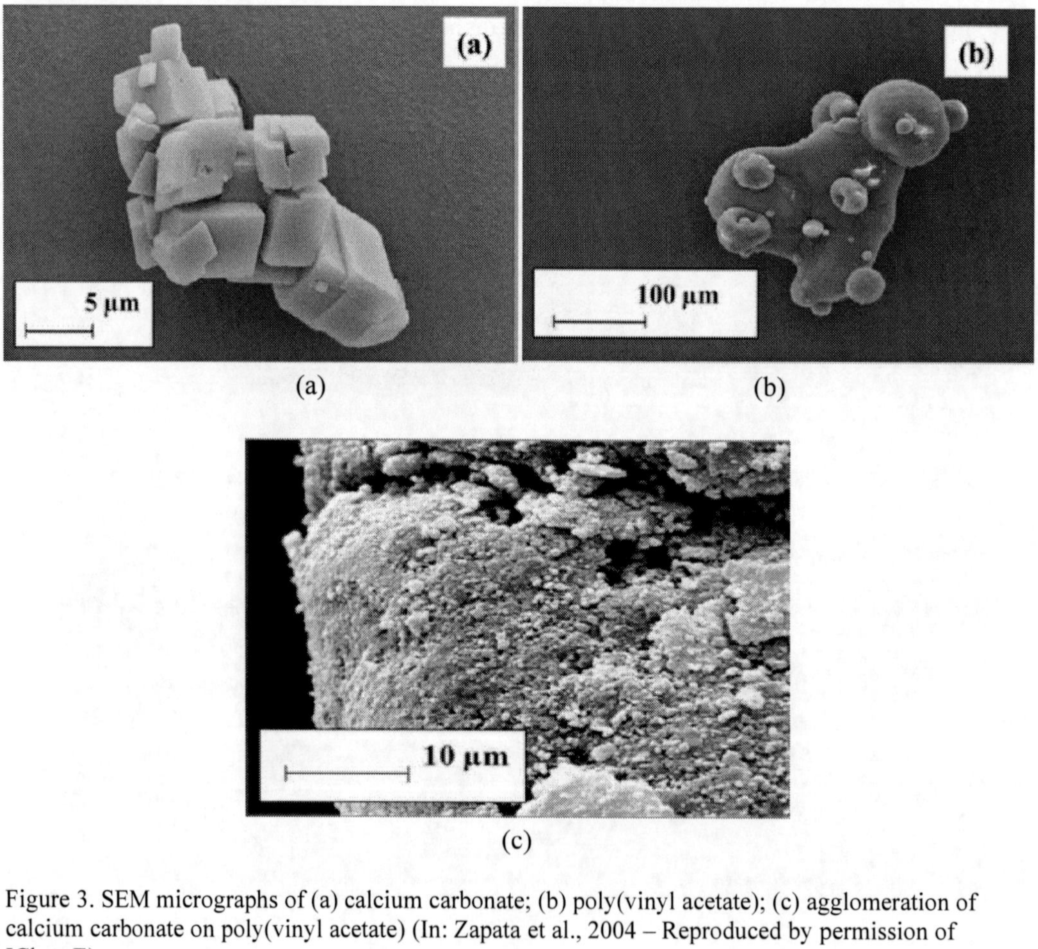

Figure 3. SEM micrographs of (a) calcium carbonate; (b) poly(vinyl acetate); (c) agglomeration of calcium carbonate on poly(vinyl acetate) (In: Zapata et al., 2004 – Reproduced by permission of IChemE)

Figure 4. An SEM image of scaffold after washing (In: Seyni et al., 2009 – Reproduced by permission of Elsevier)

The alternated fragmentation and agglomeration in co-grinding was used in the preparative step for the first time in the production of porous biodegradable composites for bone tissue engineering in a tumbling ball mill (Le Bolay et al., 2009). A polymer matrix, poly-L,D-lactic acid (PLA), is co-milled with an appropriate amount (30 to 40 wt%) of mineral filler such as nanocrystalline carbonated apatite so that it aids the fragmentation of the polymer particles and allows a satisfactory dispersion of the mineral within the matrix. Porosity is generated thereafter using a pore-forming agent such as sodium chloride particles. The resulting scaffolds (Figure 4) are shown to demonstrate sufficient mechanical resistance for implantation without changing its shape. In another development, vegetable and biodegradable fillers were used in the co-ground composite materials (Seyni et al., 2009). Co-grinding after an optimum duration offered interesting water resistant properties of the composite materials.

2.3. Co-Crystals Formation

Co-crystals synthesis using supramolecular complexes of two or more components is very attractive in the pharmaceutical industry since the stability and the dissolution of the drugs can be enhanced without changing the molecular structure of the active pharmaceutical ingredient (Myz et al., 2009). Co-crystals are generally prepared by solution crystallization and growth from melt (Lee et al., 2001; Trask et al., 2005; Braga et al., 2007). Another potential approach is the use of solid-state co-grinding or co-milling, which can remove undesired solvate formation since it required little or no solvent and hence more environmentally friendly. It is also possible to use co-grinding as a screening aid to generate novel co-crystalline phases not found in crystallization from solution (Chadwick et al., 2007; Trask et al., 2006). Kuroda and co-workers (2002) created novel crystal adducts by solid state co-grinding crystals of racemic-bis-β-napthol (rac-BN) and benzoquinone (BQ), which are structurally distinct from those produced via solution or melt. They postulated that shearing and molecular diffusion processes drive the formation of structurally distinct co-crystals.

However when the adduct crystals form I and II were exposed to air, it was found that they had reverted back to the starting rac-BN crystals. Moisture seems to have a significant effect on the formation and stability of APIs carbamazepine (CBZ) – nicotinamide (NIC) co-crystals formed by co-milling (Chieng et al., AAPS Poster, 2008). The co-crystals were stored at various conditions for 7 to 28 days and were found to be stable under high temperature (40 ^{o}C) and humidity (75%RH) conditions. At 20 ^{o}C and 33%RH, the co-crystal has reverted back to the starting material.

The success of the formation of the co-crystals depends very much on the intimate mixing of the starting materials or reactants and their specific surface areas for reaction (Braga et al., 2006). During co-milling or solid state co-grinding, the crystal lattice is destroyed and reformed through crystallization. Hydrogen bonds, π-stacking, van der Waals, ion pairing interactions and other interactions are broken and formed through the mechanochemical reaction process leading to formation of supramolecular compounds or hybrid molecular crystals. Pedireddi and co-workers (1996) discussed the need to select appropriate functional groups in controlling co-crystal formation in the solid state as well as the role of solvent inclusion in stabilizing desired supramolecular structures. For instance, ursodeoxycholic acid can only form crystalline inclusion complexes with only phenanthrene or anthrone by co-grinding in addition to co-precipitation (Oguchi et al., 2000). Co-precipitation alone, however, did not form any complexes. Chadwick and his team (2007) postulated a mechanism for grinding induced co-crystal formation. They noted that a liquid phase can form between high interfacial areas of the starting solid phases upon grinding (Figure 5). The shear induced by the grinding and the contact between the liquid phase and the individual residual solid phases would then induce the nucleation and growth of the co-crystal from the liquid phase. Once the interfacial liquid phase has crystallized, a layer of co-crystal will cover the reactant surfaces, and the reaction will cease. Fresh interfaces will form upon further attrition, hence generating more co-crystal.

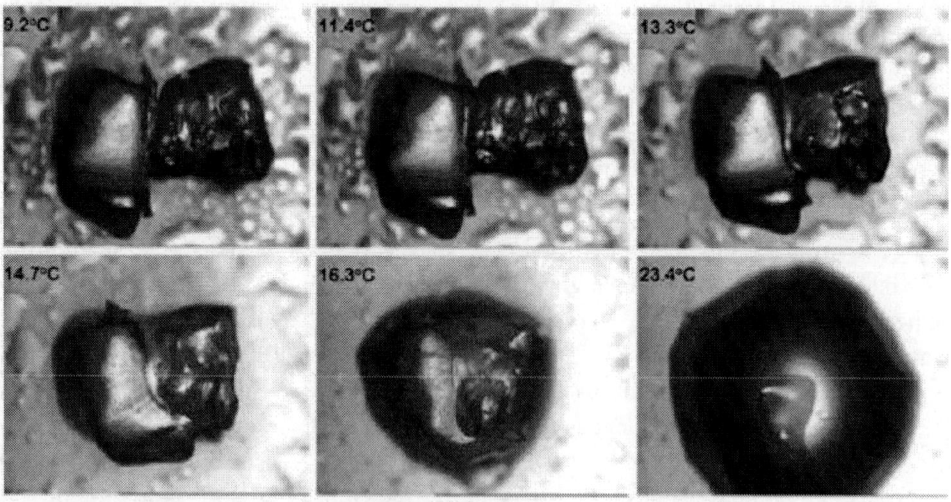

Figure 5. Effect of temperature on the solid state reaction between benzophenone (BZP) and diphenylamine (DPA). The liquid phase can be seen between the interface of the BZP and DPA crystal when the eutectic temperature of 13.3 ^{o}C is reached (In: Chadwick et al., 2007 - Reproduced by permission of the Royal Society of Chemistry)

To speed up the process of co-crystal formation and provide control over the formation of a specific polymorphic form, very small amounts of solvent can be added in a process known as solvent-drop grinding (Shan et al., 2002; Trask et al., 2004; Jayasankar et al., 2007; Myz et al., 2009). Myz and co-workers (2009) found that API meloxicam forms co-crystals with succinic and maleic acid easily by solvent drop grinding techniques. However, co-grinding with citric and ascorbic acids did not result in co-crystallization. As in the case of CBZ-NIC co-crystal formation, moisture was found to be affecting the formation of the co-crystals of meloxicam and succinic/maleic acid. In this case, however, removing adsorbed water by drying or heating before grinding actually speeds up the reaction kinetics and facilitates the co-crystallization process. Co-crystallization of aspirin and diazabicyclo[2.2.2]octane (DABCO) was also only achieved by solvent-drop grinding (Skovsgaard and Bond, 2009). It was found that the layered structure and intermolecular interactions between the acetyl groups in aspirin were disrupted when co-crystallized with DABCO.

Multi-API co-crystals can be used in combination drugs, which consist of two or more APIs produced in a single tablet (U.S. Food and Drug Adminstration, 21 CFR Part 3.2(e)). Combination drugs have become increasingly popular with advantages such as greater therapeutic effect, reduction in the number of prescriptions, and an increase in patient compliance (Werthmeier and Morisson, 2002; Bucci and Possidente, 2006; Frantz, 2006). However, their advantages are diminished due to problems in the formulation stage such as solubility, stability and chemical interactions between parent APIs. To address these problems, solid state co-grinding has been used as an alternative technique besides solution crystallization to produce stable co-crystals of various polymorphic forms. Aitipamula and co-workers (2009a) are able to obtain metastable polymorphic Form II using ethenezamide and saccharin by solid state grinding that can transform into Form I if prepared via the solution crystallization route. On the other hand, if ethenezamide and gentisic acid are used in a trimorphic co-crystal solid state grinding, metastable polymorphs can be converted to stable forms (Aitipamula et al., 2009b). The ability to form a stable co-crystal upon solid state grinding depends very much on the molecular interactions as discussed earlier. He and co-workers (2008, 2009) described a novel technique, pulsed gradient spin-echo nuclear magnetic resonance (PGSE NMR), based on intermolecular pair interactions, to screen for potential crystal pairs.

3. CRYOMILLING

Cryogenic milling (Figure 6) or cryomilling in short involves milling materials at cryogenic temperatures using cryogenic media such as liquid nitrogen (wet grinding) (Colombo et al., 2009; Suryanarayana, 2001). A cost analysis proved that cryomilling is economically feasible as materials milled under cryogenic temperatures become brittle and subsequently less energy is required for milling (Ye and Schoenung, 2004). In metallurgy, the mechanical attrition of powders within a cryogenic medium has been reported to strengthen materials by the refined grain size and the dispersed nanoscale particles (Suryanarayana, 2001). In addition, materials are prevented from thermal damage, undesirable chemical reactions or oxidations between phases are hindered (Ye and Schoenung, 2004, Milligan et al., 2009) and any potential particle aggregation is reduced (Shakhtshneider and Boldyrev, 1999).

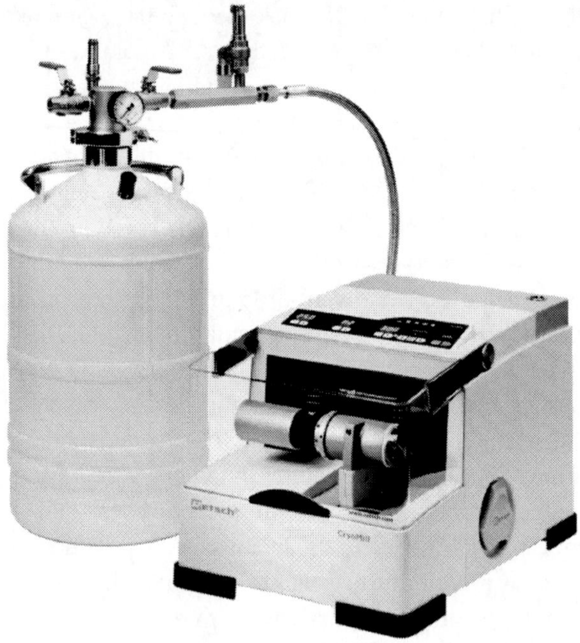

Figure 6. Retsch Cryomill (Courtesy of Retsch GmbH)

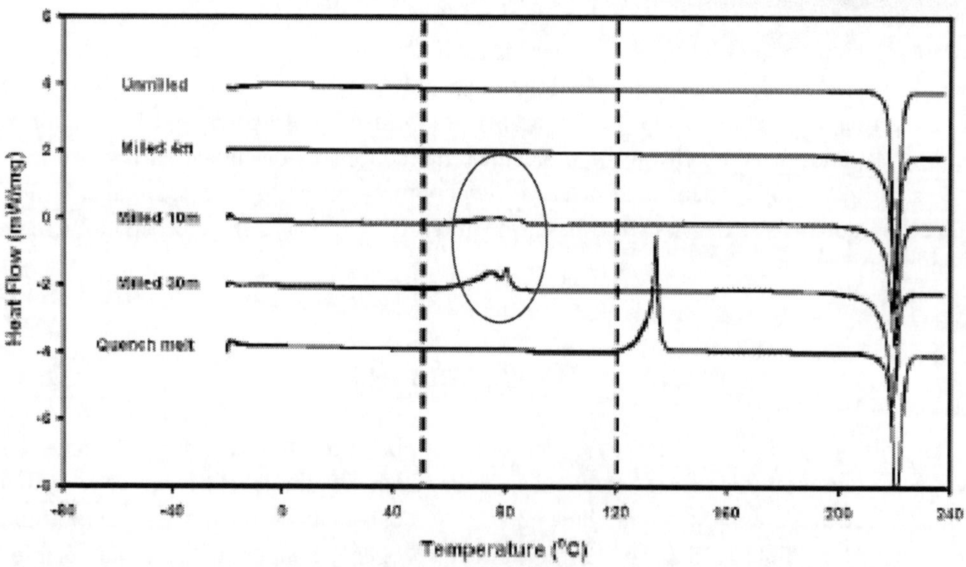

Figure 7. DSC thermograms of griseofulvin samples subjected to different cryogenic milling times. The temperature range marked by dotted lines indicates the recrystallization of defects; the T_g and the onset of recrystallization of the amorphous form. Note the exothermic event when griseofulvin is milled for 10 and 30 min respectively (circled) (In: Feng et al., 2008 – Reproduced by permission of Wiley Interscience)

3.1. Amorphous Phase Formation or Process Induced Disorder?

It was mentioned in the first section that milling could induce various degrees of disorders depending on the milling conditions and material. Feng and co-workers (2008) found that crystal defects rather than amorphous materials were obtained when they subject API griseofulvin to cryogenic milling for various durations (4 to 30 min). A characteristic exothermic event (Figure 7) was present in the differential scanning calorimetry (DSC) thermograms of the cryomilled samples instead of the endothermic glass transition event observed with the amorphous quench melt. This characteristic exothermic event was found to be associated with the rearrangement of defects towards a configuration closer to that of the original crystal and took place at temperatures below the glass transition temperatures.

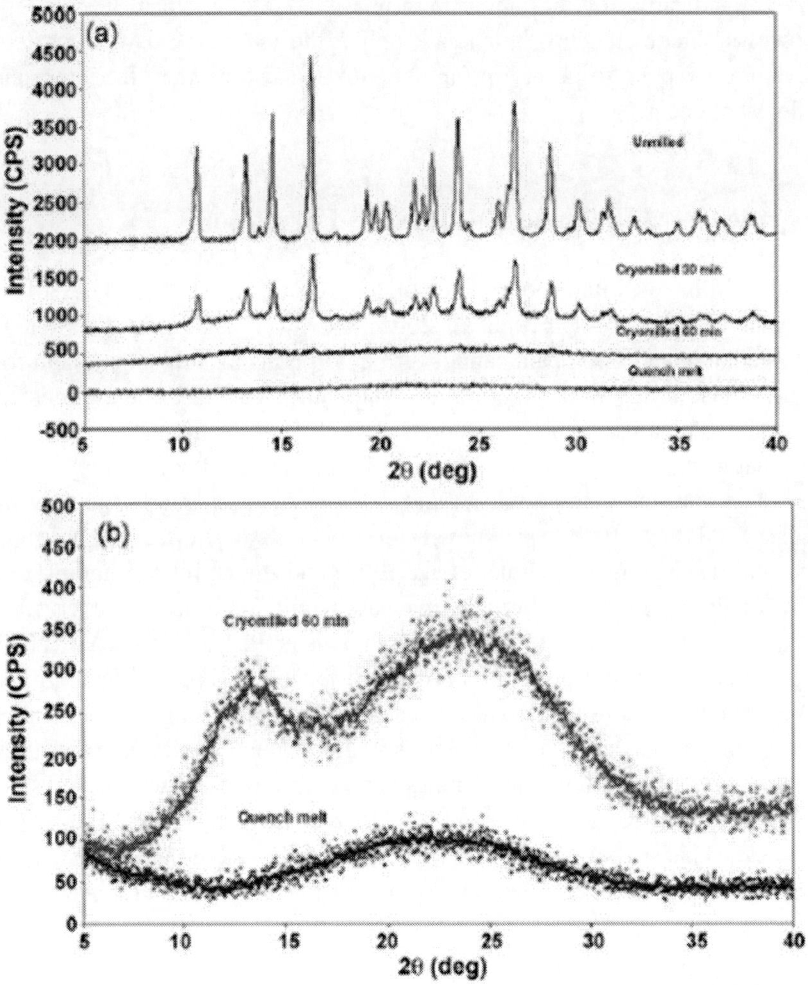

Figure 8. X-ray powder diffractograms (XRPD) of unmilled, cryo-milled 30 min, cryo-milled 60 min and quench melt griseofulvin in (a) Overall XRPD patterns; (b) Scale-up comparison between cryo-milled 60 min and quench melt griseofulvin showing the characteristic halo for the quench melt. (In: Feng et al., 2009 – Reproduced by permission of Elsevier)

Similarly, cryomilling felodipine for a total duration of 10 minutes resulted in the formation of crystal defects rather than amorphous regions (Chamarthy and Pinal, 2008). Although griseofulvin, when subject to longer cryomilling duration (60 min), appeared X-ray amorphous, close examination of the X-ray powder diffraction (XRPD) pattern revealed two major broad peaks at 13° and 24° whereas the XRPD pattern for the amorphous powder showed a single major halo at 22° (Figure 8) (Feng et al., 2009). This suggests that cryomilling is not simply a crystal-to amorphous transition but an intermediate mesophase could be present.

On the other hand, Crowley and Zografi (2002) reported that cryogenic grinding of all three polymorphs (γ, α, and δ) of indomethacin for more than 30 min resulted in the amorphous phase transformation. The occurrence of the 'halo' in the XRPD pattern of the polymorphs as well as the presence of a glass transition event in the DSC thermograms of the polymorphs are indicative of the amorphous phase after more than 30 min of cryogenic milling. However, indomethacin t-butanolate unlike the methanolate was unaffected by the grinding even after post-grind drying treatment, suggesting that the amorphous phase formation does not occur.

3.2. Cryogenic Co-Milling and Co-Crystals Formation

Cryogenic milling is often more efficient to prepare amorphous states of drugs, than grinding at room temperatures (Sheth et al., 2005). However, as described earlier, some drugs such as indomethacin and piroxicam cannot turn amorphous by subjecting them to cryogenic milling alone. These drugs would require an excipient or carrier to be cryogenically milled together in order to obtain a stable amorphous state. It was mentioned in the section on *Co-milling* that Shakhtshneider and co-workers (2007a; 2007b) used polyvinylpyrrolidone (PVP) as a carrier for cryomilling with piroxicam and indomethacin separately and they were able to obtain a glassy mixture. PVP was known for its inhibitory effect on crystallization in amorphous solid dispersions (Yoshioka et al., 1995) and the ability for complexation with a drug via hydrogen bonds to inhibit the nucleation and growth of the crystalline phase (Matsumoto and Zografi, 1999). Furthermore, Shakhtshneider and his team (2007a; 2007b) showed that the excipient to drug ratio is important for the stability of the amorphous phase. When a PVP content of more than 50% was used in the cryomilling, the glass transition was elevated with increased amorphous phase stability. Due to the increased stability in amorphous phase with co-milling under cryogenic conditions, the rate of co-crystal formation would be reduced. As demonstrated by Jayasankar and co-workers (2006), the carbamazepine-saccharin cocrystal is a thermodynamically stable phase at room temperature and storage 75% RH enhances the rate of cocrystallization through water as a potent plasticizer.

4. NANOMILLING

In both academia and industry, there has been much interest in nanoparticles because of their peculiar physicochemical properties. They have large specific surface area, high chemical reactivity and physical affinity as well as interesting optical, electrical and magnetic properties (Bilgili et al., 2006). Products of nanotechnology are expected to revolutionize many industries, especially chemicals, pharmaceuticals, ceramics as well as microelectronics, which give rise to higher homogeneity, solubility or strength. This need demands more and more suspensions of nanoparticles with high storage stability (Stenger et al., 2005). Since year 2000, US Food and Drug Administration has approved more than six nanoparticulate drug formulation products, such as Rapamune®, Emend®, Tricor®, Abraxane®, Triglide® and Megace® ES, which covers a wide range of therapies from immune suppressant, chemotherapy, cholesterol-lowering to appetite enhancement. Of the six drugs listed, five are currently produced by nanomilling in stirred media mills. An example of a nanomill is shown in the following figure.

By going down one more dimension from microrange to the nanoworld, there is a distinct increase in the surface area and related dissolution velocity. For example, when moving from a spherical 50 μm particle to micronized 5 μm particles, the total surface area enlarges by a factor of 10, moving to 500 nm nanocrystals by a factor of 100. There is an additional – but often forgotten – effect further increasing the dissolution velocity, that is, the increase in saturation solubility, when moving to sizes below 1 μm. Due to the strong curvature of the particles, they possess an increase dissolution pressure comparable to the increase vapor pressure of ultrafine aerosol particles. The increases in saturations solubility of nanocrystals reported are by a factor of about 2 to 6 (Mueller, et al., 2006). For drug delivery, most of the target sites are accessible through either microcirculation by blood capillaries or pores present at various surfaces and membranes. Most of the apertures, openings, and gates at cellular or subcellular levels are of nanometer size. Therefore, the nanoparticles are most suited to reach the subcellular level and should be smaller than 300 nm for efficient transport (Gupta, 2006).

In other applications, high energy-intensive ball milling (vibratory milling, stirred media milling, or attritors) has also been reported to be suitable for the production of submicron particles (Cho et al., 1996; Liversidge and The Cundy, 1995; Mende et al., 2003; Strazisar and Runovc, 1996) as well as nanoparticles (Bilgili et al., 2003, 2004; Czekai and Seaman, 1996; Higgins et al., 2003; Lee, 2003; Mendel et al., 1999; Wen and Chen, 1988). While searching for large-scale production of nanoparticles via milling, researchers at Eastman Kodak Company found that small polymeric media like crosslinked polystyrene beads are effective for breaking pigment agglomerates. In a subsequent series of patents (see e.g. Bishop and Czekai, 1997; Czekai and Seaman, 1996; Romano et al., 2001), it was also demonstrated that there was some improvement in the performance of inkjet inks when pigment nanoparticles as opposed to conventional submicron particles were used in ink formulations.

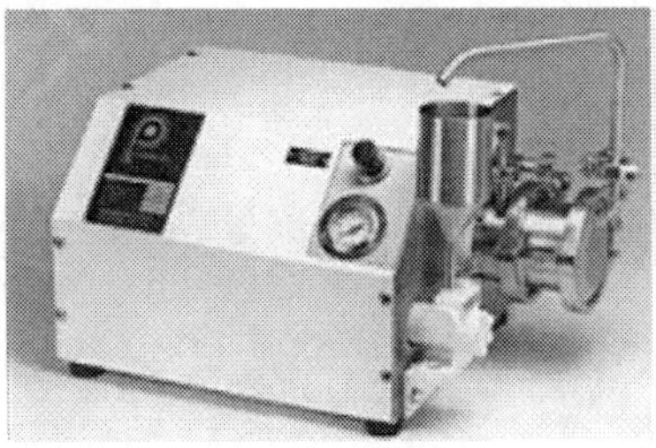

Figure 9. Pühler PHN0,3C NanoII (Courtesy of Pühler R&D)

4.1. Basic Principles and Challenges

The first comprehensive technical reports on nanomilling showed that this process depends mainly on the specific energy input which is the total energy supplied to the grinding chamber related to the product mass (Stehr, 1984; Weit, 1987). With decreasing particle size, materials exhibit increasingly plastic behavior making it more difficult to break small particles than large particles. Therefore, it is very energy intensive to go down to nanoparticles-size range. Other than size, parameters of importance are (i) toughness/brittleness (in tough materials, stress causes plastic deformation, whereas in brittle materials, cracks are propagated; hence, size reduction of brittle materials is easier than for tough materials; sometimes a material can be cooled to embrittle), and (ii) hardness, abrasiveness, particle shape and structure, heat sensitivity (only about 2% of the applied energy goes to size reduction, the rest is converted to heat; hence heat-sensitive drugs require cooling), and explodability (proper inerting may be needed) (Gupta, 2006; Way and Esterly, 2008).

For many materials, a limit in the grindability can be reached where subject to further grinding, no decrease in the particle size is observed. In the sub-micron particle size range, the behavior of the product suspensions is more and more influenced by increasing particle-particle interactions. Due to these interactions, often spontaneous agglomeration of product particles occurs and the viscosity of the product suspension increases (Mende et al., 2003; Stenger and Peukert, 2003). If product particle sizes smaller than 1 μm are reached, these interactions can lead to an interrelation between agglomeration, deagglomeration and communition, thus no further communition progress results in spite of increasing energy input. To overcome this problem, the milling suspension can be stabilized by means of electrostatic steric or electrosteric stabilization (Stenger et al., 2005).

There are a number of media mills available on the market, ranging from laboratory to industrial-scale. Nanomilling consists of a milling container filled with fine milling pearls or larger-sized balls. The container can be static and the milling material is moved by a stirrer; alternatively, the complete container is moved in a complex movement leading consequently to movement of the milling pearls. There are different milling materials available, traditionally steel, glass and zircon oxide are used. New materials are special polymers, i.e. hard polystyrene. Surfactants or stabilizers have to be added for the physical stabilization of the produced nanosuspensions. In the production of nanomedicines, the coarse drug powder is typically dispersed by high-speed stirring in a surfactant/stabilizer solution to yield a macrosuspension. The choice of surfactants and stabilizers is based on the affinity to the nanoparticle surfaces and principles of steric stabilization or a combination of steric and electrostatic stabilization as well as the route of administration for pharmaceutical applications (Mueller et al., 2006). They reduce solid surface energy by adsorption and decrease viscosity by preventing flocculation and creeping of stabilizer molecules along grain boundaries or in dislocations that occur under shear stress (Rebinder effect). Inadequate stabilizer in nanoparticle dispersions induces agglomeration or aggregation, while an excess promotes Ostwald ripening (larger crystals growing at expense of smaller crystals). To inhibit the latter and prevent recrystallization, flocculation, and phase separation, ideal nano formulation design takes advantage of synergistic interactions between certain polymers and ionic surfactants (Deng et al., 2008).

Post-milling relaxation may be essential to obtain a nanoparticle suspension at steady state. Deng and co-workers (2008) reported that after milling, a suspension of drug nanoparticles agglomerated and the nanoparticle suspension formed large clumps or a gel within one to two days. The apparent size of the particles continued to increase until reaching a maximum size. Subsequently, the gel relaxed spontaneously with an apparent size reduction, and clusters automatically dissociated into individual particles of nano size within a few days. Once the particle size decreased and big clusters dissociated into individual small particles, the system showed excellent long-term shelf stability. Milling conditions such as time, speed, and stabilizer concentration were believed to play an important role in the relaxation process that occurred in a nanoparticle dispersion. In another recent report, the operating parameters were studied to control the milling process with respect to the required milling time for reaching the minimum particle size and wear of the set-up. The milling time was found to be strongly dependent on the milling medium bead size and on the rotation speed. The process became faster when the rotation speed increased and the milling medium bead size decreased. The wear of the set-up, and therefore the contamination of the final product with heavy metals, strongly increased with the rotation speed. A similar trend was observed with an increase of milling medium bead size (Hennart et al., 2010).

Nanomilling is usually associated with high energy consumption. Depending on the product material, a typical process may require hours or even days. Heat-susceptible materials will require additional cooling to minimize product degradation. New high-power-density mills and a special modification of the process were reported with better economical values. Different mill geometries were investigated and compared with a classical stirred media mill with discs. Due to higher energy densities in the grinding chamber, these high-power-density mills are very suitable for grinding in the nanometer and submicron range (Breitung-Faes and Kwade, 2008).

Second, due to the intensive contact between milling pearls and balls, erosion may occur during the process, which can contaminate the final product. While very few published data are available, discussions after oral presentations have suggested 0.1 to 70 ppm contamination in pharmaceutical applications. It should be noted that the extent of erosion depends on the solid concentration of the macrosuspension to be processed, the hardness of drug, and based on this the required milling time and milling material. Apart from the milling material, the erosion from the container also needs to be considered (Mueller, 2006).

To achieve efficient and economic nanomilling, Kwade (2009) has recently suggested that the operating parameters should be chosen to maximize the number of stress events, so that the stress energy has an optimum value. In addition, the mill must be capable to handle very small grinding beads. The resulting suspension needs to be stabilized against agglomeration and the suspension rheology must be tailored through careful control of the particulate interfaces. It was also presented that using optimum process and formulation parameters, transparent suspensions with median particle sizes below 20 nm can be produced with hardly any product contamination.

5. Conclusion

This review summarises recent research and development on the effect of milling beyond achieving particle size reduction. The energy-intensive mechanical and thermal stresses often contribute to unintended changes in physicochemical or/and solid-state properties of the milled material such as morphology, crystallinity or amorphous content, polymorphism, glass transition, melting properties as well as stress relaxation characteristics. These may have a detrimental effect on subsequent handling, processing and product quality. Designing suitable post-milling treatments is often needed to mitigate these side-effects. On the other hand, if such effects are desired, these pitfalls can be turned into benefits by re-designing the milling process such as by co-milling with suitable additives, cryogenic milling or co-crystal formation. This may be analogous to the knowledge surge in metallurgical alloying in 1960-70s, which led to the wide range of metallic alloys that we have today. Research in altering the solid-state properties of organic materials has just been actively pursued since the last decade and is well supported by advances in powder characterisation and analyses like powder X-ray diffraction, gravimetric vapour sorption, inverse gas chromatography, isothermal calorimetry, high-sensitivity differential scanning calorimetry, Raman microscopy, etc., which provide more information about the nature of molecular interactions and spatial distributions of different species in a solid solution. This new knowledge may give rise to new milling methods to design a vast palette of "organic alloys" for different applications in the future.

Although nanomilling has overcome the previous "grinding limit" of 1 micron, it suffers from drawbacks like long milling time, residual solvents and potential degradation. Therefore, significant research efforts have been directed to pursue alternative strategies such as using supercritical fluids to directly precipitate nanoparticles, combined evaporation and high pressure homogenization methods (Moschwitzer and Muller 2006) and high-efficiency static mixing for continuous and state-up production of nanoparticles (Dong et al., 2009; 2010)

Lastly, with the rapid growth in life and biomedical science research, there is an increasing need to prepare biomaterials such as peptides, proteins, DNA, RNA as a solid phase to improve the ease of storage, formulation and delivery. Future research in milling of such materials requires more than just how to achieve desired particle sizes but better understanding on avoiding degradation, changes in solid-state and 3-D structural properties as well as stabilising the functional properties.

Acknowledgments

This work is supported by the Science and Engineering Research Council of A*STAR (Agency of Science, Technology and Research). The authors would also like to thank Mr Balani N. Prashant, Dr Shen Shoucang, Retsch GmbH and Pühler Feinmahltechnik GmbH for their permissions to use their materials and products in this chapter.

REFERENCES

[1] Aitipamula, S., Chow, P. S. & Tan, R. B. H. (2009a). *CrystEngComm, 11,* 889-895.

[2] Aitipamula, S., Chow, P. S. & Tan, R. B. H. (2009b). *CrystEngComm, 11,* 1823 – 1827.

[3] Allesø, M., Norman, C., Rehder, S., Rantanen, J., Rades, T. & Aaltonen, J. (2009). *J Control Release, 136*, 45 – 53.

[4] Arai, Y. (1996). In: Scarlett, B., Akers, R.J., Treasure, C.R.G. (Eds.), *Chemistry of Powder Production,* 1st Edition, Chapman & Hall, UK.

[5] Ashford, M. (2002). *Assessment of biopharmaceutical properties.* In: Aulton, M.E. (Ed.), *Pharmaceutics: The Science of Dosage Form Design*, 2nd Edition, Church Livingstone, Elsiever Ltd.

[6] Aso, Y., Yoshioka, S. & Kojima, S. (1996). *Chem Pharm Bull, 44,* 1065 – 1067.

[7] Bahl, D. & Bogner, R. H. (2006). *Pharm Res, 23,* 2317 – 2325.

[8] Bahl, D. & Bogner, R. H. (2008). *AAPS PharmSciTech, 9,* 146 – 153.

[9] Bahl, D., Hudak, J. & Bogner, R. H. (2008). *Pharm Dev Technol, 13,* 255 – 269.

[10] Bakker, H., Zhou, G. F. & Yang, H. (1995). *Prog Mater Sci, 39,* 159 – 241.

[11] Balani, P. N., Ng, W. K., Tan, R. B. H. & Chan, S. Y. (2010a). *J Pharm Sci, 99,* 2462 – 2474.

[12] Balani, P. N., Wong, S. Y., Ng, W. K., Widjaja, E., Chan, S. Y. & Tan, R. B. H. (2010b). *Int J Pharm,* in press.

[13] Begat, P., Young, P. M., Edge, S., Kaerger, J. S. & Price, R. (2003). *J Pharm Sci, 92,* 611 – 620.

[14] Benjamin, J. S. (1970). *Mater Trans, 1, 2394.*

[15] Bhugra, C. & Pikal, M. J. (2008). *J Pharm Sci, 97,* 1329 – 1349.

[16] Bilgili, E., Hamey, R. & Scarlett, B. (2006). *Chem Eng Sci, 61,* 149 – 157.

[17] Bilgili, E., Hamey, R. & Scarlett, B. (2004). *China Part, 2,* 93 – 100.

[18] Bilgili, E., Hamey, R., Zhupanska, O., Eskin, D. & Scarlett, B. (2003). *Proceedings of the AIChE Annual Meeting, Paper No. 30d (CD-ROM), San Francisco.*

[19] Bishop, J. F. & Czekai, D. A. (1997). *US Patent 5679138, Eastman Kodak Co.*

[20] Boldyrev, V. V., Shakhtshneider, T. P., Burleva, L. P. & Severtsev, V. A. (1994). *Drug Dev Ind Pharm, 20,* 1103-1114.

[21] Braga, D., Giaffreda, S. L., Grepioni, F., Pettersen, A., Maini, L., Curzi, M. & Polito, M. (2006). *Dalton Trans,* 1249 – 1263.

[22] Braga, D., Giaffreda, S. L., Rubini, K., Grepioni, F., Chierotti, M. R. & Gobetto, R. (2007). *CrystEngComm, 9,* 39-45.

[23] Breitung-Faes, S. & Kwade, A. (2008). *Chem Eng Res Des, 86,* 390 – 394.

[24] Brittain, H. G. (2002). *J Pharm Sci, 91,* 1573 – 1580.

[25] Brodka-Pfeiffer, K., Langguth, P., Gra, β, P. & Häusler, H. (2003). *Eur J Pharm Biopharm, 56,* 393-400.

[26] Bucci, K. K. & Possidente, C. J. (2006). *Am J Health Syst Pharm, 63,* 1654 – 1655.

[27] Burt, H. M. & Mitchell, A. G. (1981). *Int J Pharm, 9,* 137 – 152.

[28] Chadwick, K., Davey, R. & Cross, W. (2007). *CrystEngComm, 9,* 732 – 734.

[29] Chamarthy, S. P. & Pinal, R. (2008). *Colloid Surface A, 331,* 68 – 75.

[30] Chieng, N., Hubert, M., Saville, D., Rades, T. & Aaltonen, J. (2008). *AAPS Annual Meeting Poster Presentation, Atlanta, GA.*

[31] Chikhalia, V., Forbes, R. T., Storey, R. A. & Ticehurst, M. (2006). *Eur J Pharm Sci 27,* 19 – 26.

[32] Chiou, W. L. & Kyle, L. E. (1979). *J Pharm Sci, 68,* 1224 – 1229.

[33] Cho, H., Waters, M. A. & Hogg, R. (1996). *Int J Miner Process,* 44-45, 607 – 615.

[34] Choksi, R. J., Shah, N. H., Sandhu, H. K., Malick, A. W. & Zia, H. (2008). *J Pharm Sci, 97,* 2286-2298.

[35] Cirri, M., Maestrelli, F., Furlanetto, S. & Mura, P. (2004). *J Therm Anal Calorim, 77,* 413 – 422.

[36] Colombo, I., Grassi, G. & Grassi, M. (2009). *J Pharm Sci 98,* 3961-3986.

[37] Crowley, K. & Zografi, G. (2002). *J Pharm Sci 91,* 492 – 507.

[38] Czekai, D. A. & Seaman, L. P. (1996). *US Patent 5500331, Eastman Kodak Co.*

[39] Deng, Z., Xu, S. & Li, S. (2008). *Int J Pharm, 351,* 236 – 243.

[40] Descamps, M., Willart, J. F., Dudognon, E. & Caron, V. (2007). *J Pharm Sci, 96,* 1398 – 1407.

[41] Dong, Y. C., Ng, W. K., Hu, J., Shen, S. C. & Tan, R. B. H. (2010). *Int J Pharm 386,* 256 – 261.

[42] Dong, Y. C., Ng, W. K., Shen, S. C., Kim, S. & Tan, R. B. H. (2009). *Int J Pharm 375,* 84 – 88.

[43] Feng, T., Bates, S. & Teresa Caravajal, M. (2009). *Int J Pharm, 367,* 16 – 19.

[44] Feng, T., Pinal, R. & Teresa Caravajal, M. (2008). *J Pharm Sci, 97,* 3207 – 3221.

[45] Frantz, S. (2006). *Nat Rev, 5,* 881 – 882.

[46] Giry, K., Péan, J. M., Giraud, L., Marsas, S., Rolland, H. & Wüthrich, P. (2006). *Int J Pharm 321,* 162 -166.

[47] Gunawan, L., Johari, G. P. & Shaker, R. M. (2006). *Pharm Res, 23,* 967 – 979.

[48] Gupta, M. K., Vanwert, A. & Bogner, R. H. (2003). *J Pharm Sci, 92,* 536 – 550.

[49] Gupta, R. B. (2006). *Fundamentals of Drug Nanoparticles.* In: Gupta, R.B., Kompella, U.B. (Eds.), Nanoparticle technology for drug delivery, Taylor & Francis, New York.

[50] Gusseme, A. D., Neves, C., Willart, J. F., Rameau, A. & Descamps, M. (2008). *J Pharm Sci 97,* 5000-5012.

[51] He, G., Chacko, J., Guo, L. F., Chow, P. S. & Tan, R. B. H. (2008). *J Phys Chem B 112,* 9890–9895.

[52] He, G., Chow, P. S. & Tan, R. B. H. (2009). *Cryst Growth Des, 9,* 4529-4532.

[53] Heng, J. Y. Y., Thielmann, F. & Williams, D. R. (2006). *Pharm Res, 23*, 1918-1927.

[54] Hennart, S. L. A., Domingues, M. C., Wildeboer, W. J., van Hee, P., Meestersa, G. M. H. & (2010). *Powder Technol, 198,* 56-60.

[55] Higgins, J. P., Arrivo, S. M., Thurau, G., Green, R. L., Bowen, W., Lange, A., Templeton, A. C., Thomas, D. L. & Reed, R. A. (2003). *Anal Chem, 75,* 1777 – 1785.

[56] Hojjati, H. & Rohani, S. (2006). *Org Process Res Dev, 10,* 1110 – 1118.

[57] Jayasankar, A., Good, D. J. & Rodríguez-Hornedo, N. (2007). *Mol Pharm, 4,* 360 – 372.

[58] Jayasankar, A., Somwangthanaroj, A., Shao, Z. J. & Rodríguez-Hornedo, N. (2006). *Pharm Res 23,* 2381-2391.

[59] Kuroda, R., Imai, Y. & Tajima, N. (2002). *Chem Commun,* 2848 – 2849.

[60] Kwade, A. & Breitung-Faes, S. *Achema (*2009). 29th International Exhibition - Congress on Chemical Engineering, Environmental Protection and Biotechnology, Frankfurt, Germany, 11-15 May 2009.

[61] Le Bolay, N., Santran, V., Dechambre, G., Combes, C., Drouet, C., Lamure, A. & Rey, C. (2009). *Powder Technol, 190,* 89 – 94.

[62] Lee, J. (2003). *J Pharm Sci 92,* 2057 – 2068.

[63] Lee, J. C., Namura, S., Kondo, S. & Abe, A. (2001). *Polymer, 42,* 5453 – 5461.

[64] Liversidge, G. C. & Cundy, K. C. (1995). *Int J Pharm, 125,* 91 – 97.

[65] Mackin, L., Sartnurak, S., Thomas, I. & Moore, S. (2002). *Int J Pharm 231,* 213 – 226.

[66] Mahlin, D., Berggren, J., Gelius, U., Engstrom, S. & Alderborn, G. (2006). *Int J Pharm, 321,* 78 – 85.

[67] Marsac, P. J., Konno, H., Rumondor, A. C. F. & Taylor, L. S. (2008). *Pharm Res, 25,* 647 – 656.

[68] Matsumoto, T. & Zografi, G. (1999). *Pharm Res, 16,* 1722 – 1728.

[69] Mende, S., Stenger, F., Peukert, W. & Schwedes, J. (2003). *Powder Technol, 132,* 64 – 73.

[70] Mendel, J., Bugner, D. & Bermel, A. D. (1999). *J Nanopart Res, 1,* 421 – 424.

[71] Meuller, (2006). *Manufacturing of Nanoparticles by Milling and Homogenization Techniques.* In: Gupta, R.B., Kompella, U.B. (Eds), Nanoparticle technology for drug delivery, Taylor & Francis, New York.

[72] Milligan, J., Vintila, R. & Brochu, M. (2009). *Mater Sci Eng A-Struct, 508,* 43 – 49.

[73] Miyazaki, T., Yoshioka, S., Aso, Y. & Kojima, S. (2004). *J Pharm Sci, 93,* 2710 – 2717.

[74] Mura, P., Cirri, M., Faucci, M. T., Ginès-Dorado, J. M. & Bettinetti, G. P. (2002). *J Pharmaceut Biomed, 30,* 227 – 237.

[75] Myz, S. A., Shakhtshneider, P., Fucke, K., Fedotov, A. P., Boldyreva, E. V., Boldyrev, V. V. & Kuleshova, N. I. (2009). *Mendeleev Commun, 19,* 272 – 274.

[76] Ng, W. K., Kwek, J. W. & Tan, R. B. H. (2008). *Pharm Res, 25,* 1175 – 1185.

[77] Ng, W. K., Kwek, J. W., Yuen, A., Tan, C. L. & Tan, R. (2010). *AAPS PharmSciTech,* in press.

[78] Oguchi, T., Kazama, K., Yonemochi, E., Churimaworapan, S., Choi, W. S., Limmatvapirat, S. & Yamamoto, K. (2000). *Phys Chem Chem Phys 2,* 2815 – 2820.

[79] Otsuka, M. & Kaneniwa, N. (1983). *Chem Pharm Bull, 31,* 4489 – 4495.

[80] Otsuka, M. & Kaneniwa, N. (1984). *Chem Pharm Bull, 32,* 1071 – 1079.

[81] Otsuka, M., Ofusa, T. & Matsuda, Y. (1998). *Colloid Surface B, 10,* 217 – 226.

[82] Pedireddi, V. R., Jones, W., Chorlton, A. P. & Docherty, R. (1996). *Chem Commun,* 987 – 988.

[83] Révész, Á. (2005). *J Mater Sci 40,* 1643 – 1646.

[84] Réyné, M. (1995). In: Technologie des composites, 2nd Edition, Hermes, Paris, France.

[85] Romano, C. E., Bugner, D. E., VanHanehem, R. C., Bennett, J. R., Smith, D. E. & Guistina, R. A. (2001). *US Patent 6210474, Eastman Kodak Co.*

[86] Saleki-Gerhardt, A., Ahlneck, C. & Zografi, G. (1994). *Int J Pharm, 101, 237 – 247.*

[87] Sebhatu, T., Angberg, M. & Ahlneck, C. (1994). *Int J Pharm 104, 135 – 144.*

[88] Seyni, A., Le Bolay, N. & Molina-Boisseau, S. (2009). *Powder Technol, 190,* 176 – 184.

[89] Shakhtshneider, T. P. & Boldyrev, V. V. (1999). *Mechanochemical synthesis and mechanical activation of drugs.* In: Boldyreva, E., Boldyrev, V.V. (Eds.), Reactivity of molecular solids, John Wiley & Sons Ltd., Chichester, GB.

[90] Shakhtshneider, T. P., Danede, F., Capet, F., Willart J. F., Descamps, M., Paccou, L., Myz, S. A., Boldyreva, E. V. & Boldyreva, V. V. (2007a). *J Therm Anal Chem, 89,* 699 – 707.

[91] Shakhtshneider, T. P., Danede, F., Capet, F., Willart J. F., Descamps, M., Paccou, L., Surov, E. V., Boldyreva, E. V. & Boldyreva, V. V. (2007b). *J Therm Anal Chem, 89,* 709 - 715.

[92] Shan, N., Toda, F. & Jones, W. (2002). *Chem Commun,* 2372 – 2373.

[93] Shen, S. C., Ng, W. K., Chia, L., Dong, Y. C., Tan, R. B. H. (2010). *J Pharm Sci, 99,* 1997 – 2007.

[94] Sheth, A. R., Lubach, J. W., Munson, E. J., Muller, F. X. & Grant, D. J. W. (2005). *J Am Chem Soc, 127,* 6641.

[95] Skovsgaard, S. & Bond, A. D. (2009). *CrystEngComm, 11,* 444 – 453.

[96] Sonoda, R., Horibe, M., Oshima, T., Iwasaki, T. & Watano, S. (2008). *Chem Eng Bull, 56*, 1243-1247.

[97] Stehr, N. (1984). *Chem Eng Process, 18,* 73 – 83.

[98] Stenger, F., Mende, S., Schwedes, J. & Peukert, W. (2005). *Chem Eng Sci, 60,* 4557 – 4565.

[99] Stenger, F. & Peukert, W. (2003). *Chem Eng Technol, 26,* 177 – 183.

[100] Strazisar, J. & Runovc, F. (1996). *Int J Miner Process, 44 – 45,* 673 – 682.

[101] Suryanarayana, C. (2001). *Prog Mater Sci 46,* 1 – 184.

[102] Tanaka, T. & Kanda, Y. (2006). *Crushing and Grinding.* In: H., Masuda, K., Higashitani, H. Yoshida, (Eds.), *Powder Technology Handbook,* 3rd Edition, Taylor & Francis Group, CRC Press.

[103] Tang, X. C., Pikal, M. J. & Taylor, L. S. (2002). *Pharm Res, 19,* 477-483.

[104] Trask, A. V., Haynes, D. A., Motherwell, W. D. S. & Jones, W. (2006). *Chem Commun,* 51 – 53.

[105] Trask, A. V., Motherwell, W. D. S. & Jones, W. (2004). *Chem Commun,* 890 – 891.

[106] Trask, A. V., van de Streek, J., Motherwell, W. D. S. & Jones, W. (2005). *Cryst Growth Des, 5,* 2233 – 2241.

[107] U.S. Food and Drug Administration, Office of Combination Products, Website: http://www.fda.gov/oc/combination/21 CFR Part 3.2 (e).

[108] Umprayn, K. & Mendes, R. W. (1987). *Drug Dev Ind Pharm, 13,* 653 – 693.

[109] Watanabe, T., Hasegawa, S., Wakiyama, N., Kusai, A. & Senna, M. (2003). *Int J Pharm, 250,* 283 – 286.

[110] Watanabe, T., Hasegawa, S., Wakiyama, N., Kusai, A. & Senna, M. (2002). *Int J Pharm, 248,* 123-129.

[111] Watanabe, T., Wakiyama, N., Usui, F., Ikeda, M., Isobe, T. & Senna, M. (2001). *Int J Pharm 226,* 81 – 91.

[112] Way, H. & Esterly, C. (2008). *Equipment and Processing Report (PharmTech.com),* Website: http://pharmtech.findpharma.com/pharmtech/Article/The-Challenges-of-Manufacturing-Nanoparticles-thro/ArticleStandard/Article/detail/529752.

[113] Weeber, A. W. & Bakker, H. (1988). *Phys B, 153,* 93 – 135.

[114] Weit, H. (1987). *Chem Eng Technol, 10,* 318 – 319.

[115] Wen, S. B. & Chen, C. K. (1988). *Powder Technol, 55,* 11 – 17.

[116] Wertheimer, A. I. & Morisson, A. (2002). *P & T 27,* 44.

[117] Willart, J. F., Caron, V., Lefort, R., Danede, F., Prevost, D. & Descamps, M. (2004). *Solid State Commun, 132,* 693 – 696.

[118] Willart, J. F., De Gusseme, A., Hemon, S., Odou, G., Danede, F. & Descamps, M. (2001). *Solid State Commun, 119,* 501 – 505.

[119] Willart, J. F., Descamps, N., Caron, V., Capet, F., Danede, F. & Descamps, M. (2006). *Solid State Commun, 138,* 194 – 199.

[120] Willart, J. F., Lefebvre, J., Danede, F., Comini, S., Looten, P. & Descamps, M. (2005). *Solid State Commun, 135,* 519 – 524.

[121] Ye, J. & Schoenung, J. M. (2004). *Adv Eng Mater, 6,* 656 – 664.

[122] York, P. (2002). *The Design of Dosage Forms.* In: Aulton, M.E. (Ed.), Pharmaceutics: The Science of Dosage Form Design, 2nd Edition, Church Livingstone, Elsiever Ltd.

[123] Yoshioka, M., Hancock, B. C. & Zografi, G. (1995). *J Pharm Sci, 84,* 983 – 986.

[124] Yu, L. (2001). *Adv Drug Deliver Rev, 48,* 27 – 42.

[125] Zapata, C., Frances, C., Le Bolay, N. & Molina-Boisseau, S. (2004). *Chem Eng Res Des, 82,* 631 – 636.

[126] Zapata-Massot, C. & Le Bolay, N. (2007). *Part Part Syst Charact, 24,* 339 – 344.

[127] Zhang, G. G. Z., Gu, C., Zell, M. T., Burkhardt, T., Munson, E. J. & Grant, D. J. W. (2002). *J Pharm Sci, 91,* 1089 – 1100.

In: Powder Engineering, Technology and Applications
Editor: Jason M. Barker, pp.149-173
ISBN: 978-1-61761-216-9

Chapter 5

PARTICLE COLLISION HANDLING FOR HARD-SPHERE DISCRETE PARTICLE MODEL

C. L. Wu [1,] and A. S. Berrouk[2]*

[1] Engineering College, Guangdong Ocean University, Zhanjiang 524088, China
[2]Department of Chemical Engineering, The Petroleum Institute, Abu Dhabi, United Arab Emirates

ABSTRACT

The Discrete Particle Model (DPM) has been extensively used over the last decade to study the complicated flow behaviors in dense gas-solid fluidization systems relevant to a wide range of chemical and petrochemical processes. Numerical simulations of these real-life systems involve often tracking a large number of particles colliding at high frequency which results in a heavy computational burden. Thus, designing efficient collision-handling algorithms to operate collision events is paramount for its industrial use in studying flow dynamics in chemical reactors.

In this chapter we first discuss previous algorithms developed for that purpose and then present a novel, more efficient algorithm designed to handle collision events in the context of the hard-sphere DPM. The core idea is centered on the use of dynamic crosswise linked data structures. This type of data structures enables an efficient updating and managing of lists of collision events associated with particles. In addition, a chained hash table strategy is proposed to efficiently process particle collision sequence. For systems of *N* particles, this strategy allows a complexity *O(1)* time per collision regarding the event list handling operations.

INTRODUCTION

Hard-sphere Discrete Particle Modeling (DPM) coupled to a finite volume/difference description of the gas-phase dynamics has been used for a wide panoply of chemical and petrochemical applications since it was first introduced to study gas-solid flow in fluidized beds [1]. In the framework of DPM, the hydrodynamics of the gas phase is resolved at a

* Corresponding author: E-mail: chunliangwu@gmail.com, Tel: +8613828261515

larger scale than the particle size, while the particle motion is described in a Lagrangian framework applying the Newtonian kinetics for each individual particle. Thus, the particle-particle interactions are resolved at such a particle scale that many important characteristics related to the particle motion in fluidized bed reactors can be well captured [1-5].

The hard-sphere simulation technique was originated from the molecular dynamics (MD) modeling of a set of molecules interacting via hard potentials [6]. A detailed programming description of the MD simulation is available in the concerned literature [7-11]. All these developed MD algorithms use an $O(logN_p)$ event queue to handle the collision events.

In conjunction with this steady progress in MD modeling, successful attempts of combining the hard-sphere model with computational fluid dynamics to simulate gas-solid flow in fluidized beds were made [1]. In the hard-sphere DPMs, the inter-particle interactions are modeled by binary collision dynamics. Collisions are assumed instantaneous and handled one by one according to the order of the collision event occurrence. Detection and handling of potential collision events often cause a heavy computational burden for the hard-sphere DPMs. This could impede its application for real-life dense gas-particle systems if the model is not efficiently implemented.

Relatively little progress has been achieved in the development of the hard-sphere DPMs regarding the handling of the collision events compared to other aspects such as collision dynamics and fluid-particle interactions. Several optimization strategies in the context of DPM have been discussed in the literature [12], such as the delaying update of particle position [13], the local minimum strategy [10, 14] and the so-called cell method to compile the neighbor list [15, 16]. A reduction in computation time was achieved using these strategies yet the complexity is only of order $O(N_p)$ because no proper data structure was introduced to handle the event list. Also, a simple algorithm without a complex data structure that resulted in $O(N_g)$ complexity was developed [12].

In this chapter we detail an efficient approach to build a complexity $O(1)$ priority queue suitable for hard-sphere discrete particle modeling. The chained hash-table concept is introduced to speedup both insertion of collision events and location of current events with the minimum collision time. The data structure of crosswise linking event list is used to determine and delete all invalid events associated with any colliding particle pair. The basic algorithm often used in DPMs to handle collision events is reviewed then the principle features of the new strategy are detailed along with a comprehensive testing.

BASIC ALGORITHM

In DPMs, due to the weak gas-phase inertia compared to that of the dispersed phase, the complex gas-particle system is often divided in two subsystems or processes that are solved separately at each simulation time step. The solution procedure is illustrated in Figure 1. In the first process, the two-phase interactions are taken into account while the particles are assumed fixed in space. Thus, only the momentum exchange between the two phases is accounted for in this first step. The gas-phase governing equations are solved using a finite volume/difference method together with the particle momentum equation. The system of equations is solved using whether an explicitly segregated scheme [1, 12, 17] or an implicitly coupled algorithm [5, 18]. The time step should be selected at least one order of magnitude smaller than the particle response time to capture the two-phase interaction. In the second process, all possible collisions between particles are detected and the collision dynamics is computed for each occurring collision. The particle-particle interactions are assumed impulsive and do not depend on the local gas flow field. Between successive collisions, particles' motion is assumed to consist of free flights. During the second step, all collision events are handled one by one in a chronological order. This process is the core of the hard-sphere DPM yet the most time-consuming part in simulating dense gas-solid flows.

In the hard-sphere DPM, collisions are assumed binary, quasi-instantaneous, and particle contact occurs at one point. During a collision only impulse forces are considered. The relative velocity $\mathbf{v}_{12}$ at the contact point between two particles with velocities $\mathbf{v}_1$ and $\mathbf{v}_2$ is defined as follows:

$$\mathbf{v}_{12} = (\mathbf{v}_1 - \mathbf{v}_2) - (r_1\boldsymbol{\omega}_1 + r_2\boldsymbol{\omega}_2) \times \mathbf{n} \tag{1}$$

where r_1 and r_2 are the radii, and ω_1 and ω_2 are the angular velocities as shown in Figure 2.

The normal and tangential unit vectors that define the collision coordinate system are computed as follows:

$$\mathbf{n} = \frac{\mathbf{x}_1 - \mathbf{x}_2}{|\mathbf{x}_1 - \mathbf{x}_2|} \tag{2}$$

$$\mathbf{t} = \frac{\mathbf{v}_{12}^0 - (\mathbf{G}_{12}^0 \cdot \mathbf{n})\mathbf{n}}{|\mathbf{v}_{12}^0 - (\mathbf{G}_{12}^0 \cdot \mathbf{n})\mathbf{n}|} \tag{3}$$

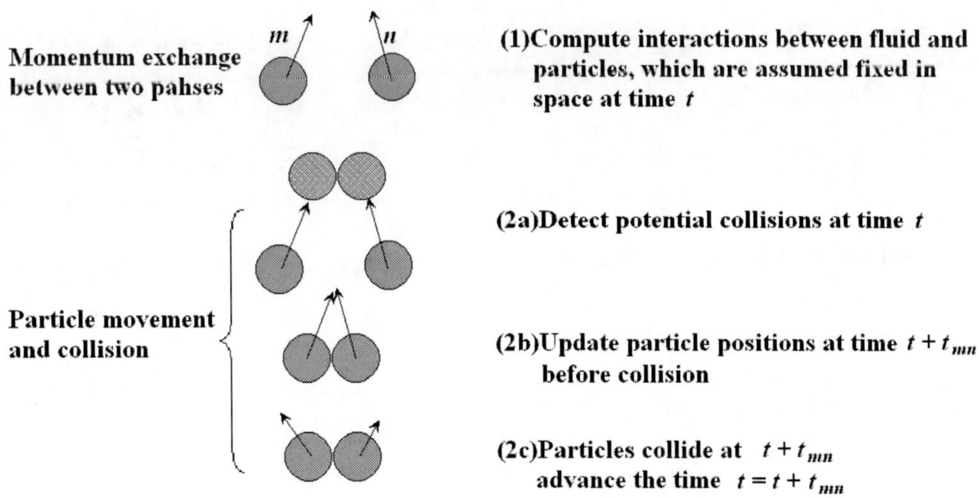

Figure 1. Two steps decoupling the hard-sphere DPM; t_{mn} is the minimum collision time in collision sequence.

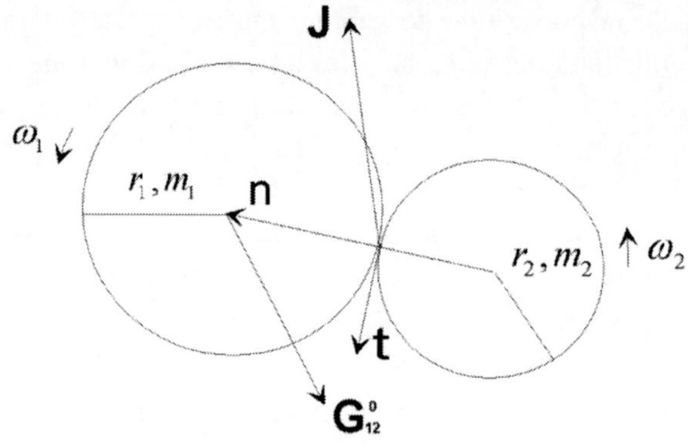

Figure 2. Illustration of collision between two particles with mass m_1 and m_2

where the superscript 0 denotes pre-collision conditions. The relative velocity of the particle centroid is given by:

$$\mathbf{G}_{12} = \mathbf{v}_1 - \mathbf{v}_2 \tag{4}$$

The equations of motions of the two colliding particles are given by the following system:

$$\begin{aligned} m_1(\mathbf{v}_1 - \mathbf{v}_1^0) &= \mathbf{J} \\ m_2(\mathbf{v}_2 - \mathbf{v}_2^0) &= -\mathbf{J} \\ \frac{I_1}{r_1}(\boldsymbol{\omega}_1 - \boldsymbol{\omega}_1^0) &= \mathbf{J} \times \mathbf{n} \\ \frac{I_2}{r_2}(\boldsymbol{\omega}_2 - \boldsymbol{\omega}_2^0) &= \mathbf{J} \times \mathbf{n} \end{aligned} \tag{5}$$

where I_1 and I_2 are the moments of inertia and $\mathbf{J}$ is the impulse vector. Using the relation $(\mathbf{n} \times \mathbf{J}) \times \mathbf{n} = \mathbf{J} - \mathbf{n}(\mathbf{J} \cdot \mathbf{n})$, one obtains:

$$\mathbf{v}_{12} - \mathbf{v}_{12}^0 = B_1 \mathbf{J} - (B_1 - B_2)(\mathbf{J} \cdot \mathbf{n})\mathbf{n} \tag{6}$$

The constants B_1 and B_2 for rigid spheres are given by:

$$\begin{aligned} B_1 &= \frac{7}{2}\left(\frac{1}{m_1} + \frac{1}{m_2}\right) \\ B_2 &= \frac{1}{m_1} + \frac{1}{m_2} \end{aligned} \tag{7}$$

Some parameters linked to the collision mechanics are required to relate the pre-collision and post-collision velocities. The first collision parameter of the model is defined as the coefficient of normal restitution, e_n:

$$\mathbf{v}_{12} \cdot \mathbf{n} = -e_n \left(\mathbf{v}_{12}^0 \cdot \mathbf{n} \right) \tag{8}$$

By combining Eq. (8) with Eq. (6), the normal component of the impulse vector is defined:

$$J_n = -(1+e_n) \frac{\mathbf{v}_{12}^0 \cdot \mathbf{n}}{B_2} \tag{9}$$

The second and the third collision parameters are the coefficient of friction μ_f and the coefficient of tangential restitution e_t. These two collision parameters represent two types of tangential collision, namely sticking and sliding collisions.

If the tangential component of the impact velocity is sufficiently high or the friction coefficient is small by comparison, i.e.:

$$\mu_f < \frac{(1+e_t)\mathbf{v}_{12}^0 \cdot \mathbf{t}}{J_n B_1} \tag{10}$$

Gross sliding occurs during the whole duration of the contact. In this case the Coulomb's law applies and the tangential component of impulse is then given by:

$$J_{t,sliding} = -\mu_f J_n \tag{11}$$

For sticking-type of tangential collisions, the relative tangential velocity between the two colliding particles becomes zero. Thus, the friction force is sufficiently high:

$$\mu_f \geq \frac{(1+e_t)\mathbf{v}_{12}^0 \cdot \mathbf{t}}{J_n B_1} \tag{12}$$

In this case, the coefficient of tangential restitution is defined as:

$$\mathbf{v}_{12} \cdot \mathbf{t} = -e_t (\mathbf{v}_{12}^0 \cdot \mathbf{t}) \tag{13}$$

and the tangential impulse for sticking collision reads:

$$J_{t,sticking} = -(1+e_t) \frac{\mathbf{v}_{12}^0 \cdot \mathbf{t}}{B_1} \tag{14}$$

With the total impulse vector known (Eqs. (9) and (11) or Eqs. (9) and (14)), the post-collision velocities can be computed using system (5).

Hereafter the basic algorithm for collision event handling is described for a system of N_P particles denoted by $P = \{0, 1, \ldots, N_p\text{-}1\}$. The system P occupies a simulation domain that is discretized into N_g finite volume grids (structured or unstructured) denoted by $Grid = \{0, 1, \ldots, N_g\text{-}1\}$. The time step used to integrate the two-phase momentum equations is Δt. Only binary collisions are considered and a collision event involving two particles m and n is denoted by $C_{m,n} = (m, n, t_{m,n})$ where $t_{m,n}$ is the collision time for the particle pair (m, n). It is defined as the time interval from the beginning of the time step to the time of their collision. It can be calculated based on the initial position and velocity vectors **x**, **v** of the particle pair (m, n):

$$t_{m,n} = \frac{-\mathbf{x}_{mn} \cdot \mathbf{v}_{mn} - \sqrt{(\mathbf{x}_{mn} \cdot \mathbf{v}_{mn})^2 - \mathbf{v}_{mn}^2[\mathbf{x}_{mn} \cdot \mathbf{x}_{mn} - (r_m + r_n)^2]}}{\mathbf{v}_{mn}^2} \qquad (15)$$

Where $\mathbf{x}_{mn} = \mathbf{x}_m - \mathbf{x}_n$ and $\mathbf{v}_{mn} = \mathbf{v}_m - \mathbf{v}_n$ $(m, n \in P)$.

A variable t_c is introduced to mark the time point at which the current collision is handled. The algorithm that handles all possible collisions in a given time step Δt is summarized as follows:

S1. Scan all particles and build up the collision event list, set $t_c = 0$;
S2. If there is no event in the event list, go to the next simulation time step; otherwise
S3. Determine the current event $C_{m,n}$ having the minimum collision time in the list;
S4. $t_c = t_c + t_{\mathrm{m,n}}$, update particle positions: $\mathbf{x}_\mathrm{p} = \mathbf{x}_\mathrm{p} + \mathbf{v}_\mathrm{p}\, t_{\mathrm{m,n}}$
S5. Update velocities of the current pair (m, n) using collision dynamics
S6. Detect future possible events for the current pair (m, n)
S7. Update the event list.

Then, steps S2 to S7 will be repeated till the end of the simulation. Similar algorithms can also be found in [1, 12, 17], where the ending condition in a time step is $t_c > \Delta t$ instead of the emptiness of the collision event list [5]. The latter condition is used in the algorithm we have developed. In this algorithm, described hereafter, only collision events with collision time less than the time step Δt are added into the event list.

If N_c is the number of collisions handled within a time step, it is evident that steps S2 to S7 will be repeated N_c times in that time step. For the same flow conditions, this frequency should be proportional to the particle number N_p, so at this stage we just assume that N_c is of the same order as N_p. It means that if millions of particles are involved in the computation, steps S2 to S7 will be computed millions of times. Thus one should pay close attention to these steps while optimizing the algorithm. Indeed, any loop in these steps must be avoided as far as possible. Though it is a time-consuming operation, all particles must be scanned when initializing the collision list in S1. We introduce the complexity in time to illustrate a rough evaluation of the looping times for a given operation. The complexity is often used in computer science to describe the algorithm efficiency. In the following discussions, the computation complexity of S1 is evaluated for the handling of one particle, while those of steps S2 to S7 are evaluated for the handling of one collision event, that is, for one collision iteration. Several scales (N_p, N_{nb}, and N_g for instance) are used in our discussions of the complexity in time of the optimized different steps of the developed algorithm and this for the clarity sake.

When building up the initial event list or redetect potential collisions in S1 and S6, it is not necessary to compute the collision time for all particle pairs that are susceptible to collide. Otherwise, the computation complexity will be $O(N_p)$ for both steps S1 and S6. Considering the fact that the time step is much smaller than the particle response time, a particle will collide only with its surrounding partners in a limited spatial range. Thus a neighbor list is constructed as follow:

$$Partner_m = \{n \mid \forall n \in P, |\mathbf{x}_n - \mathbf{x}_m| < R_s, m \neq n\}$$

This list can be compiled for each particle $m \in P$ before initializing the collision list. For a mono-disperse system, R_s can be determined by:

$$R_s = d_p + 2|\mathbf{v}|_{\max} \Delta t \tag{16}$$

$|\mathbf{v}|_{\max}$ is the maximum particle velocity, and d_p is the particle diameter. Before computing the collision time for particle m using Eq. (15), only the partner n in its neighbor list $Partner_m$ is scanned and the following condition (17) is verified. Otherwise the two particles m and n will not collide.

$$\mathbf{x}_{mn} \cdot \mathbf{v}_{mn} > 0 \tag{17}$$

If the average particle number in the neighbor list is N_{nb}, the computation complexity for S1 and S6 is $O(N_{nb})$. This greatly reduces the computation time.

Since there is no loop involved in steps S2 and S5, the complexity for these two steps is $O(1)$. By introducing a time stamp for each particle to mark the last updating time point, only the current colliding pair needs and update of their positions. So the complexity of step S4 is also $O(1)$. This strategy is the so-called delaying update of particle positions [13, 16]. When updating the event lists in Step S7, all events related to the current pair and their neighboring particles should be updated. If the events in all lists are arranged in an unsorted way, the complexity order of S7 should be $O(N_{nb})$ which adds an additional computational burden due to the handling of duplicate events. This problem is mitigated in our developed algorithm by adopting a proper data structure.

Now we turn attention to the most time-consuming part of the algorithm which is step S3. In this step the current event with the smallest collision time in the collision event list is determined. If all potential collisions are scanned, the complexity will be well beyond $O(N_c)$ since the number of the potential collisions is usually larger than that of the occurring ones. Here we call the event list having all potential events, the full-length global list GL. It is denoted by:

$$GL = \{(m,n,t_{m,n}) \mid \forall m \in P, \forall n \in Partner_m\}$$

Applying the local minimum algorithm [10], this full-length list is divided into N_p lists, in which events are individually associated with each particle.

$$LP_m = \{(m,n,t_{m,n}) \mid \forall n \in Partner_m\}$$

We call *LP* the local list of a given particle. A global list having N_p elements is composed of the event with the smallest collision time in *LP*

$$GLP = \{(p,n,t_{p,n}) \mid \forall p \in P, (p,n,t_{p,n}) \in LP_p \text{ and } \forall (p,m,t_{p,m}) \in LP_p, t_{p,n} < t_{p,m}, m \neq n\}$$

We call *GLP* the global list for all particles. This shortens the length of the event list to N_p. In some works where the finite difference grid was utilized [12], a global list for all cells was introduced:

$$GLC = \{(p,n,t_{p,n}) \mid \forall c \in Grid, \forall (p,m,t_{p,m}) \in LC_c, (p,n,t_{p,n}) \in LC_c, m \neq n \text{ and } t_{p,n} < t_{p,m}\}$$

Where the local event list for a given cell is defined as

$$LC_c = \{(p,n,t_{p,n}) \mid \forall p \text{ hosted by } c, (p,n,t_{p,n}) \in LP_p \text{ and } \forall (p,m,t_{p,m}) \in LP_p, t_{p,n} < t_{p,m}, m \neq n\}$$

After updating the local event list involved in the last collision in step S7, the local minimum event for particles and cells can be determined. Then the finite volume cells are scanned to determine the current collision. Thus the complexity of S3 is close to $O(N_{nb} + N_p/N_g + N_g)$. Usually for dense dispersed flow, the number of grid cells N_g is at least one order less than that of the particles N_p. This strategy results in a substantial cut in the CPU time dedicated for handling collision events in DPMs.

Above, a basic implementation of the hard-sphere model and the optimization strategy that has been often used to handle collision events in the context of DPM are presented. The computation complexity is summarized in Table 1. It can be seen that the most time-consuming step is S3, given a typical magnitudes of $N_c \sim N_p \sim 100\, N_g$, and $N_g \sim 100\, N_{nb}$ for 3D cases. The above-explained strategy developed to determine the current event, depends on the finite difference/volume mesh [12]. It needs to scan all cells, even cells that do not contain any particle and/or cells where no collision is taking place. These cases are often encountered in heterogeneous fluidization systems. In addition, finding the event with the smallest collision time in the cell hosting the collision pair brings additional computational overhead in the order of *O(100)* if $N_p \sim 100 N_g$, which is comparable to that of $O(\log N_p)$ algorithms. The duplicate storage of collision events also decreases the efficiency of the above algorithms as it is explained hereafter. In the next sections, the method to compile the neighbor list is introduced first. It allows the decoupling of the efficiency of the algorithm of the finite difference/volume mesh used. Then the crosswise linking strategy is proposed to eliminate duplicate storage of collision events and increase the efficiency of both event deletion and insertion operations. In addition the O(1) time- complexity hash table strategy to locate the current event is detailed.

NEIGHBOR LIST COMPILATION

It is convenient to use the finite difference/volume mesh as the background grid to compile the neighbor list in DPMs [12, 19]. Although this strategy saves a lot of CPU time, it is not efficient enough and depends on the mesh. If we suppose that the particles are distributed uniformly in the domain, the computation complexity of the neighbor list compilation is almost of the order $O(9N_p / N_g)$ and $O(27N_p / N_g)$ for 2D and 3D structured grid respectively. Besides, this compilation process becomes more complicated and more time-consuming for the case of unstructured grids since all the surrounding cells should be scanned.

In a more efficient method for 2D mono-disperse particle system, an additional structured background mesh is introduced [20]. It is designed in a way to allow every square cell to host only one particle. The cell width is taken equal to the particle radius, as shown in Fig.3a. This method needs additional memory to store the index of the particle in the fine background grid and to store the cell index (i, j) for each particle. However, it is not necessary to allocate memory space for the neighbor list of each particle. Indeed, the neighbor list can be formed by scanning all particles only one time which results in a complexity of $O(1)$. Although it is more efficient to form the neighbor list, this method may lose some of its efficiency for checking whether there is a neighboring partner in the background grid or not. Indeed, one may think that the fine grid width should be taken as the particle diameter not the radius to allow only one particle at most with its center in a cell. This is not quite correct, since particles have arbitrary configuration in space, as shown in Figure 3b. Actually the critical fine cell width Δl_p should be less than $d_p/\sqrt{2}$ for 2D case and $d_p/\sqrt{3}$ for 3D case (Figure 3c). The partner search range in a grid unit is $R_s/\Delta l_p$. In 3D numerical experiments, the background cell width can be taken equal to $d_p/1.733$.

The above method is independent of the finite volume grids. Thus it is applicable for both structured and unstructured meshes. Once the dimensions of the simulation domain are determined, the fine background grids can be constructed to locate the particles. If the memory is not an issue, the neighbor list of each particle can be built and stored for each particle by scanning the fine background grids. After compilation of the neighbor list, future collision events can be detected for each particle by scanning its partners, so the final complexity of S1 is $O(N_{nb})$.

Table 1. Complexity of a simple implementation of the DPM in steps S1 to S7

S1	S2	S3	S4	S5	S6	S7
$O(N_{nb})$	$O(1)$	$O(N_g)$	$O(1)$	$O(1)$	$O(N_{nb})$	$O(N_{nb})$

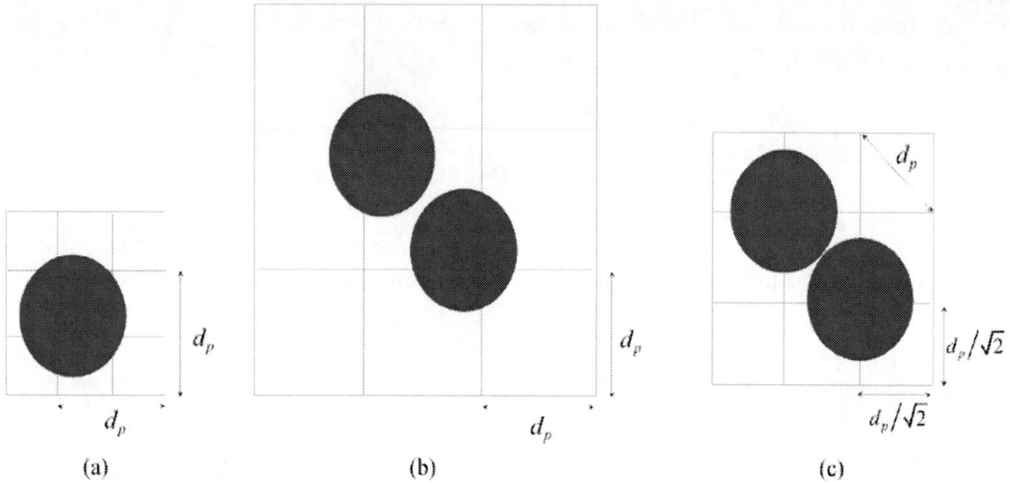

Figure 3. Searching collision partners. Cell width of background mesh: (a) $\Delta l_p = 0.5 d_p$ [20]; (b) $\Delta l_p = d_p$; (c) $\Delta l_p = d_p / \sqrt{2}$ for 2D case. d_p is the particle diameter

DATA STRUCTURE

Dynamic vs. Static

In the literature, a static data structure of the collision-handling algorithm is often used for its easy implementation [12]. Once the maximum number of the neighboring partners for a given particle is determined, two-dimensional arrays can be used to store the partner index and the collision time for the list *Partner* and *LP*. Arrays with the same dimensions of the finite volume grid can be used to store the smallest collision time and the corresponding colliding pair for the list LC_c. Additional arrays are also provided to store particles' numbers and indexes. These arrays are all statically allocated since their maximum dimensions are determined before computation. It is evident that the collision partners as well as the collision time are stored redundantly, since a particle is also a partner of its partner. This results in lists *LP* and LC_c containing unnecessary duplicates of each collision event. This feature decreases the numerical efficiency since updating individual lists of colliding pairs involves updating their partners' too. This drawback is exacerbated if a given particle and its partners are located in different finite volume cells as shown in Figure 4. In this case the elements in the corresponding lists LC_c should be updated too. Scanning the individual lists by the partners and the cells increases hugely the computational overhead.

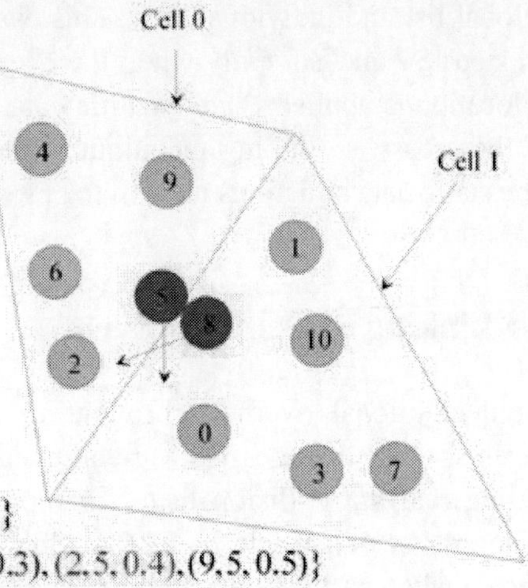

$$GLC = \{(5,8,0.1),(8,5,0.1),......\}$$
$$LC_0 = \{(5,8,0.1),(4,6,0.3),(6,4,0.3),(2,5,0.4),(9,5,0.5)\}$$
$$LC_1 = \{(8,5,0.1),(10,3,0.2),(3,10,0.2),(7,3,0.4),(1,8,0.6),(0,3,0.8)\}$$

Figure 4. Duplicate events in the local (*LC*) and global (*GLC*) lists. The events marked in red color should be updated after processing the current event (5, 8, 0.1)

To remedy to this severe shortcoming linked to static data structures, dynamic data structure can be introduced to handle the lists mentioned above. All these lists have different lengthes and they are implemented using the linear linked list of the C-programming language. This is achieved thanks to the many advanced features of the C-programming language which are appropriate for the hard-sphere modeling. These features are the pointer type, the user-defined data type and the compiler macros. Indeed, operating the pointer variable is more efficient than operating array element, since the associated memory is directly accessed without the array index transformation. User-defined data type and compiler macros make the code more readable and manageable. Some user-defined data structures are presented in Appendix 1. The collision events are stored in a doubly-linked list and the particles hosted by the finite volume cell are stored in singly-linked list. The collision event and the particle are treated as an object and the linked list is treated like a container (In fact it does not contain the objects themselves but their pointers). The pointer variable is used to mark the memory location of any object. When deleting or inserting an object to the linked list, the pointer variables are responsible of building or breaking the linkage between objects. The dynamical data structure is more natural and presents better programming logic than the static one.

It is almost impossible to determine *a priori* the number of potential collision events that can occur in one simulation time step. This number differs across different gas-solid systems and it is also time-dependent for a given system. In general, at the beginning of any time step, the number of detected potential events is usually more than N_c, but decreases to zero as $t_c \to \Delta t$. From the programming point of view, it is a good practice to allocate memories for newly detected collision events and to destroy unused memories for invalid collision events. However, the frequent allocation and release of memory in steps S6 and S7 will cost a considerable CPU time. To alleviate this computational cost, a buffering technique can be applied to reduce the memory space needed and achieve performance efficiency. This technique consists of using an object/buffer pool to buffer all invalid events; those deleted from the global list and individual lists. This event pool is recycled for new potential events detected in steps S1 and S6. Only when the event pool is empty, the corresponding dynamic memory allocation is applied. The dynamic data structure can be made more flexible through the use of the crosswise linking technique. The latter helps eliminate all the shortcomings linked to the static data structures used in the previous algorithms.

Crosswise Linking of Collision Event

To avoid additional overhead brought up by the redundant storage of the potential collision events, three linkages for a given event $C_{p,q}=(p,q,t_{p,q})$ are built up and maintained when updating relevant collision lists. As shown in Figure 5a, the first linkage is for the individual-donor list of particle p, LPD_p, which is defined as the set of potential collision events detected when particle p is the host. The second one is used for the individual-acceptor list of particle q, LPA_q, which is defined as the set of collision events detected when the partner of particle q is the host. A third list operating on the global list *GLP* or *GLC* is introduced. It is only used if there is an event that needs to be added in the particle donor list. Thus, a given event $C_{p,q}=(p,q,t_{p,q})$ is certainly contained within the lists LPD_p and LPA_q, and may be contained in the global list *GLP* or *GLC*. The collision event is crosswise linked into these lists through a doubly-linking structure (see Figure 5a). Thanks to this type of structure, the previous and the next element can be determined without scanning from the list head during the element deleting and updating operations. Because a given particle p can be a host or an acceptor, it is always associated with both the donor list LPD_p and the acceptor list LPA_p. Two pointer variables associated with the particle p are allocated to store the memory locations of the head elements of these two lists, as shown in Fig.5b. As standard practice, when operating collision events (inserting into or deleting from a list), the first particle in the pair is treated as the host and the second as the acceptor. This makes maintaining the donor and acceptor lists a straightforward operation. For instance, when initializing the individual lists in step S1, we just need to scan the particle p with its partners whose indexes are smaller than p, to build the list LPD_p

$$LPD_p = \{(p,n,t_{p,n}) \mid \forall n \in Partner_p, p > n\}$$

At this stage, the acceptor lists LPA_n of a given particle p's partners are built up too. For the partners that their indexes are larger than the one of p, we get the list LPA_p

$$LPA_p = \{(n,p,t_{p,n}) \mid \forall n \in Partner_p, p < n\}$$

By taking advantage of the cross linking structure of the collision events, any relevant particle and/or collision event can be easily identified without searching in the corresponding individual lists. Since there is only one entity of any event, there is no additional scanning to update the individual lists for the partners of the colliding pair. Moreover, by sorting the elements in the individual donor lists, only the first element (local minimum one) in the *donor list* is added to the global one so that there is no duplicate event in the global list. The additional operation on the global list is avoided too. Thus, in S7, before building up the individual lists for the colliding pair, their corresponding events in the old lists are invalidated and cleaned, resulting in an efficient update of the event lists and an $O(1)$ complexity of this step.

In summary, by introducing the donor and acceptor collision lists, we avoid redundant computation and storage. It is important to mention that although the crosswise linking storage of collision events is flexible and efficient, it improves the computational performance only for step S7. The latter improvement becomes more notable when the computation complexity in step S3 decreases to the order $O(N_{nb})$.

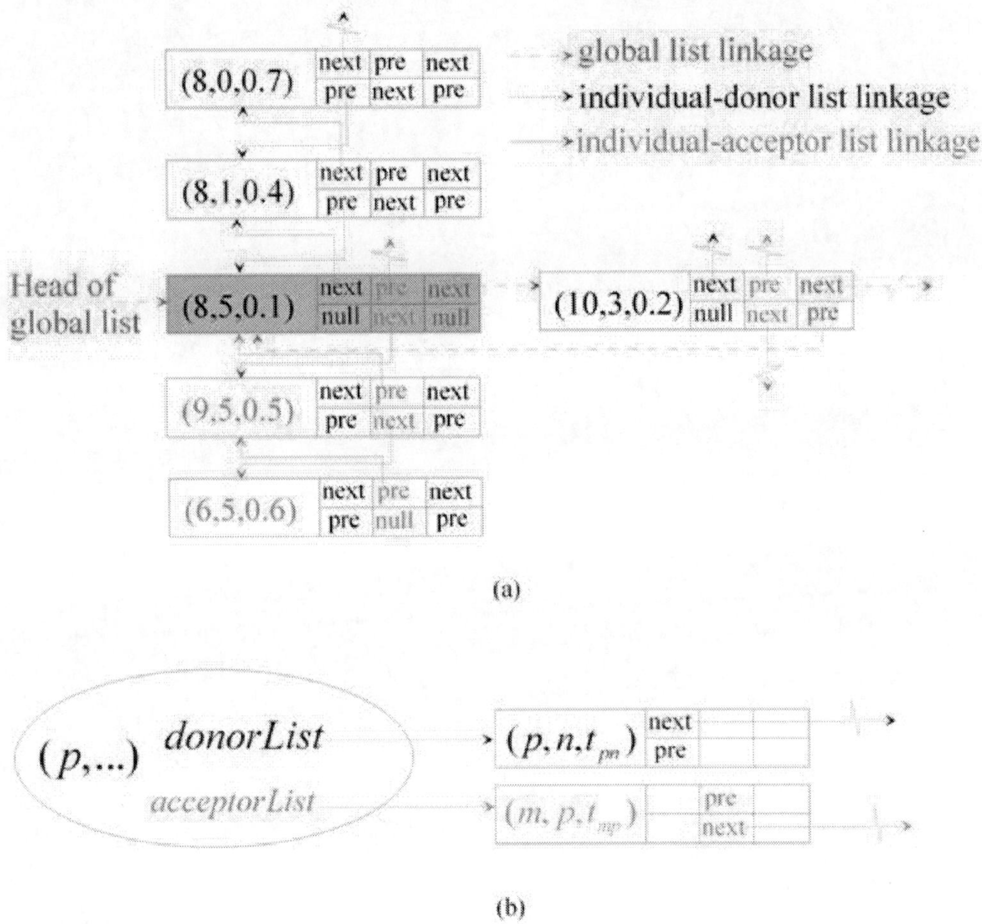

Figure 5. Illustration of the crosswise linking of collision events. (a)Three kinds of linkage are maintained for a given collision event (8, 5, 0.1). "Next" and "pre" are pointers to the next and preceding objects, respectively. "null" means a zero pointer without any object pointed to it; (b) Two pointers for a given particle p are responsible for maintaining the events in the donor and acceptor lists of particle p. If particle p becomes a partner of the current event, these events can be released from the pointers without any searching

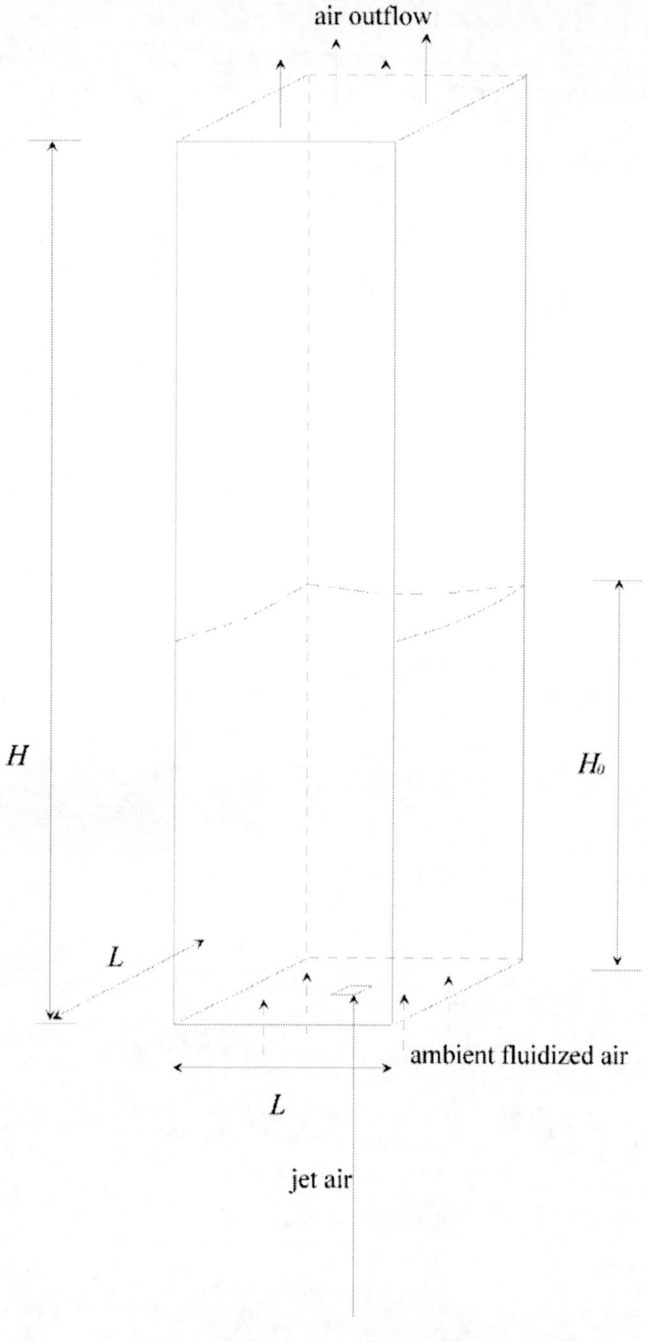

Figure 6. External sketch of a laboratory-scale bubbling fluidized bed

To illustrate the increase in the efficiency of the collision-handling algorithm brought about by the dynamic data structure and the cross-wise linking, a dense gas-solid flow in a bubbling fluidized bed was simulated. The results regarding the time efficiency of the developed algorithm is compared to those of the other existing algorithms. The numerical experiment parameters are listed in Table 2. Schematic draft of the bubbling fluidized bed is shown in Figure 6.

Table 2. Test case parameter

Domain dimensions (L x L x H)		*0.15m* x *0.15m* x *0.6m*
Grid dimensions (N_g)		30 x 30 x 45 =13,500
Particle density (kg/m^3)		2400
Normal restitution coefficient		0.97
Tangential restitution coefficient		0.3
Frictional coefficient		0.1
Ambient air velocity (m/s)		$1.2U_{mf}{}^*$
Jet air velocity (m/s)		$20U_{mf}$
d_p (mm)	H_0 (m)	N_p
3	0.015	10,000
3	0.055	40,000
3	0.21	160,000
1.5	0.05	320,000
1.5	0.2	1,280,000
1.5	0.4	2,560,000

$^*U_{mf}$ is the minimum fluidized velocity.

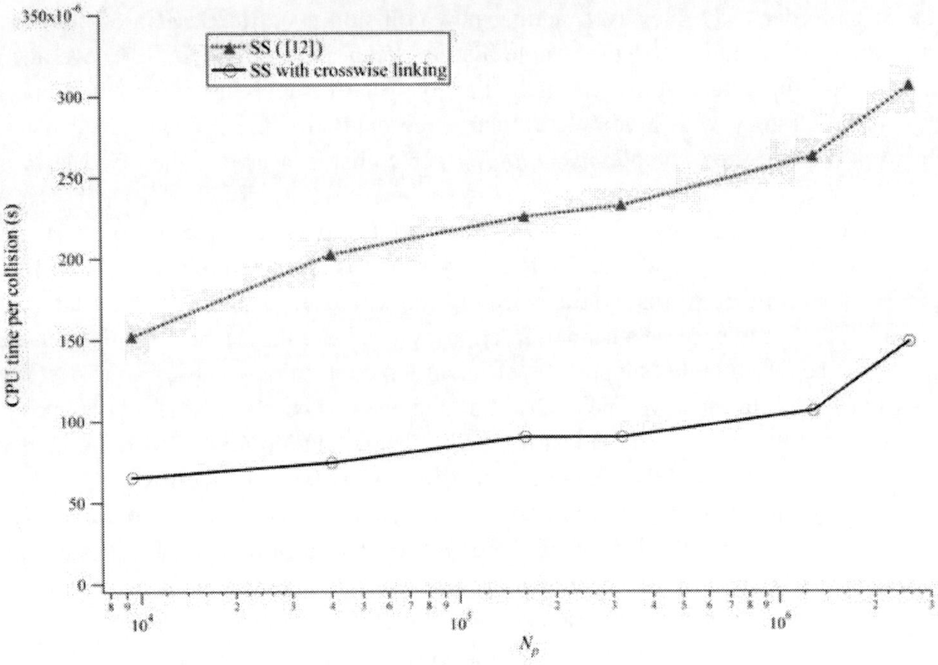

Figure 7. Comparison of the total CPU time per collision consumed in steps S1 to S7

Figure 7 shows the average CPU time per collision consumed by steps S1to S7 when the traditional method [12] (hereafter referred to as simple strategy SS) is applied with and without crosswise linking strategy. It can be seen that about 50% CPU time is saved if the crosswise linking strategy is applied. As discussed earlier, the complexity of step S7 can be reduced from $O(N_{nb})$ to $O(1)$ if the crosswise linking strategy is applied because it is not necessary to update the individual event list of each particle in the neighbor list of the colliding pair. It must be emphasized that although the complexity of step S7 is reduced to *O(1)*, this does not mean that the CPU time will cut down to $1/N_{nb}$. Simply the loop for the neighbor list in S7 becomes unnecessary.

The crosswise linking of events needs additional computation to maintain the double linkage between events. Moreover, its efficiency depends on the difference between the computation time consumed by step S7 and the one necessary to carry on operations for steps S1 to S6 (the weighted computation in S7 to that in steps S1 to S6). It is noticeable that the more particles the system contains, the more efficiency is achieved. This is due to (i) the increase in the particle number that results in more particles in the neighbor list while the number of the finite volume cells remains unvaried and, (ii) the increase in performance gained on operations of the global list since the length of the global list increases.

CHAINED HASH TABLE STRATEGY

It is well known that if all elements in the global list *GLC* are arranged in an ascending sequence (i.e. priority queue) according to their collision time, the next event is always the first element in the list. So the operation complexity to determine the current collision event will be of the order *O(1)*. However, adding new collision event to *GLC* in S7 necessitates unfortunately the scanning of the whole list to keep it ordered. On the average, the performance of applying a sorted list is higher than that of applying an unsorted one. The priority queue strategy with a complete binary tree data structure for the global list *GLP*, which gives the expected complexity $O(logN_p)$ per collision, was designed by Marin et al. [10].

In computer science, a hash table, or a hash map, is a data structure that associates keys with records. The primary operation it efficiently supports is a lookup: given a key, find the corresponding record. It works by transforming the key using a hash function into a hash code, a number that is used as an index in an array to locate the desired location where the record should be. Hash tables support the efficient insertion of new entries, in expected time complexity of *O(1)*. In the developed algorithm, the chained hash table concept is introduced for the storage of the event in the global list *GLP* as shown in Figure 8. The one-dimensional array is used to implement the hash table. The key of the hash table is the collision time $t_{m,n}$ and the record is the memory location pointing to the corresponding event in the individual-donor list. We implement the individual-donor list *LPD* as a priority queue so that only its first element is mapped/inserted into the hash table. The individual-acceptor list is implemented as an unsorted linked list since there is no need to pick events from this list (an *O(1)* complexity is achieved for insertion and deletion on this list).

To satisfy the priority queue requirement of having the event with the smallest collision time accessed first in the hash table, the hash function should be well designed to map in an orderly manner the collision times to integers or indices *I* and to locate the events' positions in the table. A simple hash function can be defined as:

$$I(t_{m,n}) = INT(t_{m,n} / \Delta t_c) \text{ with } \Delta t_c = \Delta t / N_c, \quad (18)$$

Individual donor lists

$LPD_{10} = \{(10,3,0.2),(10,7,0.4)\}$

$LPD_9 = \{(9,6,0.3),(9,5,0.5),(9,4,1.0)\}$

$LPD_8 = \{(8,5,0.1),(8,1,0.4),(8,0,0.7),(8,2,0.8)\}$

$LPD_7 = \{(7,3,0.5)\}$

$LPD_6 = \{(6,5,0.6),(6,2,0.9)\}$

$LPD_5 = \{(5,2,0.4)\}$

$LPD_4 = null$

$LPD_3 = \{(3,0,0.7)\}$

$LPD_2 = \{(2,0,0.9)\}$

$LPD_1 = null$

$LPD_0 = null$

Index:	*0*	*1*	*2*	*3*	*4*	*5*	*6*	*7*	*8*	*9*	*10*
Head of hash table →	*null*	(8, 5, 0.1)	(10, 3, 0.2)	(9, 6, 0.3)	(5, 2, 0.4)	(7, 3, 0.5)	(6, 5, 0.6)	*null*	(3, 0, 0.8)	(2, 0, 0.9)	*null*

Figure 8. Illustration of the hash table strategy. In this example, the number of particles involved equals N_p=11. N_c=10 is the number of collisions and $\Delta t = 1s$ is simulation time step

Where the function $INT(x)$ gives the integral part (lower integer) of x. If the particles collide successively in a uniform time interval, the above hash function maps only one element to a table location. When inserting a collision event, its address or index in the global list is determined by the hash function without any loop and the expected computational complexity in time will be reduced to *O(1)*.

As mentioned earlier, one cannot *a priori* determine the number of collisions in a time step. Considering the slow change of a given gas-particle system within small time steps that are often used in DPM simulations (at least one order of magnitude less than the particle response time), the number of collision occurrences for successive time steps should change very little. As a direct consequence, the length/capacity of hash table will also vary little. In fact, the number of collisions N_c in Eq. (18) is actually selected as that in the last time step N_c^0. Another fact is that the particles in gas-solid system may never collide in a uniform time interval, which may generate an identical index using the hash function Eq.(4). This conflict problem can be solved by resorting to the chaining or the re-hashing [21]. Since the re-hashing needs more tables and fails for the case when two collision events occur at exactly same collision time, the chaining of the hash table is more suitable. In the chaining methodology, the hash-table looks like a two-dimensional array with a variable length for the second dimension, or like a one-dimensional array of priority queues, as shown in Figure 9. It can be seen that deleting invalid events (in red color in Fig.9) from the hash table is straightforward. In this example, the events (8, 10, 2.4) and (10, 5, 21.6) are marked invalid since they are associated to the individual lists of the colliding pair (5, 8). Event (8, 10, 2.4) can be released from the chained hash table by re-linking its previous and next objects maintained in its third linkage pointers. Event (10, 5, 21.6) can be released directly by assigning a null pointer in the location (21) of the hash table. Because the length of the hash table is a chosen property and it is a nearly continuous function of the collision number in gas-solid system, chaining events is very rare process (in real simulations, the number of events in the chained lists is often less than 3). Thus the chained hash-table strategy gives exactly an *O(1)* time complexity for event insertion.

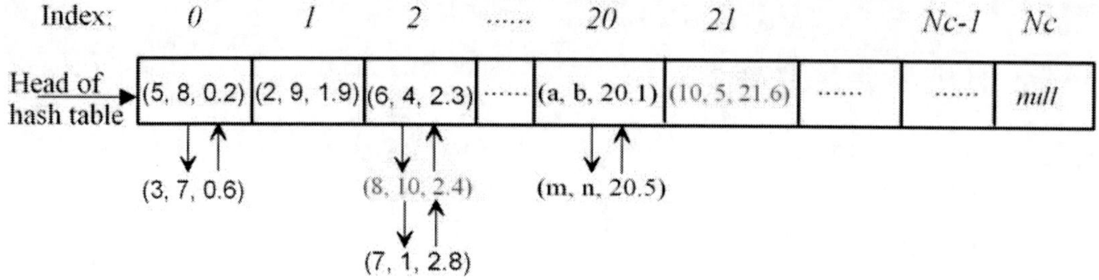

Figure 9. Logical view of event chaining. When two or more events are mapped to the same location in the hash table, they are chained in a doubly-linked list. The collision time is normalized by Δt_c

As discussed in the previous sections, there are three types of linkage to be maintained for any given event. The third linkage is used here for chaining in the hash table, as shown in Figure 5a. The event address/hash code in the hash table is stored with the corresponding event using an integer variable. When the head event $C_{a,b}$ in the donor list LPD_a is inserted to the hash table, its linkages in both the donor list LPD_a and the acceptor list LPA_b are kept. Thus, for any given event, whenever it is mapped to the hash table, its three linkages are retained. If $C_{a,b}$ becomes the current event, it will be released to the event pool while handling it, together with all other elements in LPD_a. Otherwise if there is an event $C_{b,m}$ occurring before $C_{a,b}$, $C_{a,b}$ will be released together with all other elements in LPA_b because particle b is the acceptor part of the colliding part when the event $C_{b,m}$ is processed. In the former case, there is no additional operation on the hash table since only the head event $C_{a,b}$ in LPD_a is mapped to the hash table. In the latter case, a null pointer is directly assigned to the location in the hash table indicated by the event hash code if any element in LPA_b (such as $C_{a,b}$) has been mapped to the hash table (see Figure 9 and the event data structure in Appendix 1). Thanks to the crosswise linking data structure, all events in the individual lists can be deleted without any searching. Thus the complexity for event deletion is also $O(1)$. The simulation is progressed by scanning the entire hash table from the beginning. This can be done in a simple loop. The loop indicator presents the location of the current event in the hash table. When the last recode of the hash table is null, it means there is no longer any event in this time step. Since the capacity of the hash table is chosen close to N_c, the expected complexity to locate the current event is also $O(1)$.

The increase efficiency brought about by the chained-hash-table strategy (hereafter referred to as CHTS) with that of the traditional method (SS) described in the literature [12] is demonstrated. As depicted by Figure 10, the CPU time per collision consumed handling the global list in steps S3 and S7 is essentially constant (0.00115~0.00124ms) when the chained hash table is used regardless the number of particles present in the gas-particle system. For SS, the consumed CPU time increases almost linearly as the particle number increases in the system. This is mainly due to the increasing time needed to find the local minimum event in a given cell. If the crosswise linking strategy is employed in conjunction with SS, 10-18% CPU time can be saved on the global list operations. The achieved performance is mainly the result of the crosswise linking of events in the global list which eliminates any probable duplication of events. When the particle number reaches one million, more than 98% CPU time on operations of the global list is saved applying the CHTS compared to the use of the SS.

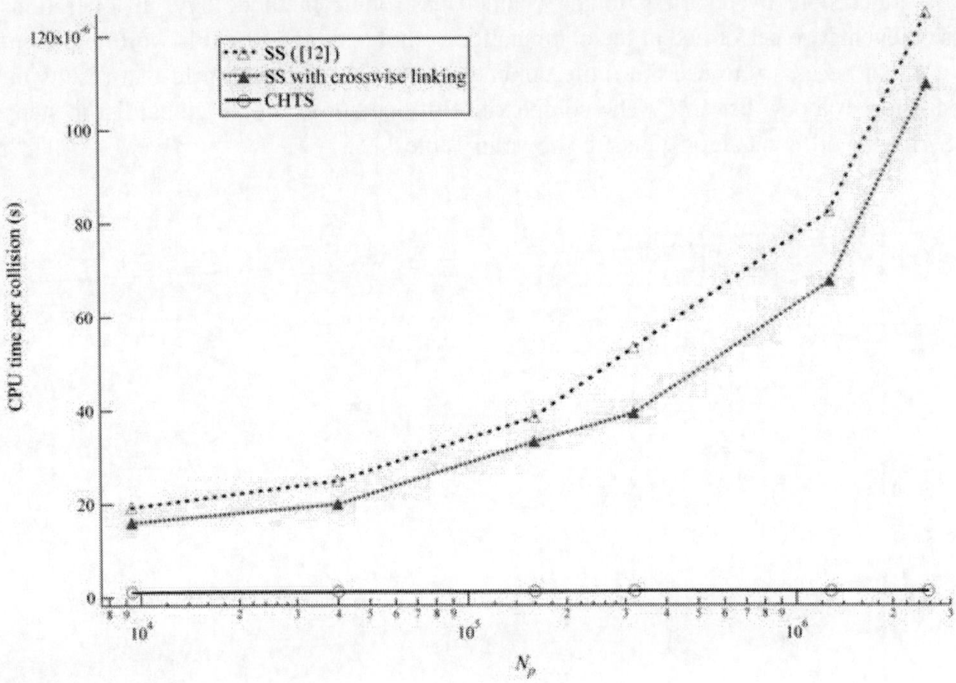

Figure 10. Comparison of consumed CPU time per collision for operations on the global event list

The efficiency of CHTS and the one of the complete binary tree strategy (hereafter referred to as CBTS, Marin et al. 1993) [10] are compared in Figures 11. The CPU time consumed by operations on the global list for the CBT algorithm increases as log(N_p) while it is almost constant for the CHTS (see Fig.11a). The present method resulted in a four-fold increase in performance compared to the CBT algorithm when there are more than 1 million particles in the fluidization system. Figure 11b and Figure 11c show the total CPU time consumed in S1 to S7 for a given number of particles in the system. More than 90% CPU time was saved by the use of CHTS compared to SS, while about 8-10% of CPU time was saved in comparison with CBTS. The denser is the regime and/or the higher is the collision frequency; the more is the CPU time that can be saved for the present method compared to CBTS. This is due to the fact that we scan the particles only once to build the initial event list in S1 and the resulted computation cost is not proportional to the collision number within a given simulation time step. Thus, the weighted CPU time for step S1 is slightly reduced compared to the one dedicated for steps S2 to S7.

Table 3. Complexity comparison of different algorithms in steps S3 and S7

Algorithms	S3	S7
Simple strategy [12]	$O(N_g)$	$O(N_{nb})$
Simple strategy with crosswise linking	$O(N_g)$	$O(1)$
CBT algorithm [10] with crosswise linking	$O(1)$	$O(\log N_p)$
Hash table algorithm with crosswise linking	$O(1)$	$O(1)$

As conclusion, by resorting to the chained hash table methodology, all operations of collision events are performed at the computation complexity of *O(1)*. It is worthy to mention that it is not necessary to use the finite volume grid as a background grid that results in the introduction of a new list *GLC*. The complexities of operations on the global list in steps S3 and S7 for the different algorithms are shown in Table 3.

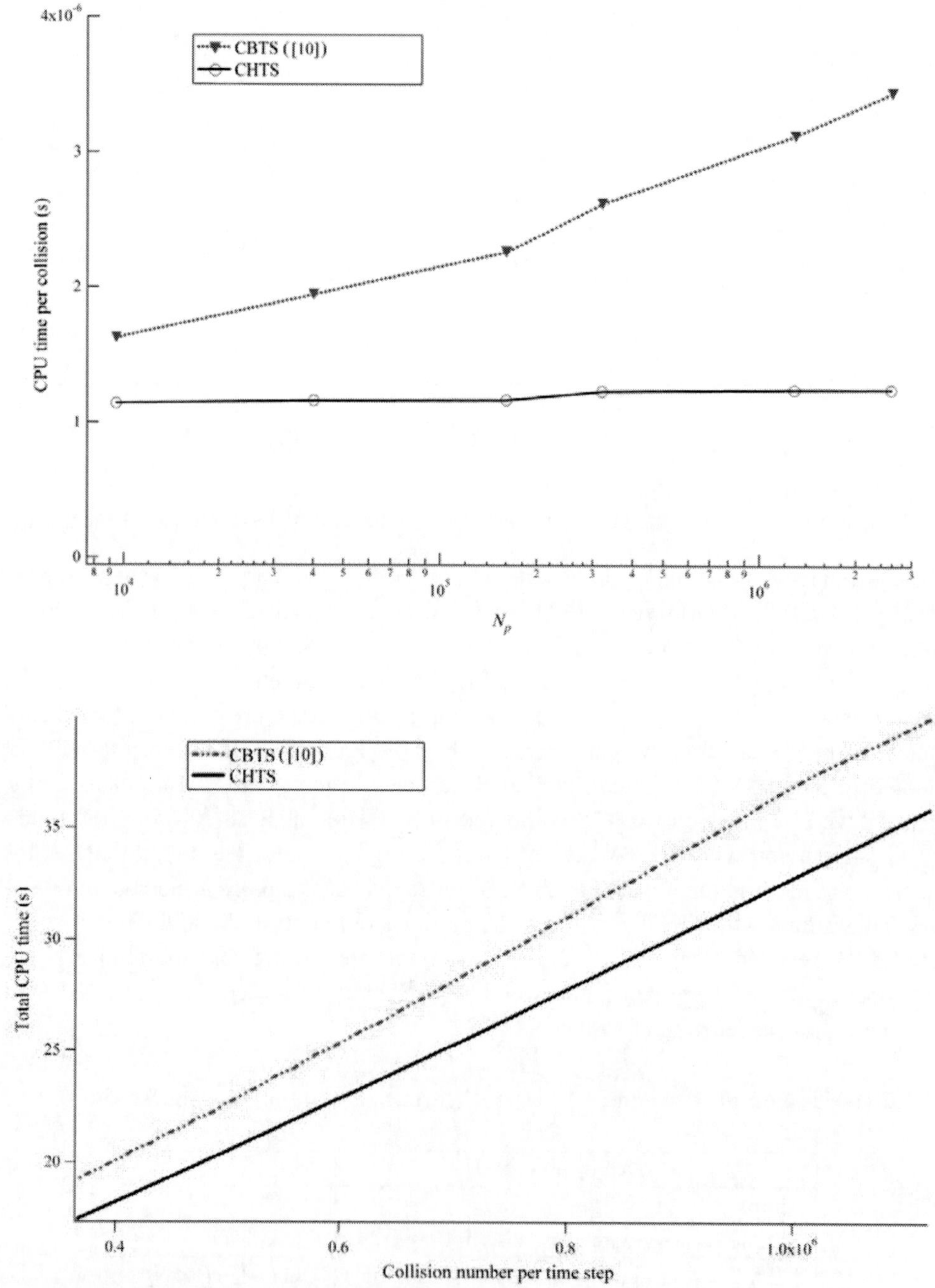

Figure 11. (Continued)

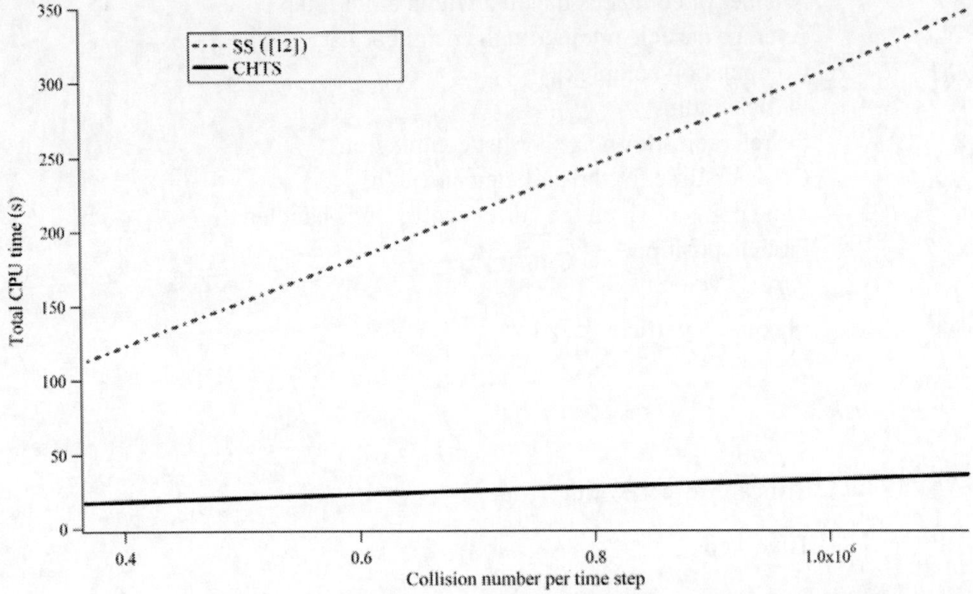

Figure 11 Comparison of the CPU time consumed by the present strategy (CHTS), the simple method (SS) and Marin et al. [10] strategy (CBTS): (a) CPU time per collision for operations on the global event list; (b) and (c) total CPU time consumed in steps S1 to S7. For (b) and (c), particle number N_p = 2,560,000

CONCLUSION

In this chapter, we detailed an efficient algorithm suitable for handling collision events in the context of Discrete Particle Modeling (DPM) of dense gas-particle systems. The new algorithm allows a complexity *O(1)* time per collision regarding the event list handling operation thanks to the use of: (i) dynamic crosswise linked data structures that enables an efficient updating and managing of lists of collision events and (ii) a chained hash table strategy to process particle collision sequence. Numerical testing showed the superiority of the developed algorithm over the traditional algorithms that have been used for DPMs.

NOMENCLATURE

Roman Letter

$C_{p,q}=(p,q,t_{p,q})$	Given collision event associating two particle p and q with a collision time $t_{p,q}$
d_p	Particle diameter
i, j	Cell indices
n, t	Normal and tangential unit vector for collision
$I(t_{m,n})$	Hash function
N_p	Number of particles
N_g	Number of finite volume grids

N_c	Number of collisions handled within a time step
N_{nb}	Average particle number in the neighbor list
O()	Computation complexity
r_p	Particle radius
R_s	Search radius for neighbor list compilation
$t_{m,n}$	Collision time for the particle pair (m, n)
t_c	Time point at which the current collision is handled.
$\mathbf{x}_p$	Particle position
$\mathbf{v}_p$	Particle velocity
$\lvert\mathbf{v}\rvert_{max}$	Maximum particle velocity

Greek Letters

Δl_p	Critical fine cell width
Δt	Time step

Acronyms

CBTS	Complete binary tree strategy
CHTS	Chained-hash-table strategy
DPM	Discrete particle model
GL	Global event list having all potential collisions
GLP	Global event list for all particles
GLC	Global event list for all cells
LC	Local event list for a given cell
LP	Local event list of a particle
LPA	Individual-acceptor list
LPD	Individual-donor list
MD	Molecular dynamics
SS	Simple strategy

APPENDIX 1 DATA STRUCTURE DEFINITIONS

```
/* -------data structure for particle properties, such as size, density etc.--------------*/
typedef struct ParticleProperty_Struct
{
    double Rp, Dp;
    double Rho,Mass;
    double Area, Volume;
    double nRest, tRest, Frict;   /*collision parameters*/
    double Umf;                   /* Minimum fluidized velocity*/
}PROPERTY;
/* -----------data structure for the particle object -----------------*/
```

```
typedef struct Particle_Struct
{
    double x[ND] , un[ND], u[ND], omega[ND];
                        /*particle position, velocity, and angular velocity*/
    int ip;             /*unique index of a given particle object */
    PROPERTY* property; /*pointer to particle property object */
    double Cd, CMagnus, CSaffman;
                        /* coefficients for two-phase momentum coupling*/

/*for voidage calculation and momentum exchange if particle is shared by several cells*/
    int* SharedCells;       /*list head to shared cells*/
    double* SharedVolumes; /*list head to volumes shared by cells*/
    int NumSharedCells;     /*number of shared cells*/

    int i, j, k;            /* particle location in the fine background grid*/
    struct Particle_Struct *Next; /* linkage to the next particle hosted in the same cell*/
    struct ParticleThread_Struct *PThread;
                        /*pointer to the thread object hosting particles in a given cell*/
    struct Event_Struct *DonorList;
                        /* pointer to the head event in the individual donor list*/
    struct Event_Struct *AcceptorList
                        /* pointer to the head event in the individual acceptor list*/
}PARTICLE;

/*----data structure for particle thread: container for particles hosted by a given cell-----*/
typedef struct ParticleThread_Struct
{
    int Number;     /*number of particles in this cell */
    PARTICLE *P;    /*pointer to the head of the particle list*/
    int Cell;       /*the index of the cell associated*/
}PTHREAD;

/*----data structure for collision event: the hash code is saved and the double linkages are
maintained when operating event lists-----*/
typedef struct Event_Struct
{
    PARTICLE *P[2]; /* colliding pair (m, n) of collision event*/
    double CT;      /*collision time*/
    int iHashCode;  /* hash code of the event mapped by the hash function*/

/* pointers to the next and previous event; three linkages maintained */
struct Event_Struct *Pre[3];
struct Event_Struct *Next[3];
}EVENT;
```

REFERENCES

[1] Hoomans, BPB; Kuipers, JAM; Briels, WJ; van Swaaij, WPM. *Chem. Eng. Sci.,* 1996, 51, 99-118. Discrete particle simulation of bubble and slug formation in a two-dimensional gas-fluidised bed: a hard-sphere approach.

[2] Bokkers, GA. van Sint Annaland, M; Kuipers, J.A.M. *Powder Technol.* 2004, 140, 176-186. Mixing and segregation in a bidisperse gas-solid fluidized bed: a numerical and experimental study.

[3] Wu, CL; Zhan, JM. *J. Hydrodynamics Ser.B,* 2007, 19, 335-341. Numerical prediction of particle mixing.

[4] Helland, E; Bournot, H; Occelli, R; Tadrist, L. *Chem. Eng. Sci.,* 2007, 62, 148-158. Drag reduction and cluster formation in a circulating fluidized bed.

[5] Wu, CL; Berrouk, AS; Nandakumar, K. *Chem. Eng. J.,* 2009, 152, 514-529. Three-dimensional discrete particle model for gas-solid fluidized beds on unstructured mesh.

[6] Alder, BJ; Wainwright, TE. *J. Chem. Phys.,* 1957, 27, 1208-1209. Phase Transition for a Hard-Sphere System.

[7] Allen, MP; Tildesley, DJ. *Computer simulations of liquids*; Oxford Science Publications: Oxford, UK, 1990, 71-108.

[8] Rapaport, DC. *J. Comput. Phys.,* 1980, 34, 184-201. The event scheduling problem in molecular dynamic simulation.

[9] Lubachevsky, BD; Stillinger, FH. *J. Statist. Phys.,* 1990, 60, 561-583. Geometric properties of random disk packings.

[10] Marin, M; Risso, D; Cordero, P. *J. Comput. Phys.* 1993, 109, 306-317. Efficient algorithms for many-body hard particle molecular dynamics.

[11] Krantz, AT. *ACM transactions on modeling and computer simulation* 1996, 6, 185-209. Analysis of an efficient algorithm for the hard-sphere problem.

[12] Hoomans, BPB. *Granular dynamics of gas-solid two-phase flows*. PhD thesis: Twente University, Netherlands, 2000.

[13] Erpenbeck, JJ; Wood, WW. *Molecular dynamics techniques for hard-core systems*. in: Statistical Mechanics B, Berne B. J; Ed; Plenum Press: New York, US, 1977, 1-40.

[14] Marin, M; Cordero P. *Comput. Phys. Commun.,* 1995, 92, 214-224. An empirical assessment of priority queues in event-driven molecular dynamics simulation.

[15] Alder, BJ; Wainwright, TE; *J. Chem. Phys.,* 1959, 31, 459-466. Studies in molecular dynamics. I. general method.

[16] Sigurgeirsson, H; Stuart, A; Wan, WL. *J. Comput. Phys.,* 2001, 172, 766-807. Algorithms for particle-field simulations with collisions.

[17] Deen, NG; Van Sint Annaland, M; Van der Hoef, MA; Kuipers, JAM. *Chem. Eng. Sci.,* 2007, 62, 28-44. Review of discrete particle modeling of fluidized beds.

[18] Wu, CL; Zhan, JM; Li, YS; Lam, KS. *Chem. Eng. Sci.,* 2006, 61, 5726-5741. Dense particulate flow model on unstructured mesh.

[19] Lu, H; Wang, S; Zhao, Y; Yang, L; Gidaspow, D; Ding, J. *Chem. Eng. Sci.,* 2005, 60, 3217–3231. Prediction of particle motion in a two-dimensional bubbling fluidized bed using discrete hard-sphere model.

[20] Ouyang, J; Li, J. *Chem. Eng. Sci.,* 1999, 54, 2077–2083. Particle-motion-resolved discrete model for simulating gas-solid fluidization.

[21] Goodrich, MT; Tamassia, R; Mount, DM. *Data structures and algorithms in C++*. Wiley & Sons Inc: Florence, US, 2004, 365-405.

In: Powder Engineering, Technology and Applications
Editor: Jason M. Barker, pp.175-196
ISBN: 978-1-61761-216-9

Chapter 6

MORPHOLOGY-CONTROLLED FABRICATION AND APPLICATIONS OF METAL OXIDE POWDERS: THE CASE OF CUPROUS OXIDE

Yunfei Si, Kunfeng Chen and Dongfeng Xue*
State Key Laboratory of Fine Chemicals, Department of Materials Science and Chemical Engineering, School of Chemical Engineering,
Dalian University of Technology, Dalian, P. R. China

ABSTRACT

The technological demands on powder-processing of functional materials have reached a particularly prosperous state. Extensive studies have been devoted to the shape-controlled production of micro/nano powders in the past few decades, and Cu_2O micro/nano powders are of particular interest owing to their potential applications. This review features the recent developments regarding wet-chemical fabrication and applications of Cu_2O micro/nano powders with well-controlled shapes. To understand how chemically derived processes control the architectures of Cu_2O powders, we examine a range of synthetic strategies that give rise to the various final shapes. For our investigation, these shapes of Cu_2O powders are broadly categorized into three themes, i.e., polyhedron, hollow structure, and dendritic morphology, and each includes several specific structures. As for these different powder-processing methods, important factors in achieving the controlled synthesis of Cu_2O micro/nanopowders are discussed. Moreover, potential applications of the Cu_2O powders with distinct morphologies in solar energy conversion, catalysis, and gas-sensing are reviewed. This work can contribute to the understanding of the control over shapes of Cu_2O powders at micro- to nano-scale in the powder technology field.

* Corresponding author: E-mail: dfxue@dlut.edu.cn

1. Introduction

In the research field of powder technology, tailoring the architecture of nano/micro powder materials has been of great research interest due to their unique shape and size-dependent properties [1-5]. Extensive studies have been devoted to the shape-controlled fabrication of powder materials from the nano- to the micro-scale and their further large-scale production [6-27]. Among them, the preparation methods for metal oxide micro/nanopowders have been a research frontier, because metal oxides (single crystal or ceramics) represent the most common, most diverse, and probably the richest class of materials in terms of physical, chemical, and structural properties [28]. As an important p-type semiconductor metal oxide with a direct band gap of 2.17 eV, cuprous oxide (Cu_2O) has been demonstrated to possess properties useful for applications in gas sensing, CO oxidation, photocatalysis, photochemical evolution of H_2 from water, organic synthesis, lithium-ion batteries, and for research on Bose–Einstein condensation [29-33]. Therefore, more and more efforts are devoted to achieve an exquisite shape control of Cu_2O micro/nano powders, which is highly desired for tailoring their properties. In particular, various Cu_2O nanostructures, including nanoplates, polyhedra, spheres, dendrites, and hollow structures, have been successfully obtained by some advanced powder preparation approaches. The combinations of such variety of properties and applications with the unique shape-related effects of micro/nano powders have made the study on morphology-controlled fabrication of Cu_2O micro/nanostructures a very important issue of research and development from both fundamental and industrial standpoints.

Among the powder preparation methods, including conventional grinding techniques, high-energy milling, chemical vapor deposition (CVD), plasma processes, solution-precipitation techniques, etc., wet chemical synthesis has been considered to be a preferred approach for controlling the shape-forming progress of the metal oxide powders [15, 25, 34, 35]. In contrast to the traditional thermal evaporation routes or CVD method, even hard template approaches, solution-based approaches can provide an alternative, convenient, lower temperature, and environmentally friendly pathway for fabrication of micro/nano powders with desirable shapes and sizes [34, 36, 37], which do not rely on drastic conditions (i.e., high temperature, high pressure etc.). Recently, it has also demonstrated remarkable performance in controlling the morphologies of metal oxide nanomaterials, such as Co_3O_4 nanorods [38] and V_2O_5-SnO_2 double-shelled nanocapsules [39]. Like other metal oxides, synthesis of Cu_2O micro/nano powders with well-controlled shapes still remains challenging but interesting. Furthermore, selective adsorption of surfactant molecules and biomimetic mineralization are typical processes for this kind of morphosynthesis [40, 41]. Before we started to work on this subject, lots of efforts had been devoted to the wet chemical synthesis of Cu_2O micro/nanomaterials, including the use of surfactants or polymers to direct the growth [42, 43]; the use of a coordinating ligand to mediate the reaction [44]; as well as the conversion of a preformed template [45]. Although the products which can be obtained by using these methods have exhibited various shapes, the control handles available for such preparation processes have been limited to the reaction temperature, the concentration of reagents or surfactants, and pH value. Moreover, the synthetic systems are always complicated and the key factors can not be easily found and manipulated to achieve shape control of the products. In order to systematically investigate the shape-property relationship and provide a guideline for the large-scale production of high-performance Cu_2O powders, it is necessary to develop a generic methodology to Cu_2O micro/nano powders with a broad range of well-defined shapes.

In this article, we review and highlight the latest development and emerging new wet chemical strategies for the fabrication of Cu_2O micro/nano powders with well-controlled shapes. Aiming to understand how chemically derived processes control the architectures of Cu_2O powders, a small number of particularly representative and instructive literature examples are chosen to investigate the important factors in achieving the controlled synthesis of Cu_2O micro/nanopowders in a pure solvent or in a mixed solvent. Moreover, investigation on their corresponding potential applications is presented.

2. Synthetic Approaches to Cu_2O Micro/ Nanostructures in Solution

Many methods have been developed to fabricate Cu_2O micro/nanostructures such as wet-chemical methods [29, 46, 47] (including solvothermal/hydrothermal method and low-temperature solution synthesis), electrodeposition approach [43], irradiation methods [48], sonochemical synthesis [49], and CVD method [50]. And most of these shape-controlled synthetic processes have utilized the capping agent/surfactant assisted or template-directed strategies to create some unique morphologies. As an *fcc* metal oxide, Cu_2O crystallizes in the cuprite structure. Furthermore, the product overall morphology is greatly related to the intrinsic crystal structure and the activity of crystal plane, when the product powder is single-crystalline [51]. Such research works have been contained in this section in details. On the other hand, numerous template-assisted synthetic approaches have been successfully designed to synthesize spherical and non-spherical hollow structures of Cu_2O powders, which may be polycrystalline phase. For our investigation on the routes to Cu_2O micro/nanostructures with various shapes, this section will mainly focus on three themes, i.e., polyhedron, branched morphology, and hollow structure.

2.1. Strategies for Achieving Polyhedral Shapes

As for the preparation of polyhedral Cu_2O micro/nanocrystals, a great deal of attention has been paid to the theoretical considerations [8, 51-55] in crystal growth in order to finely tune the morphology of Cu_2O crystal. The shape of a crystal is determined by the relative specific surface energies associated with the facets of this crystal. At equilibrium, a crystal has to be bounded by facets giving a minimum total surface energy, an argument that is known as the Wulff facets theorem [56, 57]. Confined by this requirement, the shape of a single-crystalline micro/nanostructure often reflects the intrinsic symmetry of the corresponding lattice. Belonging to the *fcc* crystal structure (Figure 1A), therefore, Cu_2O crystal prefers to be a cube or an octahedron rather than other shapes (such as a rod and a plate) [57]. Size-controllable preparation of Cu_2O nanocubes was firstly developed through reducing Cu^{2+} with sodium ascorbate and choosing CTAB as the protecting agent in an aqueous solution at 55 °C by Murphy et al. in 2003 [58]. The average edge length of the cubes which varied from 200 to 450 nm can be tuned by varying the concentration of surfactant. Almost at the same time, Qian et al. also reported that uniform crystalline Cu_2O cubes were synthesized in high yield by reducing the copper-citrate complex solution with glucose at 80 °C [59]. The shape formation process was explained as that the adsorption and de-adsorption of the citrate ions on the different planes of Cu_2O nuclei may kinetically favor the preferential crystal growth along eight {111} directions.

Since then, lots of efforts have been devoted to develop various wet chemical methods to achieve more different well-defined Cu_2O polyhedral crystals, especially with the use of various additives and surfactants. Apparently, since nanocubes represent as one of the most important structural types for Cu_2O, the preparation of Cu_2O nanocubes was further explored with the seed-mediated room temperature solution method [60], through a hydrothermal method [61], and even by reducing the precursor CuO with hydrazine hydrate [62]. It should be noted that the hydrothermal method, one of wet chemical approaches for preparing nanocrystals of controlled morphologies, has also been applied to the synthesis of Cu_2O micro/nanocrystals, and its importance and merits will be presented further in the next subsection.

Besides the cubic structure, octahedral Cu_2O crystals were synthesized by reducing copper hydroxide with hydrazine without using any surfactant [63]. (Figure 1B-i) In this synthesis, it was demonstrated that the molar ratios of the reagents (NH_3:Cu^{2+} and OH^-:Cu^{2+}) determined the morphology and size of the corresponding products via affecting the coordination between NH_3 and Cu^{2+}. Finding the key factors in determining the shapes of the micro/nanocrystals has become the most important issue for shape-controlled wet chemical synthesis of Cu_2O. For example, Xia et al. demonstrated that the introduction of an inorganic species (Cl^-) can provide a means as powerful as organic surfactants or polymers in controlling the shape of Cu_2O nanoparticles [64]. The cubic Cu_2O nanoparticles were synthesized by a polyol method using EG as the reductant (Figure 1B-ii).

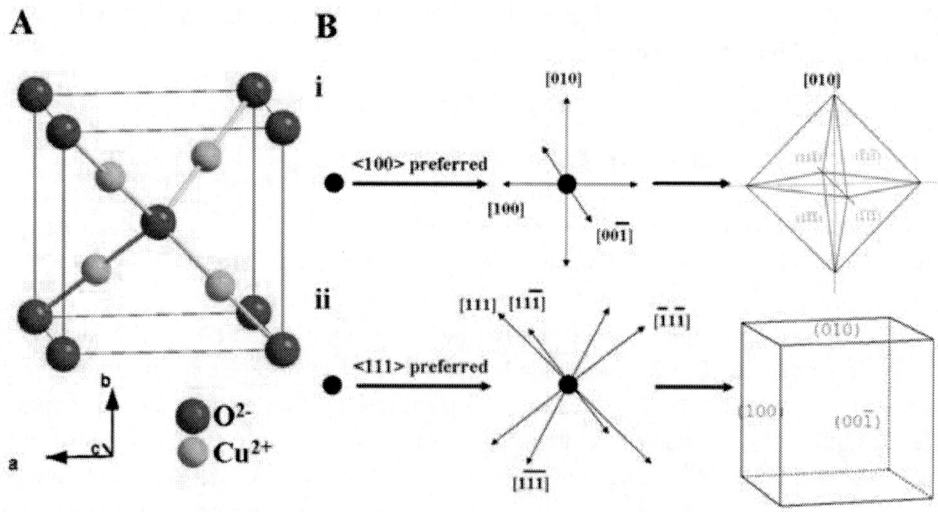

Figure 1. (A) Crystal structure of cuprous oxide, (B) Schematic illustration of the growth mode of Cu_2O (i) octahedra and (ii) cubes.

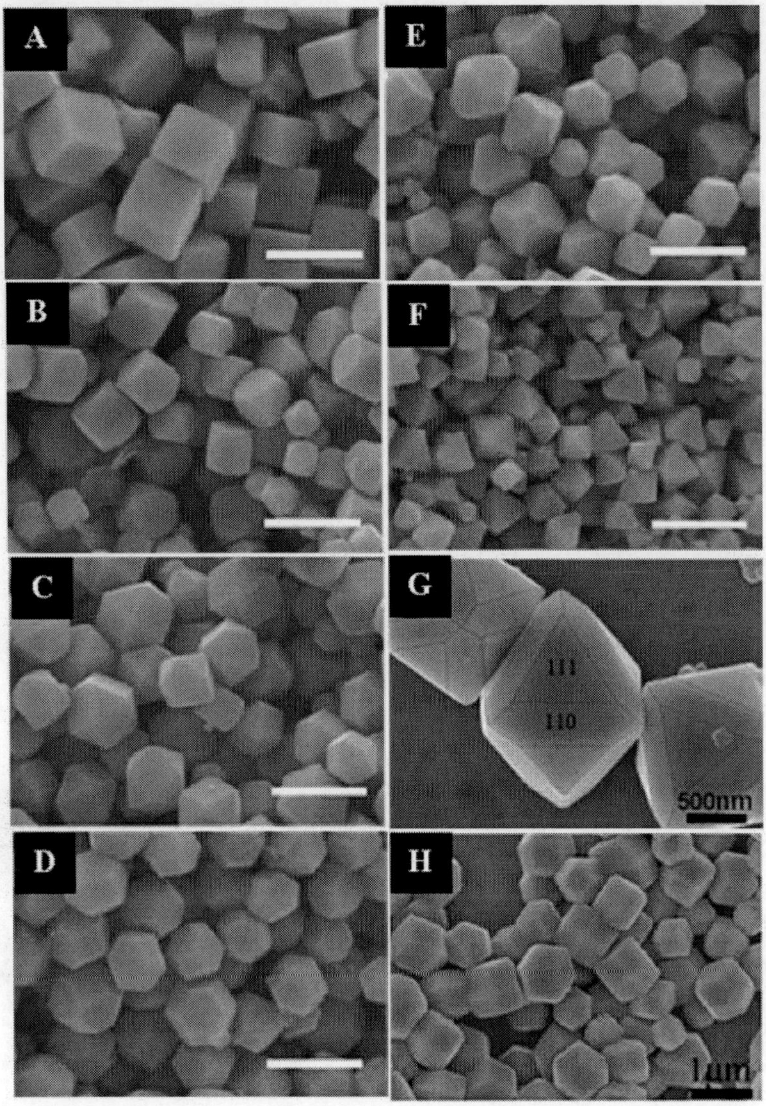

Figure 2. SEM images of the Cu_2O nanocrystals with various morphologies: (A) cubes, (B) truncated cubes, (C) cuboctahedra, (D) type I truncated octahedra, (E) type II truncated octahedra, (F) octahedra, (G) {110} truncated octahedra, and (H) rhombic dodecahedron. Scale bar = 1 μm. Reproduced with permission from [29] (A-F) and [68] (G and H). All images were reproduced with permission. Copyright: 2009 American Chemical Society (A-F); 2009 Wiley-VCH Verlag GmbH & Co. KGaA (G and H)

Despite the above success in controlling the shape of Cu_2O micro/nanocrystals, however, the major products were often restricted to cubes or octahedra due to the fast reduction associated with the strong reducing power of the reductant. Recent studies showed that systematical morphology control for Cu_2O can be achieved by taking advantage of the selective surface stabilization effect of additive or by finely tuning the use amounts of reductant in an aqueous system. Cubic, truncated cubic, cuboctahedral, truncated octahedral, and octahedral have been synthesized in an aqueous solution of $CuCl_2$, NaOH, SDS surfactant, and $NH_2OH{\cdot}HCl$ reductant by simply varying the volume of hydroxylamine added to the reaction mixture [29]. (Figure 2) The possible mechanism of this shape evolution was proposed to be that the amount of $NH_2OH{\cdot}HCl$ may influence the growth rate along the [100] direction relative to that of the [111] direction, or the value of R [65], and thus result in an evolution of final geometries. Interestingly, a kinetic model reported by Xue et al. [51] based on the calculation of surface deviation energies (defined as the bonding energy difference between bulk atoms and surface atoms) of different crystal faces has been used to study how the different molar ratio of EDTA/Cu(II) and pH values influence the morphology evolution [8]. As shown in Figure 3, the theoretical prediction has a good agreement with the current experimental observations [44], which further confirms the rationality of our analysis. Within wet-chemical synthesis methods, similar crystallographic surface control of Cu_2O nanocrystals by taking advantage of the selective interaction of polyvinylpyrrolidone (PVP) on {111} surfaces of Cu_2O powders was also achieved [66, 67]. Moreover, a novel shape evolution of Cu_2O crystals was observed via a facile solution-phase route designed by Gao et al. Interestingly, oleic acid, as a face selective adsorption additive, can subtly determine the morphology of the final Cu_2O products. Various concentrations of oleic acid can result in a series of morphologies of Cu_2O, such as cube, octahedron, {110} truncated octahedron, and microhombic dodecahedron with {110} surfaces [68].

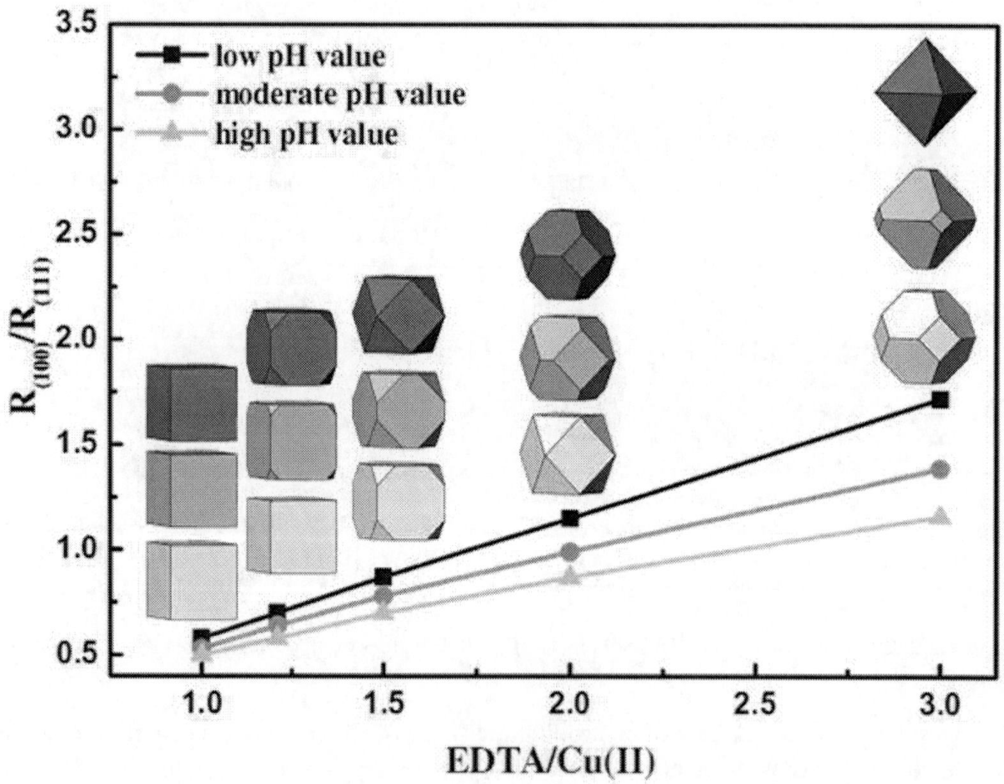

Figure 3. Simulated morphology evolution of Cu_2O crystals grown under different pH values. Three different colors (i.e., dark, dark gray and gray) of the simulated morphologies correspond to the low, moderate and high pH value, respectively. Reproduced with permission from [8]. Copyright 2008, World Scientific Publishing Co

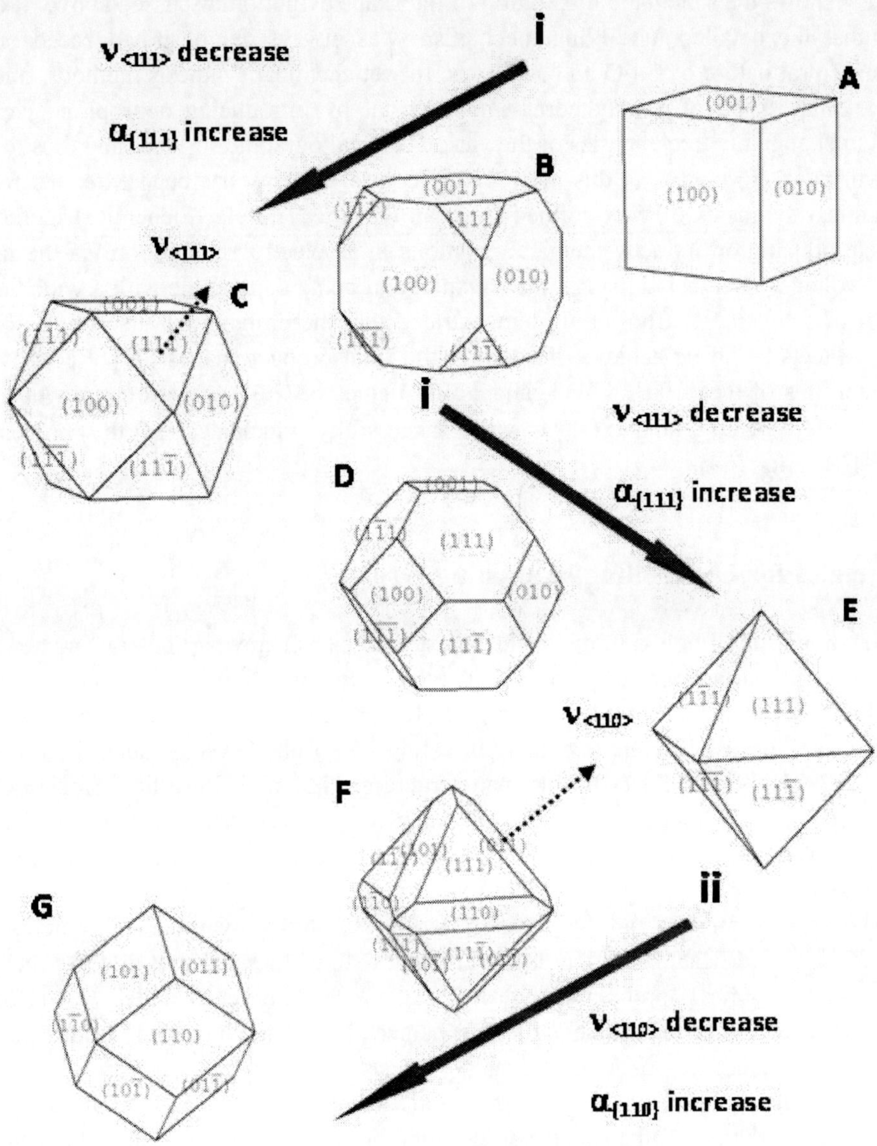

Figure 4. Cartoons showing the development of various shapes of Cu_2O crystsls, such as cube (A), truncated cube (B), cuboctahedron (C), truncated octahedron (D), octahedron (E), {110} truncated octahedron (F), and rhombic dodecahedron (G), according to the results reported [29, 68]. The schematic in process (i and ii) illustrates the effects of face-selective growth ({111} and {110} faces, respectively) on the final Cu_2O crystal shape. The rate of growth perpendicular to the {111} and {110} faces are given by $v_{<111>}$ and $v_{<110>}$ (shown by dotted-line arrows in C and F). The areas of crystallographic facets of {111} and {110} are given by $\alpha_{\{111\}}$ and $\alpha_{\{110\}}$ respectively. When the growth rates of certain faces are not changed selectively, only the particle size increases but the shape remains unchanged. If one of the faces is blocked by the adsorption of capping agents, ions, etc. or has a lower surface energy, then its perpendicular growth rate will be less than that of the other faces. In such situations, the face that is growing more slowly in its perpendicular direction will gradually appear and develop, while the other faces with faster perpendicular growth rates will gradually disappear during the growth.

Figure 4 shows the schematic illustration of the shape evolution discussed above. It could be found that the detailed modifying mechanism was always demonstrated based on the intrinsic structural nature of Cu_2O and additives. In wet chemical synthesis methods, one can control the final shape of a polyhedral Cu_2O crystal by introducing appropriate capping reagent(s) to change the free energies of the various crystallographic surfaces and thus to alter their growth rates. Furthermore, this morphology control strategy has been extended for the electrochemical synthesis of well-defined Cu_2O polyhedra. The electrochemical method of nanoparticle preparation induces chemical reactions in an electrolyte solution via the use of an applied voltage. This technique has the advantage of being compatible with a wide variety of materials. For example, Choi et al. demonstrated that the preferential adsorption of SDS onto {111} facets could be achieved to control the final product shapes [43]. Recently, the relative stabilities of the {100}, {111}, and {110} planes of Cu_2O in various growth media (i.e., Na^+, NH_4^+, NO_3^-, and SO_4^{2-}) were successfully elucidated in the process of electrocrystallization [69].

2.2. Strategies for Controlling Branching Growth

Among diverse morphologies, dendritic nanostructure has many particular characteristics, such as its large surface area, good conductivity, and the ability to provide a natural framework for the study of disordered systems [12, 70-72]. In the past few years, various branching shapes have been reported for fabricating Cu_2O nanocrystals, such as, multipod shapes [44, 73-75], double tower-tip-like nanostructures [76], and flowerlike nanostructures [77]. In recent years, various branching growth patterns have been synthesized using different methods, for example, hydrothermal reaction [44, 73], electrodeposition approach [75], and irradiation methods [48].

Hydrothermal synthesis can be defined as the synthesis by chemical reactions of substances in a sealed heated solution above ambient temperature and pressure. As one kind of wet chemical synthesis processes, hydrothermal process has been widely used for the synthesis of metal oxide micro/nanostructures, since the pioneering work from 1960s to 1980s [34]. The advantages of hydrothermal synthesis include high reactivity of reactants, easy control of solution or interface reactions, formation of metastable and unique condensed phases, less air pollution, and low energy consumption. On the other hand, the multipod branching growth can be created when the growth rate is increased to form a depletion zone of the nutrient ions around the growing crystal, and subsequently makes the apexes of a polyhedral crystal grow faster than the central part of facets. [44, 51]. Hence, hydrothermal method is an effective way to obtain branching shape. Zeng and co-workers have prepared a full range of novel multipod frameworks of Cu_2O microcrystals in hydrothermal synthesis [74].

Our group has successfully synthesized 64 kinds of Cu_2O branching morphologies by using hydrothermal method (Figure 5) [44]. In this synthetic system, the EDTA molecule is used as both chelating reagent and reductant, which is different from the previously reported reductant for the synthesis of Cu_2O. We have been proposed the five branching growth patterns, which are the full range of branching growth patterns for the cubic crystal system, as illustrated in Table 1. In particular, two kinds of high symmetric 24-pod branching growth patterns have obtained, which have not been reported previously in any synthetic systems. The branching growth patterns are mostly determined by the selective adsorption of EDTA molecules on different crystallographic planes of Cu_2O crystal. By tuning the molar ratio of EDTA/Cu(II) and pH of the initial solution, the relative growth rate of {100} and {111} facets of Cu_2O can be well controlled. Furthermore, the kinetically controlled growth process is provided by the thermodynamic and kinetic analyses [51]. Besides, we have also systematically studied the concentration of OH^- ions in influencing branching growth [73]. It has been found that various concentrations of Na_2CO_3 can result in a series of morphology evolutions of Cu_2O microcrystals, for example, from octahedron to 6-pod branching shape, from cube to 8-pod and finally to 24-pod branching shape as shown in Figure 6. Therefore, it is worth noting that the pH value is an important parameter in directing Cu_2O branching growth habit.

Figure 5. SEM images of novel Cu_2O branched microcrystal morphologies. (A) 12-pod branching growth along the <110> direction; (B) 8-pod branching growth; (C) 6-pod branching growth; (D) 24-pod branching growth, based on 12 × 2 corners of the truncated octahedron; (E) 24-pod branching growth, based on 6 × 4 corners of the truncated octahedron Scale bar = 1 μm. Reproduced with permission from [44]. Copyright 2007, Elsevier

Table 1. Five branching growth patterns of Cu_2O microcrystals under the EDTA-reduction synthetic condition

No.	Form	Corners	Description	R
1	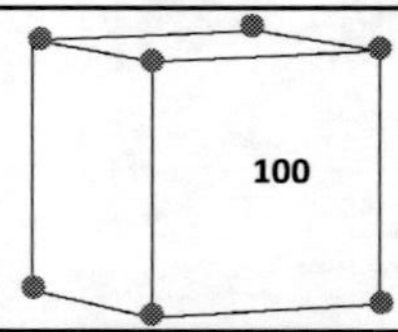	8	8-pod branching growth along the <111> direction, based on 8 corners of the cube	≤0.58
2	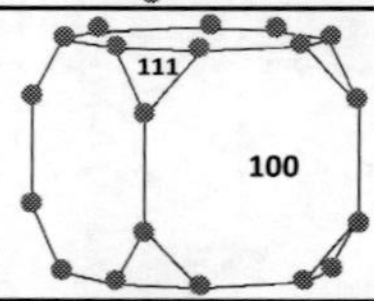	24	24-pod branching growth, based on 12 × 2 corners of the truncated cube	0.70
3	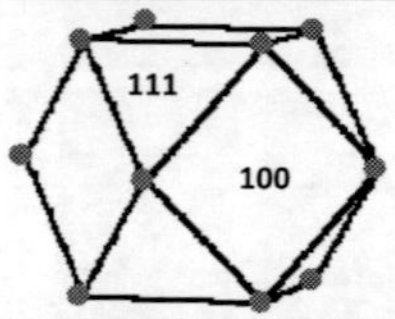	12	12-pod branching growth along the <110> direction, based on 12 corners of the cuboctahedron	0.87
4	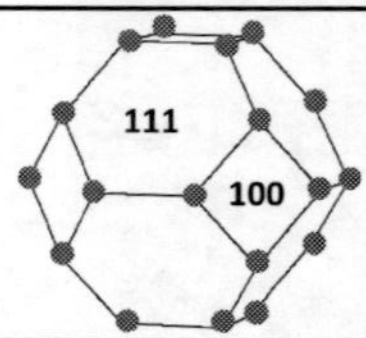	24	24-pod branching growth, based on 6 × 4 corners of the truncated octahedron	1.15
5	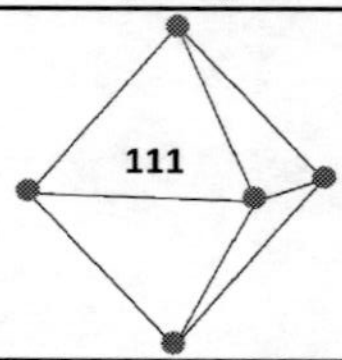	6	6-pod branching growth along the <100> direction, based on 6 corners of the octahedron	≥1.73

Note: Red spheres represent activated corners, and R is defined as the growth rate along the <100> direction to that along the <111> direction [44].

Figure 6. SEM images of Cu2O microcrystals synthesized by using different copper salts and adding different amounts of Na2CO3 solution. (A) 0.5 mmol Cu(NO3)2·3H2O. (B) 0.5 mmol CuCl2·2H2O. (C) 0.5 mmol CuSO4·5H2O. (D) 0.5 mmol Cu(Ac)2·H2O. For each row, the concentration of OH− ions increased from left to right side. scale bars = 1 μm. Reproduced with permission from [73]. Copyright 2009, World Scientific Publishing Co

Recent studies have also shown that the use of electrochemical methods in some cases has an obvious advantage for the directing branched morphologies. For example, Choi and co-workers have developed an electrodeposition approach to obtain various Cu_2O multipod shapes. The shape and size of Cu_2O can be controlled by adjusting the reaction parameters, such as, temperature, concentration of Cu^{2+}, potential and current density, additives, pH value, etc [75]. In these procedures, the deposition potential/current is the most key factor, which controls the degree of branching. So control the preferences for branching or faceting could realize by solely manipulating the deposition potential and current. Moreover, they methodically study the effect of pH value and potential on branching growth and establish a plausible mechanism for the branching growth that occurs in the low overpotential region. They found that the buffer played a key role for stabilizing dendritic branching growth, because when the pH drop during the synthesis, formation of morphologically unstable branches is prevented and faceted growth dominates.

In addition to the multipod shapes described above, Zhang and co-workers have synthesized a novel Cu_2O hierarchical double tower-tip-like nanostructure in water/oil microemulsion [76]. This kind of nanostructure is made of three-order structures. In their study, the effects of the water content, ligand (EDTA), temperature, time, and concentration of NaOH on the morphology of the products have been investigated. And, Luo et al. have obtained flowerlike Cu_2O nanoarchitectures via the solution-phase reaction [77]. In this method, the concentrations of NaOH and cetyltrimethylammonium bromide (CTAB) have been proved to play key roles in the formation of the as-synthesized nanoflowers.

2.3. Routes to Hollow Architectures

Cu_2O hollow-structured powders have attracted considerable attention because of their potentially applications in photoelectrodes, catalysis, and gas sensors. Concerning the creation of hollow for nanostructured materials, there have been two main categories of preparative strategies, i.e., template-assisted routes and template-free routes. To date, a variety of hollow nanoparticles, such as hollow spheres [78-80], hollow cubes [42, 81, 82], hollow octahedra [47], and core-shelled structures [45], have been synthesized by the well-devised wet chemical synthesis techniques. Based on this knowledge, we shall discuss various mechanisms of morphology control in hollow spherical and hollow polyhedral Cu_2O powders in the following.

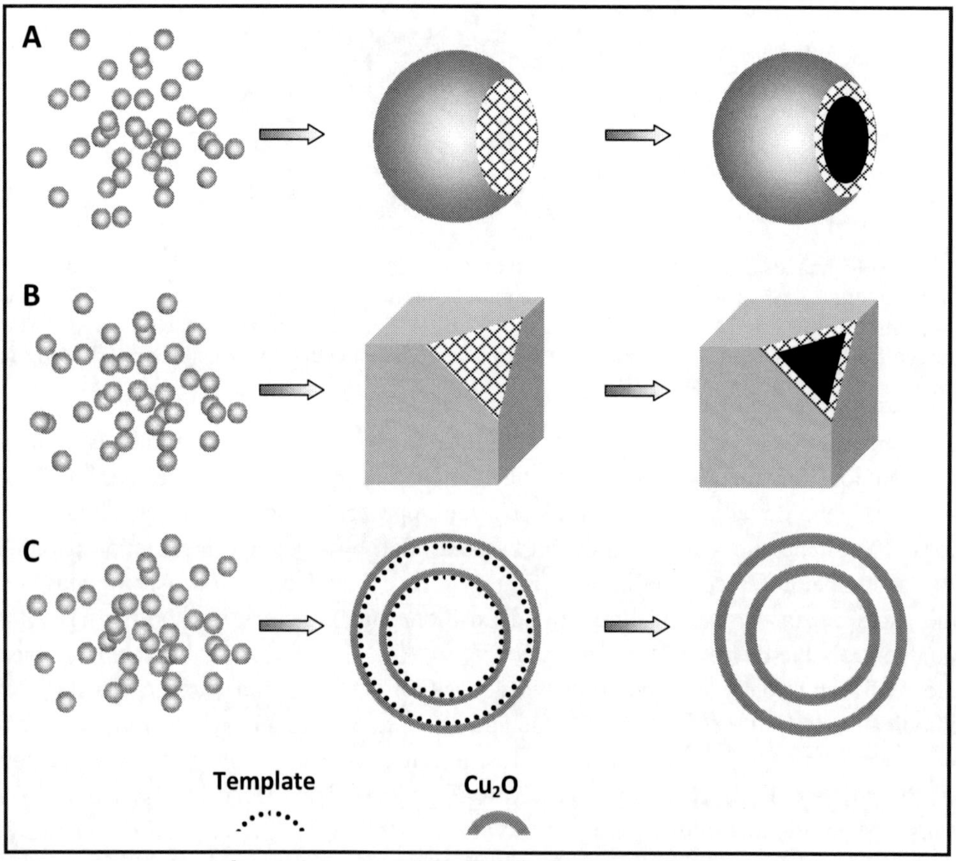

Figure 7. Three different types of synthetic methods for generation of hollow micro/nanopowders: (A) random aggregation of nanocrystallites and core hollowing via Ostwald ripening, resulting in polycrystalline spheres; (B) three-dimensional oriented attachment for solid cubes and creation of hollow interiors by Ostwald ripening; (C) the formation of multilamellar Cu_2O hollow structures in the presence of a template. Hashed areas indicate the solid parts of nanostructures, and dark areas represent interior spaces

Cu_2O spheres are one of the simplest forms for the Cu_2O powder micro/nanomaterials, yet they possess the highest symmetries. When they are prepared with hollow interiors, furthermore, the powder materials represent a class of even more attractive functional materials though the complexity of the synthesis increases. For example, Zeng and co-workers developed a DMF-assisted solvothermal technique to fabricate Cu_2O hollow nanospheres without the assistance of solid template [78]. This hollow morphology of polycrystalline Cu_2O can be stepwise controlled by three steps: forming and gathering primary CuO nanocrystallites, solid-solution redox reactions, and subsequent crystallite aging and solid evacuation. In this synthesis, the process of Ostwald ripening has been emphasized and utilized in controlling crystallite size of shell structures. Meanwhile, it should be noted that a spherical aggregate has the lowest system energy to hold the crystallites together. According to this fact, well dispersed Cu_2O hollow microspheres consisted of Cu_2O nanoparticles were quickly synthesized in aqueous solution at room temperature (25 °C) with polyvinylpyrrolidone (PVP) as surfactant [79]. In this template-free strategy, the high reduction rate resulted from the high reducing power of N_2H_4, produced a large number of Cu_2O nanoparticles, which aggregated quickly to form loose aggregations and then quickly transformed to Cu_2O hollow spheres through Ostwald ripening again (Figure 7A). Double- and even triple- and quadruple-shelled hollow spheres were firstly successfully synthesized in a solution method by using CTAB multilamellar vesicles as soft templates and ascorbic acid as the reductant at 60 °C [46]. Such particular hollow Cu_2O structures (the number of shells) could be controlled by adjusting the concentration of CTAB whilst keeping the other conditions constant. Crystal growth and solidification at the interface leads to transcriptive imprinting of spherical morphology the CTAB vesicle (template) and the formation of hollow structured Cu_2O powders, as illustrated in Figure 7C. Moreover, the formation mechanism of single-crystalline shell wall of the multishelled Cu_2O hollow spheres was further investigated [80]. It is found that the nanoparticles with a size of 2-5 nm first formed well-crystallized porous shelled hollow spheres via oriented attachment, and then the porous shells were further crystallized and became compact by Ostwald ripening to form well-crystallized hollow spheres. This understanding on the growth behavior of the single crystalline hollow spheres has been a guide for conducting the hollow-structured Cu_2O micro/nanomaterials, or even other metal oxides. Furthermore, it is clear that the physical phenomena such as Ostwald ripening and oriented attachment exist and take effect in both the template-free and template-assisted powder-processing techniques.

As one kind of template-based methods, self-sacrificial template methods have also been successfully applied to the shape-controlled synthesis of Cu_2O hollow spheres, such as hierarchical assembly of hollow microspheres [30] and core-shelled microspheres [45]. Here we will take core-shelled Cu_2O microspheres for instance to elucidate the controlling factors in constructing hollow structures via the intermediate template routes in solution. Recently, Xue and Si have developed a surfactant-free intermediate-template approach for preparing homogenous core-shell structured Cu_2O microspheres (Figure 8). In this procedure, it has been found that the addition of tartaric acid promotes the formation of monodispersed intermediate microspheres and ethanol acts as a reducing agent in current synthetic system. Furthermore, a formation mechanism has been revealed through the time-dependent experiments: (i) formation of intermediate microspheres; (ii) reductive conversion of this template to Cu_2O; (iii) inside-out Ostwald ripening of Cu_2O microspheres (Figure 9). It is worth noting that the success of this method largely depends on the effective formation of intermediate microspheres and Ostwald ripening process in the hydrothermal condition.

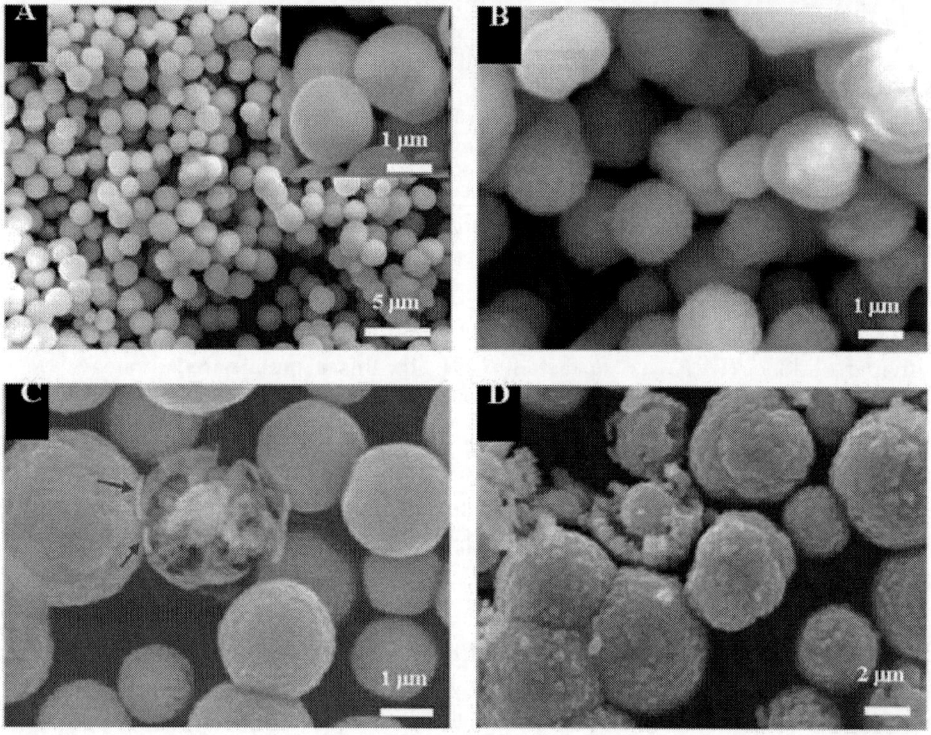

Figure 8. SEM images of particles obtained at different reaction time at 180°C, showing the evolution of particles from microspheres to core-shell structured Cu_2O particles. (A) 1.5 h (intermediate microspheres) Inset: the high-magnification SEM image. (B) 2.5 h. (C) 4 h. Red arrows show thin Cu_2O shells. (D) 14 h. Reproduced with permission from [45]. Copyright 2009, World Scientific Publishing Co

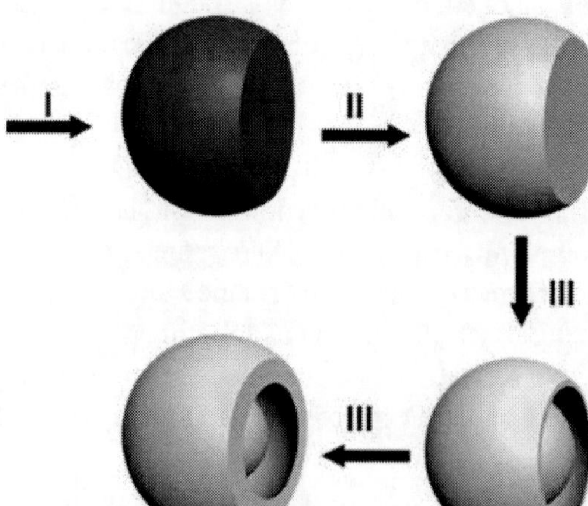

Figure 9. The schematic illustration of the formation mechanism of Cu_2O core-shell structure through the three steps proposed: (I) formation of intermediate microspheres; (II) reductive conversion of this template to Cu_2O; (III) inside-out Ostwald ripening of Cu_2O microspheres

It should be mentioned that the morphology control of Cu_2O hollow nanoproducts prepared in the template-free attempts have not been limited only to the spherical shape, single-crystalline nanocages and polyhedrons with hollow interiors have also been fabricated in solution. Among them, hollow nanocube formation has received great attention and the control over uniformity of size has been advanced, so far, by the wet-chemical methods [80-82]. For an instance, hollow nanocubes with a side length of 50-90 nm have been fabricated via a precursor hydrolysis process in aqueous solution at room temperature [42]. During the synthesis process, the addition of NaOH and PEG was the key factor responsible for the formation of well-defined single-crystalline hollow structures. Notably, Zeng et al. reported a template-free synthetic approach for generating single-crystalline hollow Cu_2O polyhedra, which is virtually a combination of both oriented attachment and Ostwald ripening [81]. In contrast to the spherical aggregates, where the nanocrystallites are randomly joined together, the Cu_2O nanocrystallites in the present case are well organized, maintaining a definite geometric shape (cubic). This formation model, chosen as an example for illustrating synthetic methods for generation of hollow nanostructures (Figure 7B), has shed light on the general formation mechanism for solution synthesis of transition-metal-oxide nanocubes with interior spaces. Besides, it has been found that the solvent composition and reducing agents can play vital roles in the achievement of well-defined hollow Cu_2O polyhedra, such as ethanol. Ethanol has been used as occupation and escape additive for formation of hollow cubes in the octylphenyl ether reduction process [82]. Moreover, Huang et al. have reported that the addition of ethanol followed by sonication of the solution promoted a selective HCl acidic etching over the {110} faces of the Cu_2O nanocages and then led to the formation of the nanoframes which have elliptical pores on the {110} faces [83].

3. Potential Applications

Compared with bulk materials, micro/nanoscale powder materials, with their larger surface-to-volume ratio and higher concentration of partially coordinated surface sites, exhibit distinct electronic, optical, chemical, and thermal properties. As discussed in the former section, Cu_2O micro/nanopowders with various well-controlled morphologies (i.e., polyhedron, branched pattern, and hollow structure) have been synthesized successfully through chemical methods. In many cases, shape control of the Cu_2O micro/nanomaterials allows one to tailor the properties with a greater versatility than can be achieved otherwise. Due to these advantages of properties, much research work has been carried out in the application of Cu_2O powders involving catalysis, photocatalysis, sensing, and photoelectrodes, some of which appear to be very shape-dependent.

3.1. Catalysis and Photocatalysis

Generally, catalytic performance of micro/nanocrystals can be finely tuned either by their composition, which mediates electronic structure, or by their shape, which determines surface atomic arrangement and coordination [84]. With recent developments in Cu_2O powder synthesis leading to the ability to control size, morphology, and reproducibility, considerable achievements in the application of Cu_2O micro/nanocrystals have been obtained such as the CO oxidation [31] and photodegradation [29, 85]. There are several important processes in heterogeneous catalysis where removal of carbon monoxide is either desired or absolutely necessary, such as in the postprocessing of Syngas to produce hydrogen as an energy source for use in fuel cells. O'Brien et al. have developed a new catalyst system comprised of Cu_2O nanoparticles supported on silica gel that carries out very efficient oxidation of CO to CO_2 [31]. The Cu_2O nanoparticles oxidized CO to CO_2 for over 144 h with no dependence on O_2 concentration before decreasing reactivity. This system is a cost-effective, highly efficient alternative to current CO oxidation systems, thus opening the doorway to a variety of applications requiring cheap one-time use or short time frame catalysts for oxidizing CO to CO_2.

Recently, Huang and co-workers successfully demonstrated that octahedra and the extended hexapods which produced exhibited good photocatalytic activity toward the photodegradation of methyl orange, while perfect cubes were practically not active [29]. The observation of this dramatic difference in the photocatalytic activity of Cu_2O crystals is only possible with the successful synthesis of sharp cubes and octahedra, as shown in Figure 10. This study also showed that hexapods, and possibly other multipods with more {111} facets, are catalytically more active and effective than the octahedra, and should be considered in future catalysis studies with Cu_2O crystals. Moreover, porous Cu_2O microcubes were found to be good photo-catalysts for the degradation of the organic dye pyronine B under irradiation with 254 nm UV light by Ni et al [85].

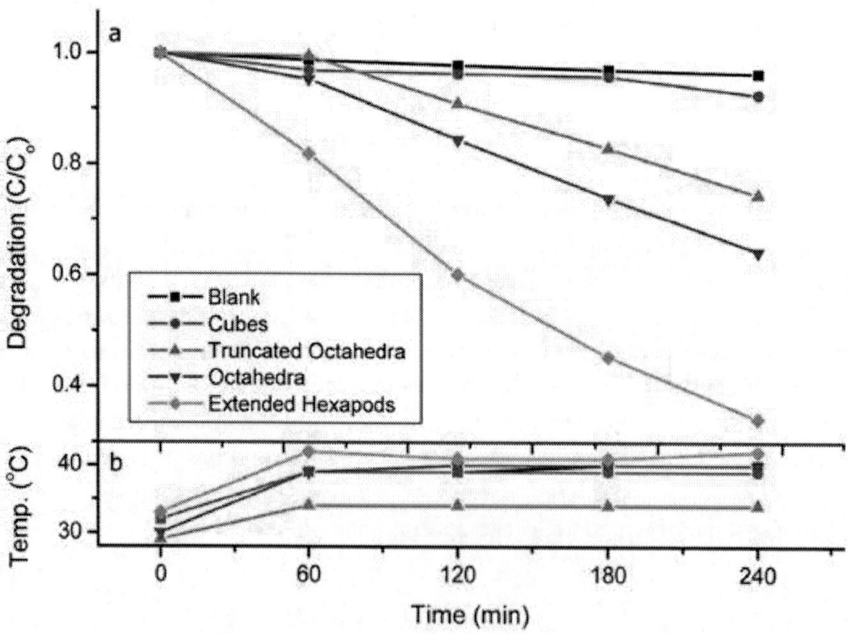

Figure 10. A plot of the extent of photodegradation of methyl orange vs. time for the various Cu_2O nanostructures is shown. The blank sample did not contain Cu_2O crystals but only the methyl orange solution. The temperature change of the solution over this time period is also given. Reproduced with permission from [29]. Copyright 2009, American Chemical Society

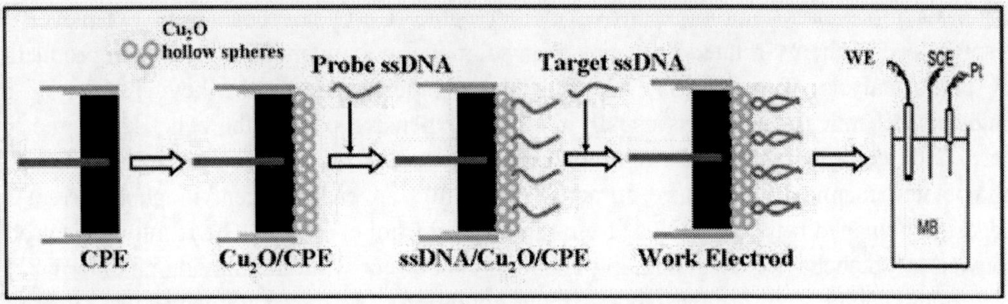

Figure 11. Schematic representation of the electrochemical detection of DNA hybridization using electrode containing synthesized Cu_2O hollow spheres. CPE is the abbreviation of carbon paste electro. Reproduced with permission from [79]. Copyright 2009, American Chemical Society

3.2. Sensors

Scientific curiosity about the behavior of Cu_2O porous capsules has led to a number of opportunities for their widespread use in gas-sensing [30, 86, 87] and biosensors [79]. Hollow micro-/nanostructures possess characteristics such as low density, high surface-to-volume ratio, and low coefficients of thermal expansion that make them suit to be sensors. For example, cuprous oxide sensors have been fabricated by depositing aqueous ink of the Cu_2O microspheres onto interdigital electrodes of a microhotplate, showing much higher sensitivity than solid Cu_2O microspheres [30]. Since cuprous oxide is a p-type semiconductor, the conductivity of the Cu_2O sensor is mainly determined by the hole carrier density of the intergranular contract region. As a reducing gas, ethanol was adsorbed on the surface of Cu_2O grains and the injected electrons from ethanol molecules neutralized holes and decreased the carrier density in the surface charge layer of p-type semiconductors. Thus, the adsorption of ethanol caused the marked increase of the resistance of Cu_2O sensors. Based on a similar mechanism, Cu_2O materials have also been made into methane gas sensors [86].

As for biosensors, electrochemical DNA biosensors maybe used in gene expression, disease diagnosis, and drug screening due to many advantages such as low cost, simple design, small dimension and high sensitivity [88]. The key element of an electrochemical DNA biosensor is the detection electrode, in which immobilized single stranded DNA (ssDNA) acts as a probe to detect the target DNA. Zhu et al. have applied their synthesized Cu_2O hollow spheres for the ssDNA probe immobilization [79]. The schematic representation of the electrochemical detection process of DNA hybridization is briefly shown in Figure 11. The results showed that the electrode modified with Cu_2O hollow spheres give a very high sensitivity for detecting the Human hepatitis B virus DNA sequences.

3.3. Photoelectrodes

Among the conventional semiconducting materials, Cu_2O is known to be one of the earliest semiconducting materials with relatively low band gap (E_g = 1.9-2.2 eV) and relatively high absorption coefficient in the visible region [89]. As a consequence of its superior optoelectronic properties, nontoxic nature, natural abundance, simplicity in the synthesis procedures, and inexpensive market value, Cu_2O has been used extensively to resolve several energy-related problems from solar energy conversion to hydrogen production by photocatalysis of water [33, 90, 91]. The photoelectrode is the key element of the phtoelectrochemical cell and is usually the most expensive part of the cell. Hence, the low cost and environmentally benign nature of copper makes developing Cu_2O-based photoelectrochemical cells more attractive. Recently, an enhancement in photocurrent for dendritic branched n-type Cu_2O has been reported by Choi et al [92]. The resulting electrodes showed n-type behavior by generating anodic photocurrent without applying an external bias (zero-bias photocurrent under short-circuit condition) in an Ar-purged 0.02 M K_2SO_4 solution. This research work provided new insight into tuning morphologies for the enhancement of desired properties of polycrystalline electrodes as well as opportunities to build various photoelectrochemical devices based on high-performance n-type Cu_2O electrodes. Sahoo and his colleges also demonstrated that the electrical devices made from nanowire networks synthesized electrochemically exhibit promising photon sensing characteristics under white light illumination and can be exploited for future applications in photodetection and photovoltaic studies at the nanoscale level [93]. Since the stability of electrodeposited p-Cu_2O in the photoelectrochemical applications is a vital factor, the crystal face dependence of p-Cu_2O stability as photocathode has also been reported by Fillinger et al [94].

4. CONCLUSION

From fundamental study through technological application, metal oxide micro/nanopowders have garnered unprecedented attention due to their strikingly different properties at nano-level dimensions as compared to the nature of its bulk counterpart, especially Cu_2O micro/nanostructures. In the past decade, considerable progress has been made in synthesis and applications of Cu_2O micro/nanopowders. This article has provided an overview on a variety of chemical methods that have been developed for generating Cu_2O micro/nanopowders with various morphologies. We have broadly divided these morphologies into three categories, in which i) solid polyhedral shapes contain cubic, truncated cubic, cuboctahedral, truncated octahedral, octahedral {110} truncated octahedron, microhombic dodecahedron and etc; ii) branched morphologies involve multipod (e.g., 6-pod, 8-pod, 12-pod, and 24-pod), flowerlike, and hierarchical double tower-tip-like structures; and iii) hollow spherical and polyhedral structures. In a solution-phase system, the success of such morphology control depends on a number of parameters that can be tuned to control the formation of Cu_2O micro/nanopowders, such as the reduction kinetics, template preformation, and surface capping. These parameters could be combined to provide an effective route to maneuver both the surface facets and morphology transformation, which are two key factors in determining the final shape of a Cu_2O micro/nanopowder. Specifically, various branching growth patterns could be controlled by introducing different amounts of reducing agents and thus manipulating the reduction kinetics. Meanwhile, a single-crystalline micro/nanopowder can further evolve into a polyhedron with one of the possible shapes as determined by the growth rates of different facets, which can be controlled with a capping agent, such as Cl^- ions, SDS, and EDTA. In most cases, the key kinetic factors in determining final shapes have common points despite the formation processes are designed and conducted through different chemical liquid-phase methods. Therefore, the methodology and design experience summarized and established for Cu_2O materials can be extended to other *fcc* metal oxides, even all the other metal oxides, although slight modification to the experimental conditions might be needed for different metal oxides.

The successes in synthesis of Cu_2O micro/nanopowders with a variety of shapes have provided opportunities to tune their optical, electrical, chemical, and other properties. These advances have in turn catalyzed exploration in a growing list of applications. However, there still remain major challenges that need to be overcome to achieve translation into practical applications. For example, the chemical/thermal/mechanical stability of Cu_2O hollow powders in the application of catalysis and sensing still need to be systematically studied and better controlled. For their application in photoelectrodes and lithium-ion batteries, some characters of branched crystal may bring negative effects, i.e., high surface area may lead to more significant side reactions, and more difficulty in maintaining interparticle contact. At the same time, achieving high yield and successful scaling up of the synthesis process are necessary for the synthesis to be commercially viable, and are expected to introduce significant challenges for size, shape, and shell thickness control. Clearly, numerous challenges remain, but the rate of advance in micro- and nanopowder science and technology over the past decade promises that this area will deliver exciting developments in the near future.

ACKNOWLEDGMENT

This work is supported by the National Natural Science Foundation of China (Grant Nos. 50872016, 20973033).

REFERENCES

[1] Cui, Y; Lieber, CM. *Science*, 2001, 291, 851-853.
[2] Sun, Y; Xia, Y. *Science*, 2002, 298, 2176-2179.
[3] Xue, D; Zou, L; Wang, L; Yan, X. *Mod. Phys. Lett. B*, 2009, 23, 3761-3768.
[4] Zhu, Y; Bando, Y; Xue, D; Golberg, D. *Adv. Mater.*, 2004, 16, 831-834.
[5] Zhu, Y; Bando, Y; Xue, D; Golberg, D. *J. Am. Chem. Soc.*, 2003, 125, 16196-16197.
[6] Liu, J; Liu, F; Gao, K; Wu, J; Xue, D. *J. Mater. Chem.*, 2009, 19, 6073-6084.
[7] Liu, J; Xue, D. *Adv. Mater.*, 2008, 20, 2622-2627.
[8] Zhao, X; Ren, X; Sun, C; Zhang, X; Si, Y; Yan, C; Xu, J; Xue, D. *Funct. Mater. Lett.*, 2008, 1, 167-172.
[9] Yan, C; Zou, L; Xu, J; Wu, J; Liu, F; Luo, C; Xue, D. *Powder Technol.*, 2008, 183, 2-9.
[10] Yan, X; Xu, D; Xue, D. *Acta Mater.*, 2007, 55, 5747-5757.
[11] Lu, P; Xue, D. *Mod. Phys. Lett. B*, 2009, 23, 3843-3849.
[12] Liu, J; Xue, D. *Mod. Phys. Lett. B*, 2009, 23, 3777-3783.
[13] Xue, D; Yan, X; Wang, L. *Powder Technol.*, 2009, 191, 98-106.
[14] Liu, M; Xue, D. *J. Phys. Chem. C*, 2008, 112, 6346-6351.

[15] Liu, F; Sun, C; Yan, C; Xue, D. *J. Mater. Sci. Technol.*, 2008, 24, 641-648.
[16] Xu, J; Xue, D. *J. Solid State Chem.*, 2007, 180, 119-126.
[17] Yan, C; Xue, D. *J. Phys. Chem. B*, 2006, 110, 7102-7106.
[18] Yan, C; Xue, D. *J. Phys. Chem. B*, 2006, 110, 25850-25855.
[19] Yan, C; Xue, D. *J. Phys. Chem. B*, 2006, 110, 11076-11080.
[20] Xu, J; Xue, D. *J. Phys. Chem. B*, 2006, 110, 7750-7756.
[21] Xu, J; Xue, D. *J. Phys. Chem. B*, 2006, 110, 11232-11236.
[22] Luo, C; Xue, D. *Langmuir*, 2006, 22, 9914-9918.
[23] Liu, M; Xue, D; Luo, C. *J. Am. Ceram. Soc.*, 2006, 89, 1551-1556.
[24] Yan, C; Xue, D; Zou, L; Yan, X; Wang, W. *J. Crys. Growth*, 2005, 282, 448-454.
[25] Yan, C; Xue, D. *J. Phys. Chem. B*, 2005, 109, 12358-12361.
[26] Xu, J; Xue, D; Yan, C. *Mater. Lett.*, 2005, 59, 2920-2922.
[27] Xu, J; Xue, D. *J. Phys. Chem. B*, 2005, 109, 17157-17161.
[28] Vayssieres, L. *Int. J. Nanotechnol.*, 2004, 1, 1-41.
[29] Ho, J. Y; Huang, M. H. *J. Phys. Chem. C*, 2009, 113, 14159-14164.
[30] Zhang, H. G; Zhu, Q. S; Zhang, Y; Wang, Y; Zhao, L; Yu, B. *Adv. Funct. Mater.*, 2007, 17, 2766-2771.
[31] White, B; Yin, M; Hall, A; Le, D; Stolbov, S; Rahman, T; Turro, N; O'Brien, S. *Nano Lett.*, 2006, 6, 2095-2098.
[32] Snoke, D. *Science*, 2002, 298, 1368-1372.
[33] De Jongh, PE; Vanmaekelbergh, D; Kelly, JJ. *Chem. Commum.*, 1999, 1069-1070.
[34] Yao, WT; Yu, SH. *Adv. Funct. Mater.*, 2008, 18, 3357-3366.
[35] Yan, C; Xue, D. *J. Alloy. Compd.*, 2007, 431, 241-245.
[36] Yoshimura, M; Suchanek, W; Byrappa, K. *MRS Bull.*, 2000, 25, 17-25.
[37] Liu, F; Xue, D. *Mod. Phys. Lett. B*, 2009, 23, 3769-3775.
[38] Xie, X; Li, Y; Liu, ZQ; Haruta, M; Shen, W. *Nature*, 2009, 458, 746-749.
[39] Liu, J; Xia, H; Xue, D; Lu, L. *J. Am. Chem. Soc.*, 2009, 131, 12086-12087.
[40] Peng, X; Manna, L; Yang, W; Wickham, J; Scher, E; Kadavanich, A; Alivisatos, AP. *Nature*, 2000, 4014, 59-61.
[41] Colfen, H; Mann, S. *Angew. Chem. Int. Ed.*, 2003, 42, 2350-2365.
[42] Xu, YY; Jiao, XL; Chen, DR. *J. Phys. Chem. C*, 2008, 112, 16769-16773.
[43] Siegfried, MJ; Choi, KS. *Adv. Mater.*, 2004, 16, 1743-1746.
[44] Xu, J; Xue, D. *Acta Mater.*, 2007, 55, 2397-2406.
[45] Si, Y; Xue, D. *Mod. Phys. Lett. B*, 2009, 23, 3851-3858.
[46] Xu, HL; Wang, WZ. *Angew. Chem. Int. Ed.*, 2007, 46, 1489-1492.
[47] Lu, CH; Qi, LM; Yang, JH; Wang, XY; Zhang, DY; Xie, JL; Ma, JM. *Adv. Mater.*, 2005, 17, 2562-2567.
[48] Liu, HR; Miao, WF; Yang, S; Zhang, ZM; Chen, JF. *Cryst. Growth Des.*, 2009, 9, 1733-1740.
[49] Xu, L; Jiang, LP; Zhu, JJ. *Nanotechnol.*, 2009, 20, 045605.
[50] Barreca, D; Gasparotto, A; Maccato, C; Tondello, E; Lebedev, OI; Tendeloo, GV. *Cryst. Growth Des.*, 2009, 9, 2470-2480.
[51] Zhao, X; Bao, Z; Sun, C; Xue, D. *J. Cryst. Growth*, 2009, 311, 711-715.
[52] Zhao, X; Si, Y; Xue, D. *J. Cryst. Phys. Chem.*, 2010, 1, 57-73.
[53] Xue, D; Si, Y; Zhao, X. *J. Cryst. Phys. Chem.*, 2010, 1, 3-11.
[54] Xu, D; Xue, D. *J. Cryst. Growth,* 2008, 310, 1385-1390.

[55] Xu, D; Xue, D. *J. Rare Earths*, 2006, 24, 228-232.
[56] Wulff, G. Z. *Kristallogr*, 1901, 34, 449.
[57] Xia, Y; Yang, P; Sun, Y; Wu, Y; Mayers, B; Gates, B; Yin, Y; Kim, F; Yan, Y. *Adv. Mater.*, 2003, 15, 353-389.
[58] Gou, LF; Murphy, CJ. *Nano Lett.*, 2003, 3, 231-234.
[59] Wang, D; Mo, M; Yu, D; Xu, L; Li, F; Qian, Y. *Cryst. Growth Des.*, 2003, 3, 717-720.
[60] Kuo, CH; Chen, CH; Huang, MH. *Adv. Funct. Mater.*, 2007, 17, 3773-3780.
[61] Zhao, HY; Wang, YF; Zeng, JH. *Cryst. Growth Des.*, 2008, 8, 3731-3734.
[62] Zhu, J; Bi, H; Wang, Y; Wang, X; Yang, X; Lu, L. *Mater. Lett.*, 2008, 62, 2081-2083.
[63] Xu, H. L; Wang, W. Z; Zhu, W. *J. Phys. Chem. B*, 2006, 110, 13829-13834.
[64] Kim, MH; Lim, B; Lee, EP; Xia, Y. *J. Mater. Chem.*, 2008, 18, 4069-4073.
[65] Wang, ZL. *J. Phys. Chem. B*, 2000, 104, 1153-1175.
[66] Zhang, DF; Zhang, H; Guo, L; Zheng, K; Han, XD; Zhang, Z. *J. Mater. Chem.*, 2009, 19, 5220-5225.
[67] Sui, Y; Fu, W; Yang, H; Zeng, Y; Zhang, Y; Zhao, Q; Li, Y; Zhou, X; Leng, Y; Li, M; Zou, G. *Cryst. Growth Des.*, 2010, 10, 99-108.
[68] Liang, XD; Gao, L; Yang, SW; Sun, J. *Adv. Mater.*, 2009, 21, 2068-2071.
[69] Siegfried, MJ; Choi, KS. *J. Am. Chem. Soc.*, 2006, 128, 10356-10357.
[70] Yan, C; Xue, D. *Cryst. Growth Des.*, 2008, 8, 1849-1854.
[71] Rothschild, WG. Fractals in Chemistry. Wiley-Interscience: New York, 1998.
[72] Liu, J; Xue, D. *Mod. Phys. Lett. B*, 2009, 23, 3817-3823.
[73] Chen, K; Si, Y; Xue, D. *Mod. Phys. Lett. B*, 2009, 23, 3753-3760.
[74] Chang, Y; Zeng, HC. *Cryst. Growth Des.*, 2004, 4, 273-278.
[75] Siegfried, MJ; Choi, KS. *Angew. Chem. Int. Ed.*, 2005, 44, 3218-3223.
[76] Zhang, H; Zhang, X; Li, H; Qu, Z; Fan, S; Ji, M. *Cryst. Growth Des.*, 2007, 7, 820-824.
[77] Luo, Y; Li, S; Ren, Q; Liu, J; Xing, L; Wang, Y; Yu, Y; Jia, Z; Li, J. *Cryst. Growth Des.*, 2007, 7, 87-92.
[78] Chang, Y; Teo, JJ; Zeng, HC. *Langmuir*, 2005, 21, 1074-1079.
[79] Zhu, H; Wang, J; Xu, G. *Cryst. Growth Des.*, 2009, 9, 633-638.
[80] Xu, H; Wang, W; Zhou, L. *Cryst. Growth Des.*, 2008, 8, 3486-3489.
[81] Teo, JJ; Chang, Y; Zeng, HC. *Langmuir*, 2006, 22, 7369-7377.
[82] Luo, F; Wu, D; Gao, L; Lian, SY; Wang, EB; Kang, ZH; Lan, Y; Xu, L. *J. Cryst. Growth*, 2005, 285, 534-540.
[83] Kuo, CH; Huang, MH. *J. Am. Chem. Soc.*, 2008, 130, 12815-12820.
[84] Tian, N; Zhou, ZY; Sun, SG; Ding, Y; Wang, ZL. *Science*, 2007, 316, 732-735.
[85] Li, H; Ni, Y; Cai, Y; Zhang, L; Zhou, J; Hong, J; Wei, X. *J. Mater. Chem.*, 2009, 19, 594-597.
[86] Jayatissa, AH; Samarasekara, P; Kun, G. *Phys. Status Solidi A*, 2009, 206, 332-337.
[87] Zhang, J; Liu, J; Peng, Q; Wang, X; Li,Y. *Chem. Mater.*, 2006, 18, 867-871.
[88] Ren, Y; Jiao, K; Xu, GY; Sun, W; Gao, HW. *Electroanal.*, 2005, 17, 2182-2189.
[89] Shen, M. Y; Yokouchi, T; Koyama, S; Goto, T. *Phys. Rev. B*, 1997, 56, 13066-13072.
[90] Yin, M; Wu, CK; Lou, Y; Burda, C; Koberstein, JT; Zhu, Y; O'Brien, S. *J. Am. Chem. Soc.*, 2005, 127, 9506–9511.
[91] Hara, M; Kondo, T; Komoda, M; Ikeda, S; Shinohara, K; Tanaka, A; Kondo, JN; Domen, K. *Chem. Commun.*, 1998, 357-358.
[92] McShane, CM; Choi, KS. *J. Am. Chem. Soc.*, 2009, 131, 2561-2569.

[93] Sahoo, S; Husale, S; Colwill, B; Lu, TM; Nayak, S; Ajayan, PM. *ACS Nano*, 2009, 3, 3935-3944.

[94] Sowers, KL; Fillinger, A. *J. Electrochem. Soc.*, 2009, 156, 80-85.

In: Powder Engineering, Technology and Applications
Editor: Jason M. Barker, pp.197-206
ISBN: 978-1-61761-216-9

Chapter 7

A COMMENTARY ON NANOSTRUCTURED FERRITE MATERIALS: SYNTHESIS, PROPERTIES AND APPLICATIONS

Narendra Nath Ghosh [*]
Chemistry Group, Birla Institute of Technology and Science,
Pilani, Zuarinagar, India

ABSTRACT

Ferrites are important class of ferromagnetic materials. These materials can broadly be classified as soft and hard ferrites, depending upon their magnetic properties. It is very interesting to note that how slight variation of composition, particle size, processing technology etc affect the properties of these materials. Traditionally they have found wide range of applications in electronic, microwave, wave absorber devices, magnetic recording industries. But recently nanostructured ferrite materials have attracted attention of scientists due to their potential applications in nanotechnology, particularly drug delivery, smart materials etc. In this paper following topics will be discussed (i) a brief history of ferrite materials, (ii) different synthesis strategies, which are adopted by several scientist for preparation of different compositions of ferrites and advantages and disadvantages associated with these methods, (iii) properties and commercial applications of different ferrites, (iv) future scope of research.

1.1. INTRODUCTION

Ferrites are magnetic ceramic materials and they are broadly categorized as soft and hard magnetic material. The word ''soft'' means temporary in the sense that the ferromagnetism emerges only when a magnetic field is applied. In contrast, hard magnets display ferromagnetism in the absence of an external field. The magnetic properties of ferrites were considerably inferior to those of ferromagnetic metallic alloys such as Alnico. However, these materials attracted worldwide attention only after new inventions like radio, television,

[*] Corresponding author: (naren70@yahoo.com).

computers and microwave devices as they had the unique combination of being magnetic and having the ability to operate at high frequencies. Structurally, ferrites can be classified as (i) cubic spinel ferrites like $NiFe_2O_4$ (ii) hexagonal ferrites like $BaFe_{12}O_{19}$ (iii) garnet ferrites such as $Y_3Fe_5O_{12}$ [1, 2].

Nanostructured materials (particle size < 100nm) are known to exhibit exotic properties as compared to those exhibited by conventional cermics (grain size > 10μm). A unique feature of these materials is that a large fraction of their atoms resides in their grain boundaries. As the particle size decreases, the interfacial volume increases. As a result, the properties of nanostructured materials are strongly governed by their surface structure and/or interface volume which represents a significant fraction of their total volume. It is thus plausible that many of the microstructure based properties such as mechanical hardness, electrical resistivity, magnetic property etc. of nanostructured ferrites would differ from those of conventional bulk materials, in which the interface volume fraction is negligible. These nanopowders, therefore, give rise to myriad of potential applications that could not have been conceived with the regular bulk microstructure [3- 6].

In the last decade especially, the field of biomedicine has witnessed a surge of interest in the use of magnetic nanomaterials in earlier diagnosis and effective treatment of some diseases such as cancer. Ferrite nanoparticles are finding novel applications in the bio-medical field in four main categories of targetted drug delivery, hyperthermia, as contrast agents in Magnetic Resonance Imaging (MRI) and in magnetic seperation of cells [3, 7, 8]. They are also currently being investigated for their applications as nanocomposites where tailoring of their properties is possible by changing their weight ratio with respect to the non-magnetic matrix [9-12].

1.2. Historical Developments

Sugimoto in his centennial feature article in Journal of the American Ceramic Society described elaborately the past, present, and future of Ferrites [13] and important references are cited there in. The first systematic study on ferrites was based upon spinel ferrites by Hilpert and it dates back to 1909. In 1932, Kato and Takei, discovered that a solid solution of magnetite and cobalt ferrite was strongly magnetized at ~300^0C. This magnet was industrialized under the name of "OP magnet". The same group experimented by mixing a normal and inverse spinel and found that the permeability of the mixed ferrite multiplied by several tens of times. This gave birth to the first commercial ferrite core of Cu-Zn ferrite. During the same period, Snoek worked with Mn-Zn and Ni-Zn ferrites to achieve high permeability and low loss materials. In 1952, the invention of hexagonal ferrites, such as barium and strontium ferrites marked an important event in the history of ferrites. Several researchers contributed significantly towards the development and commercialization of ferrites during the period from 1935 to 1970. Hogan's work on microwave ferrite devices, Albers-Schoenberg's development of Mn-Mg ferrites for magnetic memories, Dillon et al contributions to the development of magnetic memory devices and Camaras's discovery of gamma-Fe_2O_3 acicular fine particles are the most outstanding achievements [1, 13].

1.3. Common Preparation Methods for the Synthesis of Ferrites

There are several methods for the preparation of ferrites. We have critically reviewed the different patented methods in order to synthesize a wide variety of ferrites depending on the applications. The details of the methods, with specific example, have been described in this paper [14]. Herein, we will briefly describe some common methodologies that are used in order to synthesize ultrafine nanopowders.

Solid state method: [15-17]

Bulk iron oxide or ferrites have conventionally been prepared by the high temperature solid-state route that involves mechanical mixing of oxides, hydroxides, oxyhydroxides or carbonates followed by calcination at high temperatures (>1000^{0}C) (such as use of $BaCO_3$, $ZnCl_2$, $FeCl_3$ for $Ba_2Zn_2Fe_{12}O_{12}$). This method, although successful for large-scale production of bulk ceramic powders, has several limitations as far as production of nanostructured ceramics is concerned. High temperatures and long heating schedules, which are the requisites of this method, lead to coarsening of grains. Moreover, volatilization/melting of the constituent components may occur during the high temperature processing of some of the multicomponent mixed oxides systems [18]. Thus, this method is inappropriate for producing pure, nanostructured ceramic materials.

Some of the common gas phase and solution phase methods that are capable of producing nanostructured ferrites are as follows:

1.3.1. Gas phase methods

(a) Chemical vapor deposition (cvd): [19]

In the CVD process, a carrier gas stream with precursors is delivered continuously by a gas delivery system to a reaction chamber maintained under vacuum at high temperature (>900^{0}C). The CVD reactions take place in the heated reaction chamber and the products combine to form clusters or nanoparticles. The CVD process has been employed to deposit iron oxide and ferrite films at 800 to1000^{0}C.

(b) Metal organic chemical vapor deposition (mocvd): [20, 21]

In the MOCVD process, metallo organics are used as precursors that allows reactions to take place at a somewhat lower temperature (300 to 800^{0}C) and at pressures varying from 1 Torr to ambient. Iron oxide thin films have been obtained via decomposition of acetylacetonate at 400-500^0 C, and iron trifluoro-acetylacetonate at 300^{0}C in oxygen. Ferrite films have also been obtained in a similar manner.

(c) Laser ablation method: [22-24]

Well-crystallized and uniform iron oxide nanoparticles, including nanoparticles of hematite and maghemite, have been obtained in one step using laser pyrolysis. Iron pentacarbonyl is commonly used as a precursor in this method and ethylene is used as the carrier gas to transport the carbonyl vapor to the reaction zone, as ethylene does not absorb radiation at the laser wavelength. Series of Ni-Zn ferrite and Zn ferrite thin films have been obtained by laser ablation process on Si substrates.

Some limitations of gas phase methods are (i)although they are able to deliver high quality products, the yields are usually low and scale-up of the equipment is challenging. (ii) variables such as oxygen concentration, gas phase impurities, and the heating time must be controlled precisely to obtain pure products. (iii) the equipment used in these methods are also expensive and require maintenance.

1.3.2. Solution phase methods

(a) Co-precipitation method: [25-27]

Co-precipitation method is a very well established technique for preparation of iron oxides and ferrites with wide range of compositions. In this wet chemical process, ferrites with desired chemical compositions are simultaneously precipitated from aqueous solutions of metal ions. Co-precipitation occurs by the reaction of the metal ion solutions with precipitating agent, generally alkaline solution of carbonates, hydroxides, ammonia or ammines. The resulting solid precipitate contains ferrous and other metal carbonates or hydroxides, which decompose to the oxides when calcined at temperatures ranging from 500 - 800^{0}C. Final sintering at 1000 - 1200^{0}C produces the desired ferrites. Precise control over the chemical composition of the product is the main advantage of the technique in comparison with the conventional solid-state technique. The co-precipitation method, however, is associated with the limitation of producing materials that are not exceptionally homogeneous and reproducibility of the process for some cases is poor. Formation of gelatinous precipitate often causes difficulty in separation and retention of impurities in the product might cause poor magnetic properties.

(b) Sol-gel method: [28-30]

Sol-gel technique is generally employed to prepare finely divided powders that are useful particularly in the manufacture of coating compositions, intricately shaped and fine-grained ceramics, and cermets. In this technique, the organometallic compounds, namely alcoholate, carboxylate or chelate compounds of different metals are generally used as starting materials (as an example acetylacetonates of Fe, Zn and Ni were used for $Ni_{0.7}Zn_{0.3}Fe_2O_4$). Acid or base catalyzed hydrolysis followed by condensation and polycondensation reaction of starting compounds in organic solvent produce gel. The gel, upon calcination at a temperature in the range of ~300-800^{0}C, in presence of oxygen produces desired ferrites. Use of expensive metal alkoxides or complex metal compounds is the main limitation of this method.

(c) Hydrothermal method: [31-33]

Hydrothermal method is widely used for the preparation of ferrite powders, which find applications where uniform dispersion of ferrite particles is required such as application in high-density recording medium. In this method, a mixture of aqueous solutions of metal ions, and/or aqueous slurry containing metal hydroxides is mixed with alkali solution and the mixture is autoclaved between 100- 400^{0}C (e.g., alkaline suspension of α-FeO(OH), $Ba(OH)_2$ $8H_2O$ when autoclaved produced Ba- hexaferrite). This is a comparatively low temperature technique, which provides a uniform mixture of the metallic components to form ferrites with required chemical composition. Hydrothermal process is found to be superior to the conventional high temperature process because it avoids the sintering or agglomeration of ferrite grains. Although hydrothermal technique is very versatile, one of the main drawbacks of the conventional hydrothermal method is the slow reaction kinetics at any given temperature.

(d) Precursor based method: [34-36]

This method involves the preparation of a precursor compound which are complex combinations of cation and a chelating agent. The pyrolysis of the precursors gives rise to nanoscale particles of the desired mixed oxide system. Various organo-metallic complexes, metal hydroxides/oxalates/citrates/hydrazine carboxylates etc. are some of the commonly used precursor compounds. Citrate route is very commonly used for preparation of NiZn Ferrite powders [18].

(e) Combustion method: [37, 38]

This method is also known as the "self-sustaining/propagating high temperature synthesis process." This process is rapid and may approach direct conversion from the molecular mixture of the precursor solution to the final oxide product. The method utilizes a precursor that ignites at a low temperature and leads to a gas producing exothermic reactions that are self-propagating and yield voluminous fine oxides in a few minutes. The disadvantage is that sometimes the exothermic reaction might be difficult to control especially when being performed on a large-scale.

(f) Reverse micelle method: [39, 40]

The reverse micelle method leads to the controlled synthesis of ceramic materials and metals having a tailored shape, size and composition. The structure of reverse micelles is characterized by a polar core formed by hydrophilic heads of the surfactant and a non-polar shell formed by hydrophobic tails of the surfactant. The droplet size of reverse micelles can be readily modulated in the nanometer range particularly by controlling the water: surfactant molar ratios. The method can be used to synthesize monodispersed nanoparticles with various morphologies. However, this method requires a large amount of organic solvent and also the yield of production is low.

Although conventional wet chemical methods are capable of producing nano-structured iron oxide and ferrite powders but they are also associated with some inherent limitations such as (i) use of expensive metal alkoxides or complex metal compounds, (ii) use of delicate reagents which are difficult to handle, (iii) frequent use of strong acids/bases and organic solvents, (iv) formation of heterogeneous/undesirable crystalline phases (v) requirement of elaborate experimental setup and time consuming processes.

A synthesis method becomes highly attractive if it is versatile and has the capability to produce single phase, nanopowders at a low temperature in a cost-effective and time-efficient manner without using any elaborate experimental set up. The importance of single phase material should not be undermined as the electrical and magnetic properties of materials are highly sensitive to the purity of phase of the synthesized material. Therefore, development of a technically simple but cost-effective synthetic methodology for preparation of pure and homogeneous ferrite nanopowders is a major challenge till now.

In view of above, we have recently developed three simple, aqueous solution based chemical methodologies to synthesize single phase, iron oxide and ferrite nanoparticles [41-47]. Advantages offered by these developed aqueous solution based methods are (i) metal alkoxides or complex metal compounds, which are expensive, difficult to handle, synthesize, and sometimes toxic have not been used in the developed methods.(ii) strong acid, base or organic solvents have not been used in any of the methods.(iii) use of simple and cheap metal nitrates as starting materials and water as solvent helps in reducing the processing cost as compared to other reported wet chemical methods. (iv) No elaborate experimental setup is required for the synthesis of nanopowders by these methods (v) Unlike other reported methods, Zn loss has not been observed during high temperature sintering of pellets, prepared by the synthesized nanopowders. This ensures the maintenance of stoichiometry of the final product, which govern the electrical and magnetic properties of the materials.

1.4. Applications of Ferrites

Ferrites possess a plethora of applications depending on their composition and property [1, 14, 48-50]. Some of the important applications, from technological point of view, are summarized in Table 1.

Table 1. Different compositions of ferrites along with their properties and applications

Sl No.	Ferrite composition	Property	Application
1.	Mn-Zn ferrites, Ni-Zn ferrites	High resistivity and saturation magnetization	Ferrite cores
2.	Ni-Cu-Zn ferrite and Ba-Co-Zn ferrite	Low losses at high frequencies and high permeability	Multilayer Ferrite Chip Inductors
3.	Yttrium iron garnet (YIG), Hexagonal ferrites $MFe_{12}O_{19}$ (M = Ba, Sr), Ni-Zn ferrites, Mn-Mg-Zn, Cu-Zn and Li ferrites	Low losses at high frequencies and high permeability	Microwave Ferrites for applications such as non-reciprocal circuit elements, telecommunication circuitry uses, e.g. as channel filters and for switched-mode power supplies in power applications for electronic control and computer uses. Ferrite Wave Absorbers
4.	Ba hexaferrite	High magnetization and high coercivity	high-density perpendicular magnetic media
5.	Hexagonal ferrites $MFe_{12}O_{19}$ (M = Ba, Sr)	High magnetization and high coercivity	Permanent Magnets
6.	XFe_2O_4 where X is either Mg, Zn and Sn.	Optical Property	Yellow pigment capable of withstanding elevated temperatures such as 200^0C.
7.	$(M(II)_xFe(II)_{1-x}Fe(III)_2O_4$ and $M(II)_x\ Fe(III)_{3-x}O_{4.5-0.5x}$ where M is Mg, Mn, Co, Ni, Cu and/or Zn and $0<x<1$.	High saturation magnetization, high permeability	Magnetic heads, magnetic liquids and inks and for the preparation of plastoferrites.
8.	$Zn_xFe_{1-x}O.Fe_2O_3$ where x is from 0.05 to 0.62	High saturation magnetization	Magnetic cards, magnetic coating.
9.	Magnetite particles	Optical property	Magnetic pigment for paints for vehicles, magnetic toner material for electrostatic reproduction.

In addition to the traditional applications listed in Table 1, magnetic nanoparticles of ferrites are finding new applications in the bio-medical field [7, 8]. Some of them are described below:

1. Contrast agents: Superparamagnetic nanoparticles are potential candidates for the development of new smart contrast agents for Magnetic Resonance Imaging (MRI) allowing for an early detection of several diseases. These are smart as they are designed to sense and detect molecular interactions in biological media.
2. Bioseperation: The use of magnetic nanoparticles for in vitro protein or cell separation is simple and can be performed without using liquid chromatography systems.
3. Targetted drug delivery: Nanoparticles with diameter between 10-100nm are optimal for drug delivery. Magnetic drug delivery is especially promising for cancer treatment as it avoids the side-effects of conventional chemotherapy.
4. Hyperthermia: This medical treatment is based on the fact that tumor cells are more sensitive to temperature increase than healthy ones. Superparamagnetic nanoparticles can absorb energy of an oscillating magnetic field and convert it into heat. This property is used to destroy unhealthy cells in the body.

Magnetic nanocomposites [9-12], is another area where ferrite nanoparticles are embedded in an immiscible nonmagnetic matrix. They are of current interest as these materials have tremendous potential to exhibit certain remarkable properties. For example, they are capable of making the ferrite devices light weight, less brittle and more durable. It is also possible to tailor their physical properties (magnetic, electrical, optical, mechanical, etc.) by controlling the particle size of the magnetic nanoparticles and their weight ratio with respect to the non-magnetic matrix. Moreover, these materials are expected to demonstrate multifunctionality in terms of performance and applications based on the properties of the individual constituents [9-12].

1.5. Future Scope of Work

Though, ferrites have been in use for the past hundred years, the discovery of nanoferrites has widened the horizon of their applications in diverse fields.

(i) Miniaturization of electronic components, potential bio-medical applications and the advantages offered by magnetic nanocomposites, has led to an increased demand on the synthesis and characterization of nanostructured ferrites. Hence, the synthesis of magnetic nanoparticles, with different compositions and tailor made particle sizes, has made considerable progress over the past decade. Although there is a spectrum of applications for these nanomaterials, it is not easy to produce ferrites with good homogeneity and desirable properties. Ferrites are known to be highly structure-sensitive materials and their properties greatly depend on the synthesis routes. Therefore, large scale and cost effective synthesis of high quality magnetic nanoparticles, having uniform particle sizes and desirable properties will still remain a challenge in the coming years.

(ii) Another critical issue to be addressed is the inherent tendency of nanoparticles to agglomerate over long periods of time. Therefore, it is crucial to develop optimized coatings to improve their stability. For example: silica doped Ni-Zn ferrite nanoparticles have been prepared by Stoia et al [9].

(iii) Last but not the least, safety and long term bio-compatibilty of the synthesized nanoparticles, should be carried out before utilizing them for bio-medical applications.

Acknowledgments

Dr. N. N. Ghosh gratefully acknowledges financial support from DRDO (Project No. ERIP/ER/0904500/M/01/1204).

References

[1] Smit, J; Wijn, HP. J. Ferrites: *Physical properties of ferromagnetic oxides in relation to their technical applications;* Cleaver-Hume Press Ltd.: London, KY, 1959.
[2] Kingery, WD; Bowen, HK; Uhlmann, DR. Introduction to Ceramics; John Wiley and Sons (Asia) Pte. Ltd: *Singapore*, RS, 2004.
[3] Faraji, M; Yamini Y; Rezaee M. *J. Iran. Chem. Soc*., 2010, 7, 1-37.
[4] Koch, CC. *J. Mater. Sci*., 2007, 42, 1403–1414.
[5] Gryaznov, VG; Trusov, LI. *Prog. Mater. Sci*., 1993, 37, 289-401.
[6] Pathak, A; Pramanik, P. *Proc. Ind. Natl. Sci. Acad*. 2001, 67, 47-70.
[7] Laurent, S; Forge, D; Port, M; Roch, A; Robic, C; Elst, LV; Muller, RN. *Chem. Rev*. 2008, 108, 2064-2110.
[8] Lee, SW; Kim, CS. *J. Magn. Magn. Mater*., 2006, 304, e418-e420.
[9] Stoia, M; Caizer, C; Stefanescu, M; Barvinschi, P; Julean, I. *J. Therm. Anal. Calorim*. 2007, 88, 193-200.
[10] Sarangi, PP; Naik, B; Vadera, SR; Patra, MK; Prakash, C; Ghosh, NN. Mater. Technol. *Adv. Perform. Mater*., 2009, DOI 10.1179/106678509X12519671504700 (In Press).
[11] Ghosh, NN; Sarangi, PP; Naik BD. *World J. Eng*., 2008, 5, 622-623.
[12] Naik, B; Sarangi, PP; Hui D; Ghosh, NN. *Plastic Masses,* (Russian Version). 2008, 12, 13- 17.
[13] Sugimoto, M. *J. Am. Ceram. Soc*., 1999, 82, 269-280.
[14] Pant, P; Bhuvaneswari, S; Ghosh, NN. *Recent Patents Nanotechnol*., 2008, 2, 8-18.
[15] Kaiser, M. *J. Alloys Compd*., 2009, 468, 15-21.
[16] Bhame, SD; Joy, PA. *J. Am. Ceram. Soc*. 2008, 91,1976–1980.
[17] Sakellari, D; Tsakaloudi, V; Polychroniadis, EK; Zaspalisw, V. *J. Am. Ceram. Soc*. 2008, 91, 366-371.
[18] Verma, A; Chatterjee, R. *J. Magn. Magn. Mater*., 2006, 306, 313-20.
[19] Maruyama, T; Kanagawa, T. *J. Electrochem. Soc*., 1996, 143, 1675-1677.
[20] Itoh, H; Uemura, T; Yamaguchi, H; Naka, S. *J. Mater. Sci*., 1989, 24, 3549-3552.
[21] Fitzgeralda, AG; Engin, R. *Thin Solid Films*., 1974, 20, 317-327.

[22] Alexandrescu, R; Morjan, I; Voicu, I; Dumitrache, F; Albu, L; Soare, I; Prodan, G. *Appl. Surf. Sci.*, 2005, 248, 138-146.
[23] Bomati-Miguel, O; Mazeina, L; Navrotsky, A; Veintemillas-Verdaguer, S. Chem. *Mater.*, 2008, 20, 591-598.
[24] Sorescu, M; Diamandescu, L; Swaminathan, R; McHenry, ME; Feder, M. *J. Appl. Phys.,* 2005, 97, 10G105-10G107.
[25] Zhang, H; Wang, WW; Li, H; Meng, S; Li, D. *Mater. Lett.*, 2008, 62, 1230-1233.
[26] Raming, TP; Winnubst, AJA; van Kats, CM; Philipse, AP. *J. Colloid Interface Sci.*, 2002, 249, 346-350.
[27] Zhang, HE; Zhang, BF; Wang, GF; Dong, XH; Gao, Y. *J. Magn. Magn. Mater.*, 2007, 312, 126-130.
[28] Suri, K; Annapoorni, S; Tandon, RP. *Bull. Mater. Sci.*, 2001, 24, 563-567.
[29] Roy, PK; Bera J. *J. Mater. Process. Technol.*, 2008, 1 9 7, 279–283
[30] Cao, M; Liu, T; Gao, S; Sun, G; Wu, X; Hu, C; Wang, ZL. *Angew. Chem. Int. Edit.*, 2005, 44, 2-6.
[31] Wang, HW; Kung, SC. *J. Magn. Magn. Mater.*, 2004, 270, 230–236.
[32] Li, X; Li, Q; Xia, Z; Yan, W. *J. Alloys Compd.*, 2008, 458, 558–563.
[33] Satyanarayana, L; Reddy, KM; Manorama, SV. *Mater. Chem. Phys*.**,** 2003, 82, 21-26.
[34] Chikate, RC; Jun, KW; Rode, CV. *Polyhedron.*, 2008, 27, 933-938.
[35] Ghodake, SA; Ghodake, UR; Sawant, SR; Suryavanshi, SS; Bakare, PP. *J. Magn. Magn. Mater.*, 2006, 305, 110-119.
[36] Verma, A; Thakur, OP; Prakash, C; Goel, TC; Mendiratta, RG. *Mater. Sci. Eng. B***.** 2005, 116, 1-6.
[37] Costa, ACFM; Leite, AMD; Ferreira, HS; Kiminami, RHGA; Cavac, S; Gama, L. *J. Eur. Ceram. Soc.*, 2008, 28, 2033-2037.
[38] Mangalaraja, RV; Lee, ST; Ananthakumar, S; Manohar, P; Camurri, CP. *Mater. Sci. Eng. A*.**,** 2008, 476, 234-239.
[39] Morrison, SA; Cahill, CL; Carpenter, EE; Calvin, S; Swaminathan, R; McHenry, ME; Harris, VG. *J. Appl. Phys*., 2004, 95, 6392-6395.
[40] Gubbala, S; Nathani, H; Koizol, K; Mishra, RDK. *Physica B*, 2004, 348, 317-328.
[41] Sarangi, PP; Naik, B; Ghosh, NN. *Powder Technol.*, 2008, 192, 245–249.
[42] Ghosh, NN., Sarangi, PP. In Nanomaterials: Properties, Preparation and Processes; Editor, Cabral, V; Silva, R; Nova Publishers: Hauppauge, NY, 2010. (In press).
[43] Pant, P; Naik, BD; Ghosh, NN. *Mater. Technol.: Adv. Perform. Mater.*, 2009, 24, 213-216.
[44] Sarangi, PP; Naik, B; Ghosh, NN. *J. Amer. Ceram. Soc.*, 2008, 91, 4145–4147.
[45] Sarangi, PP; Vadera, SR; Patra, MK; Prakash, C; Ghosh, NN. *J. Amer. Ceram. Soc.* , 2009, 92, 2425-2428.
[46] Ghosh, NN; Sarangi, PP. In Combustion Synthesis: Novel Routes to Novel Materials; Editor, Lacner, M; Bentham *Science Publishers Ltd.: Bussum*, NL, 2010. (Accepted).
[47] Sarangi, PP; Naik, BD; Vadera, SR; Patra, MK; Prakash, C; Ghosh, NN. *Mater. Technol.: Adv. Perform. Mater.*, 2009, 24, 97-99.
[48] Snelling, EC. Soft Ferrites-Properties and Applications; ILIFFE Books Ltd.: London, KY, 1969.

[49] O¨ zgu¨r, U; Alivov, Y; Morkoc, H. Microwave ferrites, part 1: fundamental properties. *J. Mater. Sci. Mater. Electron.*, 2009, 20, 789-834.

[50] O¨ zgu¨r, U; Alivov, Y; Morkoc, H. Microwave ferrites, part 2: passive components and electrical tuning. *J. Mater. Sci.: Mater. Electron.*, 2009, 20, 911-952.

In: Powder Engineering, Technology and Applications
Editor: Jason M. Barker, pp.207-215
ISBN: 978-1-61761-216-9

Chapter 8

DEVELOPMENT OF A SIMPLE BUT NOVEL CHEMICAL METHOD FOR PREPARATION OF AG NANOPARTICLE INCORPORATED MESOPOROUS SILICATE MCM-41

Narendra Nath Ghosh and Bhanudas Naik
Birla Institute of Technology and Science, Pilani, Zuarinagar, Goa, India

ABSTRACT

Preparation of homogeneous dispersion of nanostructured catalytically active sites (metal species) within the porous, high surface area supports (such as SiO_2 or Al_2O_3) has been a goal of catalyst scientists for many years. Mesoporous materials (e.g. SBA-15, MCM-41 etc) have exhibited their potential as catalysts, catalyst support, adsorbents for large organic molecules, guest-host chemical supports etc. Silver nanoparticles have myriad applications in diverse fields due to their catalytical, antibacterial, and biosensing properties. Several methodologies have been reported for preparation of nanosized Ag particles, such as colloidochemical, sonoelectrochemical, thermal decomposition, UV irradiation, γ-irradiation, protection-reduction, cation exchange, use of biosurfactants as reducing agent etc. However, most of these synthesis methods are complex in nature. Thus, development of a simple synthetic methodology for preparation of homogeneously dispersed Ag nanoparticles on mesoporous silica matrices is still a major challenge.

A novel chemical synthetic technique has been developed for the preparation of Ag nanoparticle loaded mesoporous silica, such as MCM-41. MCM-41 silicate material was synthesized by using a liquid crystal templating method. Here, in this synthesis process tetraethyl orthosilicate (TEOS) was used as an inorganic silica source and cetyltrimethyl ammonium bromide (CTAB) as pore structure directing template. Incorporation of Ag nanoparticle within hexagonal mesosructure was achieved through wet impregnation of $AgNO_3$, followed by in situ chemical reduction of Ag^+ ions by trisodium citrate. The synthesized materials were characterized by using room temperature powder XRD, BET surface area and porosity analysis using N_2 sorption- desorption study, high resolution TEM. X-ray analysis and TEM micrographs confirmed the formation of Ag nanoparticles in the MCM-41 matrix. BET surface area of the synthesized materials was in the range of 1000 m^2/ gm with narrow pore size distribution. A decrease in surface area and pore volume was observed due to impregnation for the Ag nanoparticle within porous silicate structure of MCM-41.

1. INTRODUCTION

The synthesis of mesoporous materials with liquid crystal templating method was initiated by the pioneering work of the scientist of Mobil Oil Corporation. They discovered the novel family of molecular sieves called M41S [1-3]. These mesoporous silicate materials possess ordered pores and narrow pore sizes between 15-100 Å [4-5]. The high surface areas (>1000 m^2/g) and tailoring of pore sizes are among the many desirable properties that have made such materials the focus of great interest [1-6]. Mesoporous materials (e.g. MCM-41) have exhibited their potential as catalysts, catalyst support, adsorbents for large organic molecules, guest-host chemical supports etc. MCM-41 is the most important and principle M41S material, which has hexagonally packed open cylindrical structure [4-6]. The extremely high specific surface areas are conducive to high catalytic activity and adsorption of ions [4]. MCM-41 has been the main focus of catalytic studies compared to the other M41S mesoporous structures because of its ease of preparation and large pore sizes [4]. However, other M41S structures are either thermally unstable or difficult to obtain [7]. The unique physical properties of MCM-41 have made these materials highly desirable for metal nanoparticle incorporation for specific applications [4]. Silver nanoparticles have myriad applications in diverse fields due to their catalytical, antibacterial, and biosensing properties [8-15]. As catalyst, use of silver nanoparticles was investigated in a variety of reactions, such as NO_X reduction, CO and hydrocarbon oxidation, electroxidation of polypyrrole, photocatalysis reactions etc [16-18]. Claus et al. synthesized Ag/SiO_2 catalysts, which exhibited a superior performance for the selective hydrogenation of α, β-unsaturated aldehydes to allylic alcohols [19]. Shin et al. synthesized Ag-deposited Fe_2O_3 nanoparticles and their application for the catalytic reduction of 4-nitrophenol to 4-aminophenol by $NaBH_4$ was studied [20].

Impregnation technique for incorporation of metal nano particles, which serve as active catalyst site, on/ within the porous support (silica, alumina etc) is a common strategy for preparation of heterogeneous catalysts [21-24]. However, most of the time porous structure of support gets destroyed during impregnation and nanostructured metal particles form agglomeration. Several methodologies have been reported for preparation of nanosized Ag particles, such as colloidochemical, sonoelectrochemical, thermal decomposition and UV irradiation, γ-irradiation, protection-reduction, cation exchange, use of biosurfactants as reducing agent etc [25-27]. However, most of these synthesis methods are complex in nature. In this paper, we have reported a simple but cost effective aqueous solution based chemical technique for synthesis of Ag nanoparticle containing MCM-41, where Ag particles are deposited on mesoporous silica matrix. Impregnation of Ag nanoparticles within the porous structure of MCM-41 was carried out by simultaneous wet impregnation and chemical reduction of silver nitrate with tri-sodium citrate within the porous structure.

2. Experimental

2.1. Synthesis of Materials

In a typical synthesis of MCM-41, 1.922 g of CTAB (Cetyl triethyl ammonium bromide) Molychem, India) was dissolved in 89.2 g of water with 4.8 g of ~30 % ammonia solution (Merck, India). The mixture was stirred till the formation of clear solution. 6.9 g of tetraethyl orthosilicate (98%, Merck, Germany) was added to this solution and stirred for 4 h. The resulting mixture was then aged for 24 h for condensation at 30^0C. The precursor formed was then filtered and washed with 150 ml of distilled water, dried in the oven at 100^0C for 24 h. It was then calcined at 550^0C for 3 h in air atmosphere to obtain MCM-41. Ag nanoparticles were impregnated on the porous matrix of MCM-41 by using the reduction of $AgNO_3$ with trisodium citrate. To prepare MCM-41 with 5wt.% Ag nanoparticles, 23.4 ml aqueous solution of 0.02M $AgNO_3$ (99.9% Merck, India) was first heated at 80^0C for 20 min in a round bottom flask. To this warm solution 0.95 g of MCM-41 was dispersed, followed by drop wise addition of 23.4 ml 0.2M trisodium citrate dehydrate (99%, Merck, India) aqueous solution. After 1 h, the reaction mixture was cooled to room temperature. The silver impregnated MCM-41 was filtered and washed with 100 ml of warm distilled water at least five times and dried at 60^0C for 60 h in a vacuum oven. The grayish colored product obtained is now onwards referred to as "Ag-MCM-41".

2.2. Sample Characterization

Thermogravimetric analysis (TGA) and differential thermal analysis (DTA) analysis were carried out for the precursor and Ag-MCM-41 by using Shimadzu DTG-60 (Japan) in air atmosphere at a heating rate of 10^0C/ min. Room temperature powder X-ray diffraction patterns were obtained with a Rigaku powder X-Ray diffractometer (Mini Flex II) with Cu Kα radiation at a scanning speed of 2°/min respectively. Nitrogen adsorption-desorption isotherms were obtained with a Micromeritics Tristar 3000 surface area and porosity analyzer. Samples were prepared for surface area and porosity analysis by drying at 180^0C for 12 h with constant nitrogen gas flow in a degassing unit. Surface areas were determined according to Brunauer-Emmett-Teller (BET) method. Pore size distributions were determined by the BJH method applied to desorption branch of the N_2 isotherm. The High Resolution Transmission Electron Microscope images of samples were obtained using FEI HRTEM (Tecnai G2 30S-Twin) with an accelerating voltage of 300 kV.

3. Result and Discussion

3.1. TGA-DTA Analysis

To investigate the oxidative decomposition behavior MCM-41 precursor, TGA and DTA analysis was employed in the air atmosphere at a heating rate of 10^0C/min (Figure 1a,b). It was observed that, ~54% weight loss occurred in the TGA thermogram in the temperature range of 40-350^{0}C. An Exothermic peak at 295^{0}C was appeared in DTA. This may be due decomposition of organic surfactant part of hybrid organic-inorganic structure of MCM-41 gel. Heating the sample beyond 350^{0}C resulted in neither the weight loss in TGA nor formation of any new peak in DTA. Therefore it was confirmed that, full decomposition of precursor gel into silicate (MCM-41) occurred at 350^{0}C.

Figure 1c shows the TGA of Ag-MCM-41 in the temperature range of 30-550^{0}C. Initial weight loss of ~ 8 wt%, in the region of 40 to 200^{0}C might be due to the removal of water, which was entrapped during impregnation of Ag nanoparticles in the pores. Heating the sample beyond 200^{0}C resulted in no further weight loss, which confirmed the stability and purity of the synthesized Ag-MCM-41.

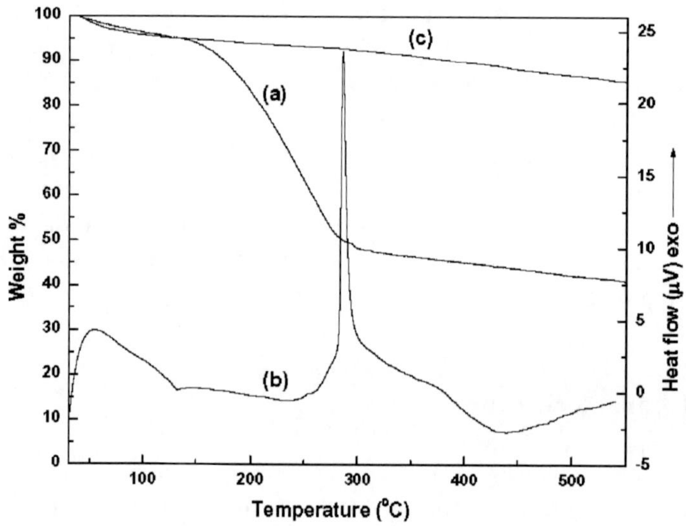

Figure 1.Thermogravimetric (a) and Differential thermal analysis curve (b) of MCM-41 precursor gel and Thermogravimetric curve of (c) Ag-MCM-41

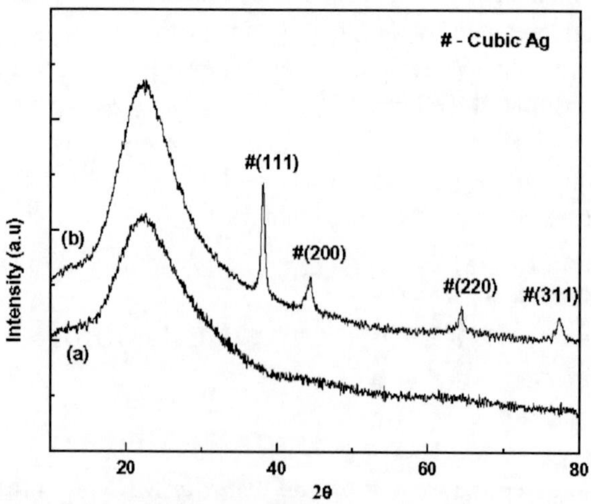

Figure 2. Room temperature XRD patterns of (a) synthesized MCM-41 and (b) Ag- MCM-41

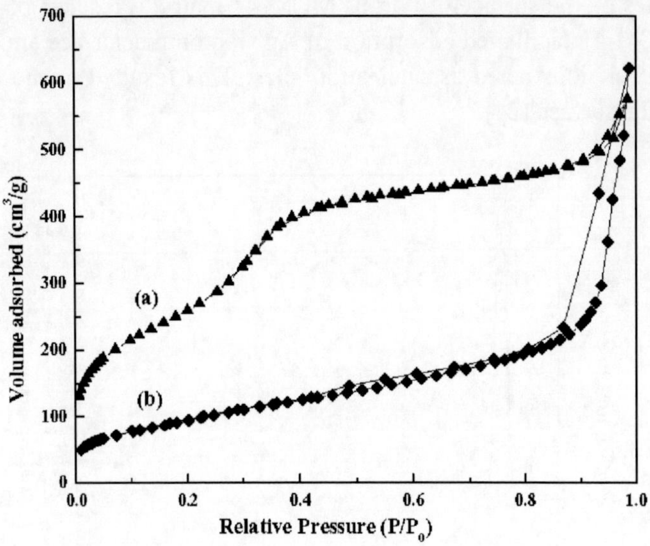

Figure 3. N_2 adsorption- desorption isotherms of (a) MCM-41, (b) Ag-MCM-41

3.3. N_2 Adsorption and Desorption Analysis

N_2 adsorption-desorption analysis of MCM-41 (Figure 3a) showed typical type IV physisorption isotherm with no hysteresis loop, where adsorption and desorption loops overlap with each other even in the pore filling region [28]. This is due the fact that equilibrium exists in the adsorption-desorption steps in the small regular pore of MCM-41[29]. Steep adsorption/desorption steps in the isotherm indicate that MCM-41 possess regular cylindrical mesopores with narrow pore size distribution. At a relative pressure ~ 0.37, sudden steep increase of adsorption volume was observed. This was due to the capillary condensation of N_2 inside the pores of MCM-41 [29]. Pore filling pressure range of MCM-41 was very small between $P/P_0 \approx$ 0.34-0.40, which indicated that the pores associated in the adsorption process are of equal size [29]. Pore size distribution of MCM-41(Figure 4a) showed a narrow distribution centered at 2.5 nm with BJH desorption volume of 0.595cm^3/g. In the isotherm of Ag-MCM-41(Figure 3b) a hysteresis loop in the higher pressure range was observed, isotherm upto $P/P0 \approx 0.8$ was observed to be similar to MCM-41. This indicated the presence of considerable number of small regular size pores along with some bigger pores. Presence of hysteresis loop may be attributed to the presence of few bigger nanoparticles of Ag, which lead to partial destruction few pores [24].

In the pore size distribution plot Ag-MCM-41 (Figure 4b) the decrease of intensity of peak height (centered at ~2 nm) was observed compared to MCM-41.This might be due to the partial filling of pores of MCM-41 with Ag nanoparticle. The pore volume of Ag-MCM-41 obtained was 0.205 cm^3/g which is smaller then pure MCM-41. The decreased pore volume of the Ag loaded MCM-41 thus confirmed the formation of Ag nanoparticles within the porous matrix of MCM-41. Multiple point BET surface area of MCM-41 and Ag-MCM-41 was found to be 1008.5 and 350.3 m^2/g respectively. These significant decrease in surface area and total pore volume of Ag-MCM-41 were due to partial filling of pores by deposition of Ag nanoparticles on the surface of porous MCM-41 matrix. Presence of silanol groups on the surface of MCM-41 facilitated adsorption of Ag^+ ion on the surface and during reduction of $AgNO_3$, these Ag^+ ions acted as nucleation sites. This resulted in the deposition of Ag nanoparticles on the surface [24].

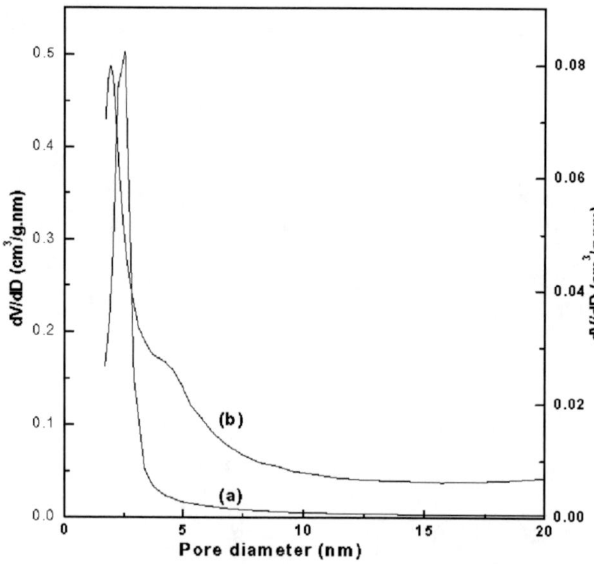

Figure 4. Pore size distribution of synthesized (a) MCM-41 and (b) Ag-MCM-41

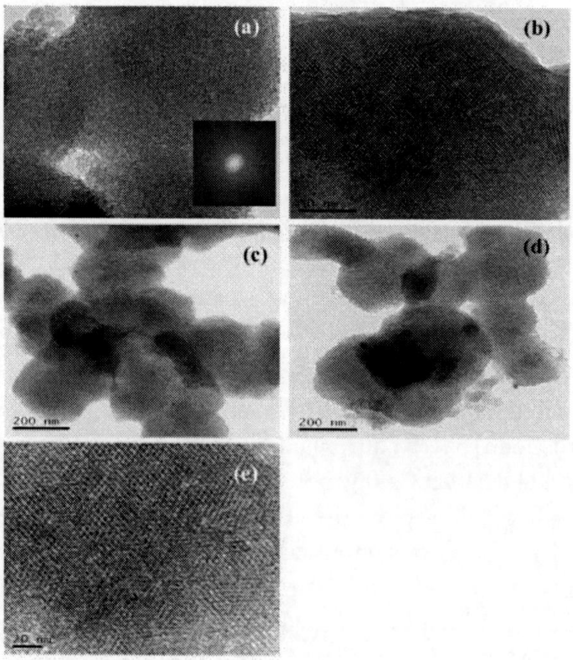

Figure 5. HRTEM micrograph of synthesized (a, b) MCM-41, (c, d) Ag-MCM-41 showing the deposition of Ag nanoparticles on silica surface , (e) presence of mesostructured pores (like MCM-41) in the Ag-MCM-41 sample even after deposition of Ag

3.4. HRTEM Analysis

HRTEM micrographs of MCM-41 (Figure 5a,b) showed that, synthesized MCM-41, possessed uniform hexagonal mesostructured pores, with long range ordering and long open channels. Average diameter of the pores was ~3 nm. In TEM images of Ag-MCM-41 (Figure 5c,d) spherical Ag nanoparticles, were seen to be dispersed in to the surface of MCM-41. TEM images clearly showed the infiltrated Ag nanoparticles in the porous matrix of MCM-41. In Ag-MCM-41, presence of some open pores and long channels (like MCM-41) was also observed (Figure 5e), even though the loading of Ag was as high as 5 wt.%.

4. Conclusion

A simple chemical method has been developed for preparation of nanosized Ag-particle incorporated MCM-41. Important features of this method are as follows: (i) This process is very simple and auto clave, which is conventionally used for mesoporous materials preparation, was not used (ii) Ag-MCM-41was synthesized in mild condition, which helped to maintain the porous structure of MCM-41 even after loading of Ag nanoparticles, (iii) Ag nanoparticles were synthesized by using simple reduction of $AgNO_3$ in aqueous medium. No capping agent was used during synthesis but agglomeration of Ag nanoparticles was not observed in HRTEM micrographs. (iv) reaction was carried in aqueous medium. No organic solvent was used which made the process cost-effective and simple, (v) even after high loading of Ag (5 wt.%), certain amount of mesoporous structure with long range ordering of pores of MCM-41, was preserved in Ag- MCM-41. However, BET surface area and total pore volume of MCM-41 decreased after deposition of Ag nanoparticles. This was due to the deposition of Ag particles on silica surface. Even though, N_2 adsorption-desorption investigation showed that Ag-MCM-41 possess regular cylindrical mesopores with narrow pore size distribution. The surface area of the synthesized Ag-MCM-41 was considerably high (350.3 m^2/g) for catalysis applications. In all, the novelty of this method lies in its simplicity, cost effectiveness and the synthesized material, Ag-MCM-41, has capability to be used as efficient heterogeneous catalyst for various reactions.

Acknowledgments

Authors gratefully acknowledge financial support from the Department of Science and Technology, New Delhi India (Project no: SR/S1/IC-39/2006). We also express our thanks to Dr. V.S. Prasad, NIIST, Thiruvananthapuram India for recording TEM micrograph of the samples.

References

[1] Kresge, CT; Leonowicz, ME; Roth, WJ; Vartuli, JC; Beck, JS. *Nature*, 1992, 359, 710-712.

[2] Beck, JS; Vartuli, JC; Roth, WJ; Leonowicz, ME; Kresge, CT; Schmitt, KD; Chu, CTW; Olson, DH; Sheppard, EW. *J. Am. Chem. Soc.*, 1992, 114, 10834-10843.

[3] Kresge, CT; Leonowicz, ME; Roth, WJ; Vartuli, JC. inventor; Mobil Oil Corp., assignee. *Synthetic mesoporous crystalline material* United States Patent US 5098684, 1992, March 24.

[4] Ying, JY; Mehnert, CP; Wong, MS. *Angew. Chem. Int. Ed.*, 1999, 38, 56-77.

[5] Zhao, XS; Lu, GQ; Millar, G. *J. Ind. Eng. Chem. Res.*, 1996, 35, 2075-2090.

[6] Naik, B; Ghosh, NN. *Recent Pat. Nanotechnol.*, 2009, 3, 213-224.

[7] Vartuli, JC; Schmitt, KD; Kresge, CT; Roth, WJ; Leonowicz, ME; McCullen, SB; Hellring, SD; Beck, JS; Schlenker, JL; Olson, DH; Sheppard, EW. *Chem. Mater.*, 1994, 6, 2317- 2326.

[8] Akhavan, O; Azimirad, R; Moshfegh, AZ; *J. Phys. D Appl. Phys.*, 2008, 41, 195305.

[9] Breen, JP; Burch, R; Hardacre, C; Hill, CJ. *J. Phys. Chem.*, B 2005, 109, 4805-4807.

[10] Liu, JH; Wang, AQ; Chi, YS; Lin, HP; Mou, CY. *J. Phys. Chem.*, B 2005, 109, 40-43.

[11] Shimizu, K; Satsuma, A. *Phys. Chem. Chem. Phys.*, 2006, 8, 2677-2695.

[12] Shashikala, V; Sivakumar, V; Padmasri, AH; David, BR; Venkata, MS; Sarma, PN; Rama Rao, KS. *J. Mol. Catal. A-Chem.*, 2007, 268, 95-100.

[13] Sivakumar, V; Nagaraja, BM; Shashikala, V; Padmasri, AH; Madhavendra, SS; Raju, BD; Rama Rao, KS. *J. Mol. Catal. A: Chem.*, 2004, 223, 313-319.

[14] Ren, X; Meng, X; Tang, F. *Sensor. Actuat. B-Chem.*, 2005,110, 358-363.

[15] Dorjnamjin, D; Ariunaa, M; Shim, YK. *Int. J. Mol. Sci.*, 2008, 9, 807-820.

[16] Bera, P; Patil, KC; Hegde, MS. *Phys. Chem. Chem. Phys.*, 2000, 2, 3715-3719.

[17] Lee, H; Liu, Y. *Polymer*, 2005, 46, 10727-10732.

[18] Rupa, AV; Manikandan, D; Divakar, D; Sivakumar, T. *J. Hazard. Mater.*, 2007, 147, 906-913.

[19] Claus, P; Crozier, PA; Druska, P. Fresen. *J. Anal. Chem.*, 1998, 361, 677-679.

[20] Shin, KS; Choi, J; Park, CS; Jang, HJ; Kim, K. *Catal. Lett.* 2009, 133, 1-7.

[21] Kim, P; Kim, Y; Kim, H; Song, IK; Yi, J. *J. Mol. Catal. A: Chem.*, 2004, 219, 87-95.

[22] Hornebecq, V; Antonietti, M; Cardinal, T; Treguer-Delapierre, M. *Chem. Mater.*, 2003, 15, 1993-1999.

[23] Liu, Q; Wang, A; Wang, X; Gao, P; Wang, X; Zhang, T. *Microporous Mesoporous Mater.* 2008, 111, 323-333.

[24] Naik, B; Prasad, VS; Ghosh, NN. *Powder Technol.*, 2010, 199, 197-201.

[25] Quanfa, Z; Jianchun, B; Zheng, X. *Sci. China. Ser.*, B 2002, 45, 416-420.

[26] Hornebecq, V; Antonietti, M; Cardinal, T; Treguer-Delapierre, M. *Chem. Mater.*, 2003, 15, 1993-1999.

[27] Kolesnik, IV; Eliseev, AA; Garshev, AV; Lukashin, AV; Tret´yakov, YD. Russ. Chem. Bull. 2004, 53, 2496-2498.

[28] Sing, KSW; Everett, DH; Haul, RAW; Moscou, L; Pierotti, RA; Rouquerol, J; Siemieniewska, T. *Pure Appl. Chem*., 1985, 57, 603-619.

[29] Northcott, K; Oshima, S; Perera, J; Komatsu, Y; Stevens, G. *Advanced Powder Technol*., 2007, 18, 751-762.

In: Powder Engineering, Technology and Applications ISBN: 978-1-61761-216-9
Editor: Jason M. Barker, pp. 217-255

Chapter 9

STOCHASTIC MODELING TECHNIQUES IN POWDER TECHNOLOGY

Alex C. Hoffmann
Dept. of Physics and Technology,
The University of Bergen, Norway

Abstract

In stochastic modeling the focus is on one particle, namely on the time-evolution of the probability distribution of the particle's state (e.g. spatial position or physical or chemical state), given its present state. This in contrast to the standard modeling approach in process technology, which is based on formulating and solving differential balance equations. Stochastic modeling is useful throughout process technology but is especially relevant in powder technology.

Stochastic modeling is found in the powder technology literature in various forms and guises. This chapter gives an up-to-date discussion of stochastic modeling in powder technology, showing through some chosen examples how the different forms of stochastic modeling are really variations of the same basic method, and bringing the literature into one unified notation for clarity.

After a brief introduction to the foundations of stochastic modeling and its strengths and limitations when applied within powder technology, some recent literature will be reviewed and specific examples of the main techniques will be analyzed in detail. The relation of stochastic modeling to other techniques, such as 'discrete element modeling' and 'population balance modeling' will be discussed.

1 Foundations of Stochastic Modeling

Research into modeling of processes involving powders has intensified during the last few decades, as industry has increasingly been developing processes involving finely dispersed solids.

Most of these processes come under the heading of transport processes, namely transport of momentum, heat or mass, which may be exchanged between the particles themselves, the particles and the carrier fluid and the particles or carrier fluid with the solid surfaces of the processing vessel. Mass transfer between the particles and the carrier fluid may be associated with a chemical reaction at the surface of the particles, such as in heterogeneous catalysis or combustion.

Different modeling approaches are being used, the most popular of which are:

- Modeling by balance equations for momentum, energy or mass. These equations are differential in their final mathematical form, but are developed on basis of balances over volumes that are large compared with the individual particles while still small in comparison with the process as a whole. modeling of particle flows and (often dense) fluid/particle flows by constitutive equations fall in this category. A numerical modeling technique consistent with this approach is Eulerian-Eulerian simulation.

- Modeling by tracking individual particles. While this approach is not finding much application in theoretical modeling, it is widely used in Eulerian-Lagrangian simulations of fluid/particle flows and 'finite element' modeling of particle flows.

The former of these approaches is mainly useful on a large scale, but nevertheless demands intimate knowledge of the physics of particle-particle, particle-fluid and particle-wall interactions for formulation of realistic balance equations. The latter approach is, in numerical simulations, due to its computational intensity, mainly useful on a small scale. This technique also requires knowledge of the physics of the interactions, but the details of the physics can more easily and directly be implemented in the models.

This chapter will describe an alternative method of modeling, one which can be used for both theoretical and numerical modeling of processes, and which can be advantageous both in the model formulation and the model solution stages. This is the method of stochastic modeling.

In stochastic modeling the focus is on one material particle, which may be a molecule, a solid particle or a fluid element, and the time-evolution of the probability distribution of its 'state', which may be its spatial position or chemical or physical state.

The first section will introduce the basic tools and terminology for stochastic modeling.

2 Basics of Stochastic Modeling

A particle may at any given time, t, take on any state, x, in the 'state space', S, where S, for example, could be the three-dimensional state space constituted by the volume of a reactor. The focus here will be on processes that are discretized in both time and state space, so that any time interval can be denoted by an integer index, say $n \in I$, and the state also by an integer index, say h, i, j or $k \in I$.

It is further assumed that a set of subsequent observations of the state, x_{i_n}, of the particle can be expressed by a series of discrete random variables, X_n.

2.1 Markov Chains

A discrete stochastic process constitutes a 'Markov chain' if it satisfies the Markovian property that the probability of any given state at time n depends on the state at time $n-1$ only, and not on the previous states:

$$P(X_n = x_{i_n}|X_1 = x_{i_1}, \ldots, X_{n-1} = x_{i_{n-1}}) = P(X_n = x_{i_n}|X_{n-1} = x_{i_{n-1}})$$

where, e.g. $P(X_n = x_{i_n}|X_{n-1} = x_{i_{n-1}})$ denotes a 'conditional probability', namely the probability that the state is x_{i_n} in the n'th time interval under the condition that it was $x_{i_{n-1}}$ in the $(n-1)$'th time interval.

The distribution of a Markov chain is completely specified by two pieces of information, namely:

1. the distribution, $\mathbf{p}^{(0)}$, of X_0, i.e. the initial distribution over the states.
2. the transition probabilities $p_{ij} \equiv P(X_{n+1} = x_j|X_n = x_i)$.

where the probability distribution over the discrete state space has been denoted by the vector $\mathbf{p}$.

Markov chains possess some properties that are very useful when applying them to stochastic modeling of processes in powder technology. Denoting the probability of reaching state k in time-step n after having been in state i at time-step l by $p_{ij}^{(ln)}$, the Chapman-Kolmogorov equations for a Markov chain state that:

$$p_{ik}^{(ln)} = \sum_{j \in S} p_{ij}^{(lm)} p_{jk}^{(mn)} \quad \text{for } l \leq m \leq n.$$

Thus by using the Chapman-Kolmogorov equations all transition probabilities can be calculated from the one-step transition probabilities $p_{ij}^{(n,n+1)} = P(X_{n+1} = j|X_n = i)$.

The Markov chain $(X_n)_{n \geq 1}$ is 'time homogeneous' if the one-step transition probabilities $p_{ij}^{(n,n+1)}$ do not depend on n. In many, but not all, models in process technology this assumption can be made, and one can delete the index denoting time

The matrix $\mathbf{P} = p_{ij}$ is the 'transition matrix' of the Markov chain. Transition matrices are characterized by the properties that $p_{ij} \geq 0$ and $\sum_{j \in S} p_{ij} = 1$.

Some remarks on notation The notation used in this chapter is often used in mathematical texts. This does, however, involve a lot of p's, so it is useful to briefly review some aspects of the notation at this stage:

- Probabilities are denoted as normal: $P(\,\cdot\,)$, and conditional probabilities as $P(\,\cdot\,|\,\cdot\,)$.
- The vector of the probability distribution over the finite number of possible states is $\mathbf{p}$ with single-index components p_i.
- The matrix of transition probabilities is $\mathbf{P}$ with double-index components p_{ij}.

Two more remarks on notation are that indices are separated by comma's when necessary for clarity and that the time step is denoted by superscripts in parentheses to distinguish the time-index from the state-index (subscripts) on the one hand and from exponents on the other.

The m-step transition probabilities of a time-homogeneous Markov chain are defined by:

$$p_{ij}^{(m)} \equiv P(X_{n+m} = x_j|X_n = x_i),$$

with the right-hand-side independent of n. The m-step transition matrix is then $\mathbf{P}^{(m)} \equiv p_{ij}^{(m)}$.

The Chapman-Kolmogorov equations lead to:

$$p_{ik}^{(m)} = \sum_{j \in S} p_{ij}^{(m-1)} p_{jk},$$

or in matrix notation: $\mathbf{P}^{(m)} = \mathbf{P}^{(m-1)}\,\mathbf{P}$. Iterated application of this leads to:

$$\mathbf{P}^{(m)} = \mathbf{P}^{m}, \tag{1}$$

i.e. the m-step transition matrix equals the m-th power of the one-step transition matrix $\mathbf{P}$.

The probability distribution of X_n, i.e. the probability distribution over the states for the particle at time-step n, can then, for a time-homogeneous Markov chain, be found from the transition matrix and the initial distribution by matrix multiplication as follows.

If $p_i^{(n)}$ is the probability for the particle to be in state x_i at time-step n, the Chapman-Kolmogorov equations imply that:

$$\begin{aligned} p_j^{(n)}\,(= P(X_n = j)) &= \sum_{i=1}^{N} P(X_{n-1} = i)\, P(X_n = j | X_{n-1} = i) \\ &= \sum_{i=1}^{N} p_i^{(n-1)}\, p_{ij}, \end{aligned}$$

which in vector-matrix notation is:

$$\mathbf{p}^{(n)} = \mathbf{p}^{(n-1)}\,\mathbf{P}. \tag{2}$$

Iterating this equation leads to:

$$\mathbf{p}^{(n)} = \mathbf{p}^{(0)}\,\mathbf{P}^{n}. \tag{3}$$

It is thus possible to find the probability distribution at time n either directly by multiplying the initial probability distribution $\mathbf{p}^{(0)}$ from the right by $\mathbf{P}^n$ or recursively, by multiplying $\mathbf{p}^{(n-1)}$ by $\mathbf{P}$. Numerically, the computations involved in determining $\mathbf{p}^{(n)}$ can be carried out rather quickly using a compiled programming language such as Fortran, or a matrix-oriented package such as MATLAB.

If x_i and x_j are two states in S, then x_j is said to be 'accessible' from x_i, if there is an n such that:

$$p_{ij}^{(n)} > 0,$$

i.e. if it is possible to reach x_j from x_i in a finite number of steps. The symbol $i \rightarrow j$ will expresses that x_j is accessible from x_i.

The states x_i and x_j are said to 'communicate' with each other, denoted by $i \leftrightarrow j$, if $i \rightarrow j$ and $j \rightarrow i$ hold simultaneously.

The relation $\leftrightarrow$ is an 'equivalence relation' on the state space S, and as such has the properties of reflexivity: $i \leftrightarrow i$, symmetry: $i \leftrightarrow j$ if and only if $j \leftrightarrow i$ and transitivity: $i \leftrightarrow j$ and $j \leftrightarrow k$ imply $i \leftrightarrow k$. The first two of these properties are obvious, and the third one can easily be proven.

The state space S of a Markov chain can thus be split into disjoint classes $S_1, S_2, \ldots$ in such a way that any two states from the same class communicate with each other and that states from two different equivalence classes do not communicate.

A state x_i is called an 'absorbing state', if

$$p_{ii} = 1,$$

and $p_{ij} = 0$ for all $j \neq i$.

Two main types of Markov chains can be distinguished when modeling particle transport processes:

- In batch processes, all states communicate with each other, i.e. there is just a single class of states. The particle distribution in such a process will typically converge to a steady state.
- In continuous processes, the exit state, or absorbing state, plays a special role as it does not communicate with the interior states; all particles eventually leave the reactor by entering the absorbing state. A charge, or feed, of particles can be implemented as shown by example later in this chapter.

2.2 An Example

A specific type of Markov chain, which is often useful when applying stochastic modeling to processes involving solid particles, is the so-called 'birth-death process'. Figure 1 illustrates the possible transitions in such a process. A birth-death Markov chain is characterized by the fact that only three possible one-step transitions are allowed, namely

- remaining in the same state,
- moving one state ahead or
- moving one step back

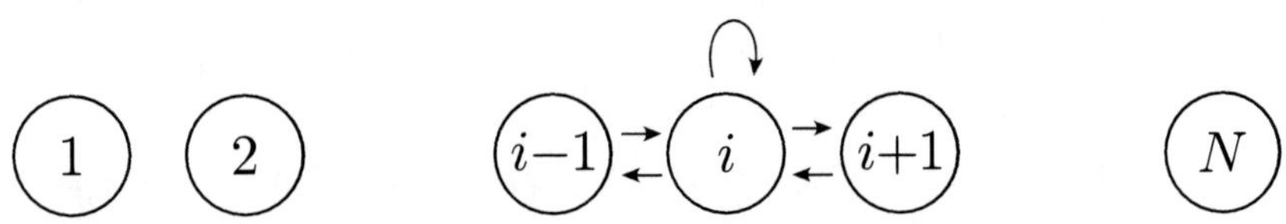

Figure 1: Possible transitions to and from the state labelled i in a birth-death Markov chain with state space $S = \{1, \ldots, N\}$.

For simplicity the state, which until now was denoted x_i, will be denoted only by its index, e.g. i, in this section and this will also be done where ever convenient in the rest of this chapter. Letting i label the present state of the process, the probabilities of the three possible transitions may be denoted by: α_i, β_i and δ_i, respectively, where, as mentioned before, $\alpha_i + \beta_i + \delta_i = 1$. Setting $\delta_1 = \beta_N = 0$ at the boundaries yields the transition probabilities:

$$p_{ij} = \begin{cases} \delta_i & j = i-1, i \neq 0 \\ \alpha_i & j = i \\ \beta_i & j = i+1, i \neq N \\ 0 & \text{otherwise} \end{cases}$$

and the resulting transition matrix becomes:

$$\mathbf{P} = \begin{pmatrix} \alpha_1 & \beta_1 & 0 & 0 & \cdots & 0 & 0 & 0 & 0 \\ \delta_2 & \alpha_2 & \beta_2 & 0 & \cdots & 0 & 0 & 0 & 0 \\ 0 & \delta_3 & \alpha_3 & \beta_3 & \cdots & 0 & 0 & 0 & 0 \\ 0 & 0 & \delta_4 & \alpha_4 & \cdots & 0 & 0 & 0 & 0 \\ \vdots & \vdots & \vdots & \vdots & \ddots & \vdots & \vdots & \vdots & \vdots \\ 0 & 0 & 0 & 0 & \cdots & \alpha_{N-3} & \beta_{N-3} & 0 & 0 \\ 0 & 0 & 0 & 0 & \cdots & \delta_{N-2} & \alpha_{N-2} & \beta_{N-2} & 0 \\ 0 & 0 & 0 & 0 & \cdots & 0 & \delta_{N-1} & \alpha_{N-1} & \beta_{N-1} \\ 0 & 0 & 0 & 0 & \cdots & 0 & 0 & \delta_N & \alpha_N \end{pmatrix}.$$

Originally, birth-death chains were invented as models for the development of the size of a population, hence their name. In the period between two transitions, one death or one birth may take place or there may be no change; no change can also be the net result of one birth and one death occurring simultaneously. In particle transport applications, birth-death chains constitute a flexible class of discrete models for plug-flow with dispersion.

When modeling continuous processes which allow inflow and outflow of material, the exterior can be added as an extra state, say $N + 1$. If a particle cannot return after leaving the reactor:

$$\alpha_{N+1} = 1 \quad \text{and} \quad \delta_{N+1} = \beta_{N+1} = 0.$$

In this model the exit state is thus an absorbing state, and the transition matrix becomes:

$$\mathbf{P} = \begin{pmatrix} \alpha_1 & \beta_1 & 0 & 0 & \cdots & 0 & 0 & 0 & 0 \\ \delta_2 & \alpha_2 & \beta_2 & 0 & \cdots & 0 & 0 & 0 & 0 \\ 0 & \delta_3 & \alpha_3 & \beta_3 & \cdots & 0 & 0 & 0 & 0 \\ 0 & 0 & \delta_4 & \alpha_4 & \cdots & 0 & 0 & 0 & 0 \\ \vdots & \vdots & \vdots & \vdots & \ddots & \vdots & \vdots & \vdots & \vdots \\ 0 & 0 & 0 & 0 & \cdots & \alpha_{N-2} & \beta_{N-2} & 0 & 0 \\ 0 & 0 & 0 & 0 & \cdots & \delta_{N-1} & \alpha_{N-1} & \beta_{N-1} & 0 \\ 0 & 0 & 0 & 0 & \cdots & 0 & \delta_N & \alpha_N & \beta_N \\ 0 & 0 & 0 & 0 & \cdots & 0 & 0 & 0 & 1 \end{pmatrix}$$

3 Fluidized Beds

Stochastic models have been applied to the particle transport in various types of fluidized beds. The particle transport processes in fluidized beds are quite complex, making traditional modeling based on differential balance equations difficult in some cases. Stochastic modeling techniques may make both the formulation and solution of models for many types of fluidized beds substantially easier.

3.1 Slugging Fluidized Beds

Slugging fluidized beds is a good example of stochastic modeling of the particle transport in fluidized beds for two reasons:

- it illustrates stochastic modeling of a batch process and
- it constitutes a case where no other modeling technique is practicable

Remark Fluidized beds generally (but not always) contain so many particles that their behaviour becomes deterministic in nature. This, however, does not detract from the advantages of using a stochastic modeling approach in both model formulation and model evaluation, something that this example illustrates.

Slugging fluidization normally occurs in beds with a high height-to-diameter ratio. A slug is simply a fluidization bubble, the diameter of which is nearly equal to that of the bed itself. Even so, slugs influence the bed material in a different way than smaller bubbles do.

Some empirical information about the nature of slugging fluidized beds is available [1–4]. Figure 2 illustrates two modes of slugging observed in fluidized beds. The focus in this section is on axisymmetric slugging, although the modeling approach equally well can be used for fluidized beds with wall slugs.

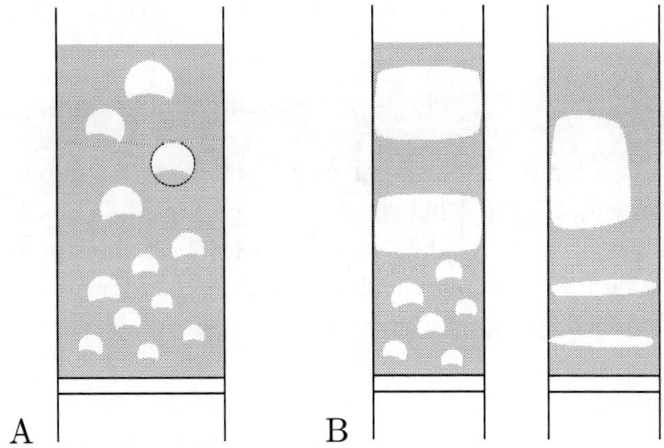

Figure 2: Fluidized beds. A: bubbling bed, the approximately spherical region enclosing the bubble and its associated wake is indicated in one case. B: Modes of slugging; left: axisymmetric, right: wall slugs [5]

The bed to be modeled contains a binary mixture of particles in which one of the two components segregates to the bottom in competition with mixing caused by rising slugs. In binary particle mixtures the larger and/or denser fraction will segregate to the bottom of the bed, the density being the dominant factor.

It can be noted at this point that segregation in *bubbling* beds has been studied extensively, pioneered by Rowe and coworkers [6, 7]. The fundamental particle transport processes proposed by these workers were: particle transport upward in the wakes of fluidization bubbles, a compensating transport downward in the bulk, and dispersion of particles in the bulk due to the motion of fluidization bubbles. A stochastic model for segregation in bubbling beds was proposed by Dehling and Hoffmann [8]. In this model a detour was taken around defining continuous parameters to describe the particle transport processes rather than formulating the model directly from the physical processes. This was done to utilize the empirical information available in the literature about the particle transport processes, which is given in terms of continuous transport parameters. This work also demonstrated the effect of particle 'gulf streaming' in increasing the solids circulation beyond that attributable to transport only in bubble wakes.

3.1.1 Model Formulation

The motion of a single particle is modeled by a stochastic, Markovian process (X_n), where $(X_n)_{n\geq 0}$ gives the axial location of the particle at time-step n.

Space is discretized by subdividing the fluidized bed in the axial direction into a finite number of cells labelled $1 \leq i \leq N$ obtaining the state space $\{x_1, \ldots, x_N\}$. As before, the transition probabilities are denoted $p_{ij}^{(n)} \equiv P(X_{n+1} = j|X_n = i)$.

Given an initial distribution vector $\mathbf{p}^{(0)}$ and the transition matrices $\mathbf{P}^{(n)} \equiv p_{ij}^{(n)}$ for all $n \geq 0$ the probability distribution for the position of the particle at some time n_0 can, according to Eq. (3) be calculated from

$$P(X_{n_0} = x_i) = \mathbf{p}^{(0)}\mathbf{P}^{(1)} \cdot \ldots \cdot \mathbf{P}^{(n_0)}. \tag{4}$$

3.1.2 Particle Transition Probabilities in a Slugging Fluidized Bed

Figure 3 shows the development of a slugging bed with height h_b at rest.

As shown in reference [1] and indicated in the left plate of Fig. 2B, slugs are formed by coalescing bubbles at some height, h_0, in the bed. The slugs grow in size until they have reached a certain height, h_s, at which time they 'detach'[1] and start rising as a whole toward the top of the bed. When the apex of the slug has reached the top of the bed, the slug collapses.

Two separate phases in the process can thus be distinguished:

Slug formation: The slug forms at a height h_0 above the bottom of the bed and grows in vertical size at some rate v from 0 to h_s. In this phase, all particles above the slug are pushed upward, and as a result the bed height also rises above its original level, h_b. Assuming that the slugs nearly fill up the cross-section, the bed rises at the same speed, v, as the slug to its maximum height $H \equiv h_b + h_s$ (see Figure 3).

Slug rise: When the slug has reached its final height, it starts rising toward the top of the bed. In this phase the height of the slug, and consequently also the height of the bed, remain constant. At the end of the slug rise phase, the apex of the slug has reached the top

[1] At 'detachment' a new slug starts to form underneath the slug in question, but in this model concept it is assumed that there is no interaction between succeeding slugs, see further on

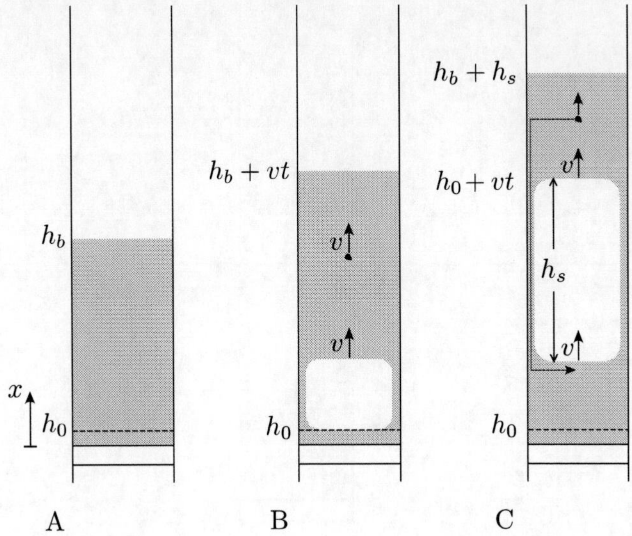

Figure 3: Slugging fluidized bed in different states of development: **A** Bed at rest, **B** Slug formation phase and **C** Slug rise phase.(The dot represents a single particle)

of the bed. At this point the slug ceases to exist, and the bed collapses back to its original height.

As a result of slug formation and rise, particles in the bed are displaced. They are either pushed upward or fall through the annulus around the slug and are deposited directly underneath the slug. The probability of the particle moving to the annulus and down below the slug likely depends on the vertical distance of the particle from the slug apex, x, and its *rate*[2] is a function, $\lambda_h(x)$, of h where h denotes the position of the slug apex and $h \leq x \leq H$. The flowpattern to the annulus indicated by the study of Davidson and Harrison [1] is sketched in Figure 4.

Remarks on the modeling strategy For simplicity a bed of unit cross-sectional area will be considered so that superficial velocities and volumetric flows are equal; since only axial particle transport is considered, this does not lead to loss of generality. The modeling strategy is as follows. First the effect of one single slug is modeled, and the effects of subsequent slugs are then modeled as superposition of the effect of the individual slugs, even if more slugs are present in the bed at the same time. To model the details of the effect of one slug, it is necessary to operate with a transfer probability matrix that varies in time, but once the transfer probability matrix for the complete effect of a slug has been found, this can then be used in a time-homogeneous Markovian model for the effect of subsequent slugs. First the displacement of bulk material by a slug will be modeled, and then the segregation of larger/denser particles.

As mentioned, according to Davidson and Harrison [1] the rate function $\lambda_h(x)$ is a decreasing function of x. The slug is assumed to move upward at a constant speed v. The

[2] Basically the probability of moving to the annulus during one second

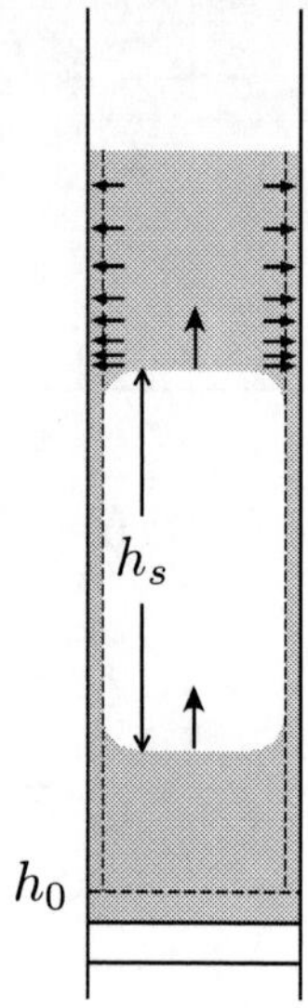

Figure 4: A sketch indicating the flow pattern of particles to the annulus in front of the rising slug, consistent with Davidson and Harrison [1]. Particles close to the slug apex are more likely to move to the annulus

total flow around the slug must equal the speed at which the slug rises giving the continuity condition

$$\int_h^H \lambda_h(x)dx = v, \tag{5}$$

for any $h(= h_0 + vt)$ in the interval $[h_s + h_0, H]$, where $h_s + h_0$ is the axial position of the apex at the start of the rise phase and $H(= h_b + h_s)$ is its position at the end of the rise phase. In the following it is assumed that $h_0 = 0$ and that $\lambda_h(x)$ follows a power law: $\lambda_h(x) = c(H + h - x)^r$ with $r \geq 1$, where the constant c is determined by the continuity condition (5)

$$\lambda_h(x) = \frac{v(r+1)}{H^{r+1} - h^{r+1}}(H + h - x)^r, \tag{6}$$

for $h_s \leq h \leq H$ and $h \leq x \leq H$.

The exponent r still remains to be determined, e.g. from experimental observations.

Particles above the slug that do not flow into the annulus rise at a speed $v_h(x)$ which depends on their vertical location. It is calculated as the difference between the speed of the slug apex and the flow of particles into the annulus below height x, that is

$$v_h(x) = v - \int_h^x \lambda_h(y)dy.$$

Using equation (5), this can also be written as

$$v_h(x) = \int_x^H \lambda_h(y)dy, \tag{7}$$

which can be interpreted as equating the upward flow of the rising particles to the downward flow in the annulus.

To obtain a discrete Markov chain approximation to the continuous system the bed is divided into horizontal slices of width Δ and time is discretized into steps ε such that $\frac{\Delta}{\varepsilon} = v$. In this way, one step per unit of time corresponds to the speed v. Furthermore it is assumed that h_s and h_b are integer multiples of Δ, i.e., $h_s = i_s\Delta$ and $h_b = i_b\Delta$. The horizontal slices (cells) are numbered by $i = 1, ..., N = i_s + i_b$. The discrete Markov chain model thus has state space $\{x_1, \ldots, x_N\}$ and transition probabilities $p_{ij}^{(n)}$, which, as mentioned, depend on time and are defined below. Since $v = \frac{\Delta}{\varepsilon}$ it takes i_b time units for the formed slug (with its apex at height i_s already) to reach the top of the bed. Since in the slug formation phase all particles above the forming slug move upward with velocity v the process can be simplified by assuming that the slug has already formed by shifting the initial distribution, $\mathbf{p}$, i_s cells upward to $\bar{\mathbf{p}} = \mathbf{pS}$ via the matrix $\mathbf{S} = (s_{ij})_{1 \leq i,j \leq N}$ defined by

$$s_{ij} \equiv \begin{cases} 1 & if\, 1 \leq i \leq i_b \text{ and } j = i + i_s \,, \\ 0 & \text{otherwise.} \end{cases} \tag{8}$$

Therefore time starts after the slug has already formed, such that time 0 becomes the point of time just before the slug starts to rise. There are no transitions by particles below the slug apex so that $p_{i,i}^{(n)}$ can be set $= 1$ for all $i \leq n + i_s$ because at time n the slug apex is in cell $n + i_s$.

For a particle in a cell i above cell $n + i_s$ at time $n \leq i_b - 1$, i.e., $n + i_s \leq i \leq N$, there are three different possible transitions (see Figure 5)

- move to what will be the first cell below the slug after the time step, i.e., to $j = n$,
- move one cell up, i.e., to $j = i + 1$,
- stay in the same cell.

The probabilities for these three transitions can be found using the rate function $\lambda_h(x)$. The probability of a particle moving from a cell into the annular region and into cell n is given as the average probability rate for removal from that cell multiplied by the length of one time unit ε. The probability of upward movement to cell $i + 1$ can be found from equation (7). This gives:

$$\begin{aligned} p_{i,n}^{(n)} &= \varepsilon \frac{1}{\Delta} \int_{(i-1)\Delta}^{i\Delta} \lambda_{(n-1+i_s)\Delta}(x)dx \\ &= \frac{(H + (n - i + i_s)\Delta)^{r+1} - (H + (n - i - 1 + i_s)\Delta)^{r+1}}{H^{r+1} - ((n - 1 + i_s)\Delta)^{r+1}} \end{aligned} \tag{9}$$

$$\begin{aligned} p_{i,i+1}^{(n)} &= \varepsilon \frac{1}{\Delta} \int_{i\Delta}^{H} \lambda_{(n-1+i_s)\Delta}(x)dx \\ &= \frac{(H + (n - i - 1 + i_s)\Delta)^{r+1} - ((n - 1 + i_s)\Delta)^{r+1}}{H^{r+1} - ((n - 1 + i_s)\Delta)^{r+1}} \end{aligned} \tag{10}$$

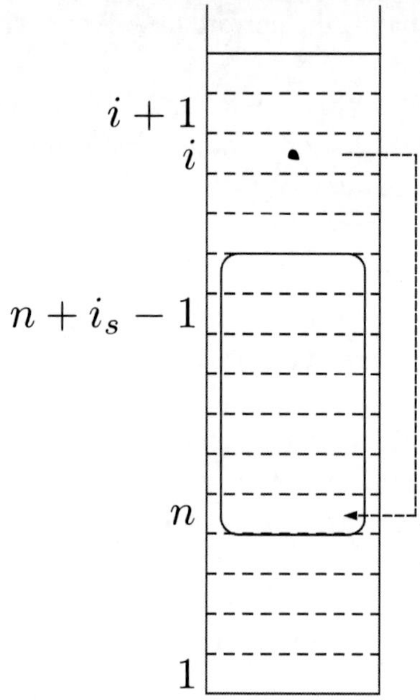

Figure 5: The discrete Markov chain model for the n-th transition. Just before the transition the slug occupies the cells $n, \ldots, n+i_s-1$. The possible transitions are up into cell $i+1$, staying in cell i or moving to cell n.

$$\begin{aligned} p_{i,i}^{(n)} &= 1 - p_{i,n}(n) - p_{i,i+1}(n) \\ &= \varepsilon \frac{1}{\Delta} \int_{(n-1+i_s)\Delta}^{(i-1)\Delta} \lambda_{(n-1+i_s)\Delta}(x)dx \\ &= \frac{H^{r+1} - (H + (n-i+i_s)\Delta)^{r+1}}{H^{r+1} - ((n-1+i_s)\Delta)^{r+1}} \end{aligned} \tag{11}$$

for $1 \le n \le i_b$ and $n + i_s \le i \le N$. Thus the transition probability matrices $\mathbf{P}^{(n)} \equiv p_{ij}^{(n)}$ are defined for all $1 \le n \le i_b$. The probability distribution at some time $0 \le n_0 \le i_b$ is now given by equation (42) as

$$\mathbf{p}^{(n_0)} = \bar{\mathbf{p}}\mathbf{P}^{(1)} \cdot \ldots \cdot \mathbf{P}^{(n_0)}, \tag{12}$$

where $\mathbf{p}^{(n_0)} = (P(X_{n_0} = 1), \ldots, P(X_{n_0} = N))$. Considering $n_0 = i_b$ captures the total effect of one slug formation and rise. The probability distribution at time $n_0 = i_b$ is equal to the distribution after one slug of the axial positions of an infinite number of particles with initial distribution $\mathbf{p}$.

So far the model describes the particle behavior in a bed containing particles of uniform properties. A segregation process can be introduced based on the idea that 'jetsam' particles—the heavier and/or larger particles that tend to sink in the mixture—sink when slugs disturb the system, in the same way as particle segregation in bubbling beds only takes place when fluidization bubbles disturb the region in question.

Segregation of jetsam and flotsam particles manifests itself as a difference in the downward drifts. The mass balance equation (5) only holds on average, with $\int_h^H \lambda_h(x)dx > v$ for jetsam particles and $\int_h^H \lambda_h(x)dx < v$ for flotsam particles. The segregation effect is simulated by increasing the probability of a down-transition for the jetsam particles, giving an extra downward drift. To account for this the transition probabilities (9), (10) and (11) can be modified to:

$$\bar{p}_{i,n}^{(n)} = \xi + (1-\xi)p_{i,n}^{(n)}, \tag{13}$$

$$\bar{p}_{i,i+1}^{(n)} = (1-\xi)p_{i,i+1}^{(n)}, \tag{14}$$

$$\bar{p}_{i,i}^{(n)} = (1-\xi)p_{i,i}^{(n)} \tag{15}$$

for all $1 \leq n \leq i_b$ and $n + i_s \leq i \leq N$. Here $0 \leq \xi \leq 1$ denotes a constant defining the strength of the segregation effects. When calculating the probability distributions for jetsam particles the transition probabilities (13), (14) and (15) have to be used for $\mathbf{P}^{(\cdot)}$ in equation (12).

As mentioned, the effect of several slugs can now be calculated via superposition. The probability distribution of the particle's position after s slugs, given its initial probability distribution $\mathbf{p}$, is:

$$(P(X_s = i))_{1 \leq i \leq N} = \mathbf{p}(\mathbf{S}\mathbf{P}^{(1)} \cdot \ldots \cdot \mathbf{P}^{(i_b)})^s\,, \tag{16}$$

with X_s denoting the jetsam particle's position after s slugs. $\mathbf{P}^{(i)}$ for $1 \leq i \leq i_b$ is given by equations (13), (14), (15) and the shift matrix $\mathbf{S}$ given by equation (8). The matrix product in the parentheses on the right-hand-side of the equation can be said to constitute the transition probability matrix of a time-homogeneous Markov process.

3.1.3 Comparison with Experiments

In this section results from the model are briefly compared with some of the experiments of Abanades and Atarés [4]. Abanades and Atarés arranged a layer of white jetsam particles on a bed of red flotsam particles (the minimal fluidization velocity, U_{mf} of the mixture was 1 m/s), and fluidized this suddenly with a pre-set flowrate, video-recording the process.

Equation (16) was implemented with number of cells, N, between 15 and 17, of which $N - 10$ were occupied by the slug, and otherwise using the same experimental conditions as Abanades and Atarés given in the figures below. It was assumed that $h_0 = 0$. The slug frequency f was, following Abanades and Atarés [4], calculated from the empirical correlation:

$$f = 0.32\frac{U^{-0.15}}{h_b}$$

where U is the superficial fluidization velocity. The height of a stable slug, h_s, was calculated by solving [9]:

$$\frac{h_s}{D_{bed}} - 0.495\sqrt{\frac{h_s}{D_{bed}}}\left(1 - \frac{U - U_{mf}}{0.35\sqrt{gD_{bed}}}\right) + 0.061 - \frac{1.939(U - U_{mf})}{0.35\sqrt{gD_{bed}}} = 0$$

with D_{bed} the bed diameter and g the gravitational acceleration.

The experimental and the model results are not in the same form. The experimental results are given as the volume fraction of bed at a given height that consists of tracer, say $V_{fr,i}$. The p_i of the model represent, as mentioned earlier, the fraction (e.g. volume fraction) of the tracer that is in cell i. If there are N equi-sized cells in the bed, the latter can, since the volume fraction of each cell is $1/N$, be transformed to the former by:

$$V_{fr,i} = \frac{V_{tr} p_i}{\frac{1}{N}}$$

where V_{tr} is the total volume of tracer in the bed.

Remark The effect of the jetsam particles filling up a given cell is not accounted for here. This would impose a maximum value for p_i of $p_{max} \equiv p_{i,max} = 1/(NV_{tr})$. However, to take full account of these effects in the model, the change of the segregation behavior of the jetsam particles in response to the change in the local jetsam concentration will also have to be accounted for. In other words an assumption of this model is that the jetsam is infinitely dilute so that all the jetsam particles can collect in one cell (i.e. no interaction between particles in the sense that they can fill a cell) and the segregation rate is not influenced by the presence of other jetsam particles (the local environment is not influenced by the concentration of jetsam particles). The issue of interfering particles is discussed briefly in Dehling et al. [10] and is an aspect open for further investigation.

The two model parameters r, determining the flow profile of particles to the annulus in front of the slug, and ξ, determining the segregation, are functions of the experimental conditions and particle properties. Their quantification awaits more empirical information and they are used as adjustable parameters here.

Figure 6 shows plots of distribution of jetsam volume fraction as a function of height in the bed after 20 seconds for different bed heights and superficial gas velocities. The adjustable parameters are set to $r = 1$ and $\xi = 0.17$ in all cases to give acceptable agreements between experimental results and model predictions.

Figure 6 shows good agreement between the predicted and experimental data. There is one weak effect in the experimental data that is not accounted for in the model predictions. The tracer fraction in the upper part of the bed (high cell index) increases slightly with bed height while in the model predictions the tracer fraction is a decreasing function of bed height also in this region. It should be possible to achieve even better predictions with more knowledge on how to choose r and ξ. The model is compared with more experimental results in [11].

The process discussed in this section is an example of a process which does not at all lend itself to modeling by differential balance equations formulated in terms of continuum-flow parameters. An added attraction of the Markovian model is that it is very easily evaluated by matrix operations. These advantages are not dependent on the system exhibiting significant stochastic behaviour, but are, as mentioned before, equally valid for a process containing so many particles that the system behaviour is essentially deterministic in practice. If, however, a slugging system (or any system containing particles) contains so few particles that there is a significant stochastic element in the system results, this aspect can easily and directly be accounted for in this type of model.

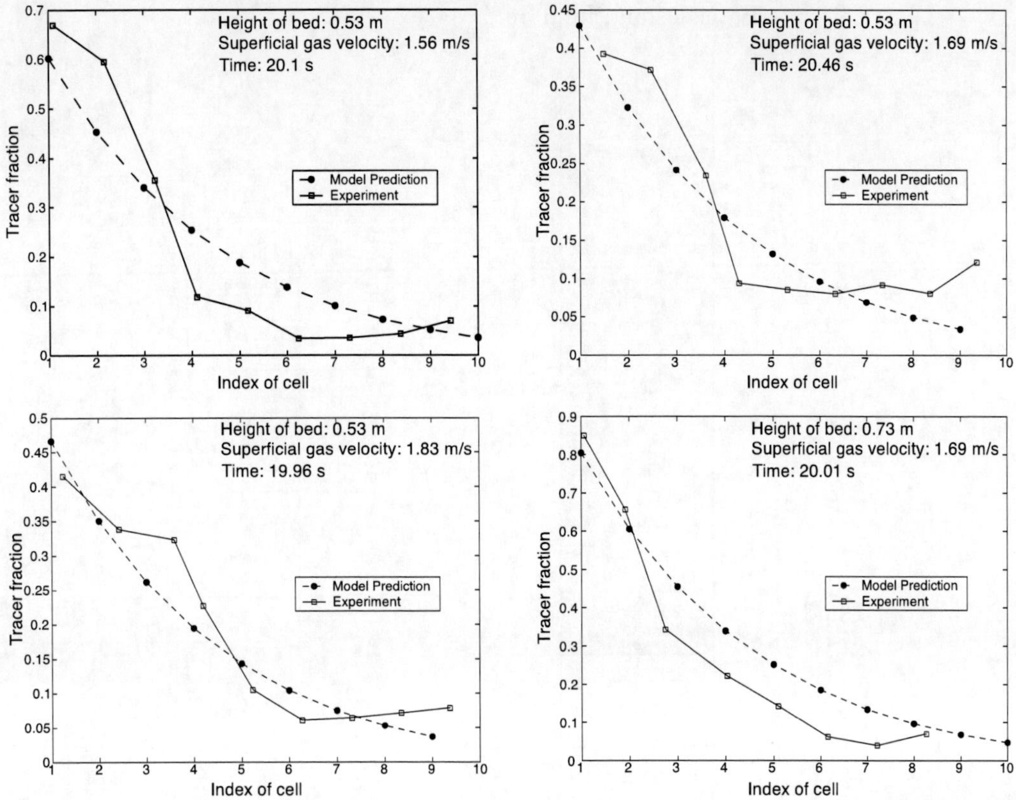

Figure 6: Distribution of jetsam volume fraction with $\xi = 0.17$ and $r = 1$ compared with the experimental results of Abanades and Atarés [4].

3.2 Fluidized Bed Riser

Another example of stochastic modeling of fluidized beds is the stochastic model for a fluidized bed riser proposed by Harris et al. [12]. The modeling concepts build on a quite general stochastic model proposed by Fan et al. [13].

3.2.1 Physical Phenomena

Fluidized bed risers are beds operated in the transport regime, as indicated in Figure 7 A. Riser reactors are used, for example, for the combustion of coal or for fast reactions involving the contact of gaseous reactants with a solid catalyst, such as in the fluidized bed catalytic cracking (FCC) process. In fact, in the FCC process, the catalyst is circulated between two fluidized beds, one (the reactor) operated as a riser and the other (the catalyst regenerator) operated as a bubbling/turbulent fluidized bed.

For such processes the particle residence time distribution (RTD) in the fluidized bed is crucial but very difficult to model, something that has impeded the development of this advanced type of reactor for other applications [15]. Harris et al. [16] give a thorough discussion of these issues, and the development of CFB reactors in general.

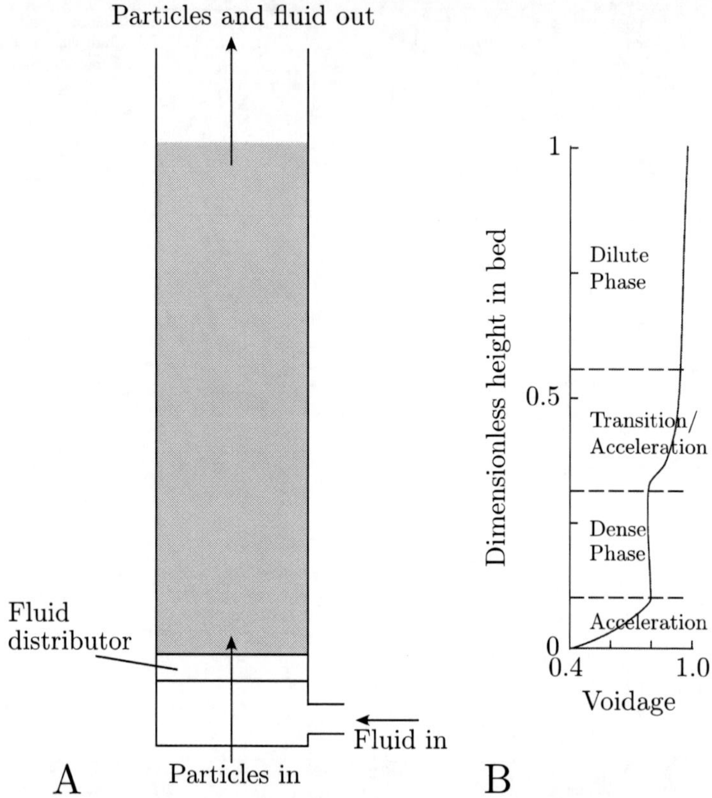

Figure 7: A: Flow of fluid and particles through a fluidized bed riser. B: axial distribution of the average voidage fraction in a fast fluidized bed riser according to Yang [14] and Harris et al. [12]

Yang [14] proposes three sub-flowregimes for fluidized bed risers: dilute phase pneumatic transport, fast fluidized bed and dense phase transport. The model of Harris et al. is concerned with the second of these three. The axial profile of voidage, $\bar{\epsilon} = 1 - \bar{\epsilon}_s$ where $\bar{\epsilon}_s$ is the cross-sectionally averaged volume fraction taken up by the solids, is, for this flowregime, given in Figure 7 B.

Note that ϵ_s is not homogeneous in the bed, neither axially or, as shown below, cross-sectionally. This complicates the modeling of the solids flow in the bed.

Particle transport in such beds is not uniform over the cross-section. In the transition and dilute phase regions a 'core-annular' flow regime exists with upflow of particles in a central core, occupying most of the cross-sectional area, downflow in an annular region on the vessel wall. An important physical parameter when modeling such a process is the thickness, δ_f, of this annular region. The correlation used for this by Harris et al is one proposed by Bi et al. [17]:

$$\frac{\delta_f}{D_v} = 0.5\left[1 - \sqrt{1.34 - 1.30(\bar{\epsilon}_s)^{0.2} + (\bar{\epsilon}_s)^{1.4}}\right] \text{ for } 0.0015 \leq \bar{\epsilon}_s \leq 0.20 \qquad (17)$$

where D_v is the vessel diameter.

The material reaching the reactor top through the core splits in two streams, one exiting the reactor and another refluxing through the annular region. Another crucial parameter is the fractional split between these two streams. van der Meer et al. [18] define the 'reflux ratio' as:

$$k_m = \frac{\text{material flow refluxing through the annular film}}{\text{material flow exiting the reactor}},$$

and give empirical values for k_m for various configurations for the riser top, ranging from 0.11 for a short, but smooth, exit section to 2.72 for a larger, abrupt exit.

Harris et al. operate with material fluxes between the two countercurrent flows in the core and in the annulus. Very little information is available in the literature for the quantification of these fluxes, and Harris et al. varied them in their paper. It can be gleaned that they in most cases fixed the annulus→core flux as 0.6 of the core→annulus flux.

3.3 Model

The stochastic model of Harris et al. builds, as mentioned on work by Fan et al. [13] who again built on pioneering work of Gibilaro et al. [19].

The method of Fan et al. essentially builds on the concept of a number of CSTR compartments with material exchange between them. This method is suitable for modeling flows between actual vessels in a process, but also flows within vessels, which are then conceptually divided into subregions, each of them modeled by a CSTR. Fan et al. even used this concept for modeling the flow through a plug-flow reactor by coupling a large number of such CSTRs in series. A brief discussion of the principles behind their modeling approach is give in the following paragraphs, a more thorough discussion is given in [10].

In accordance with the well-known theory for CSTRs, the probability of a particle staying in, say, the i'th CSTR, p_{ii} depends on the length of the time interval chosen, Δt, and the mean residence time of particles in the CSTR, τ_i, which, in accordance with the classical theory of Danckwerts, equals the volume of the CSTR divided by the volumetric flow through it:

$$p_{ii} = e^{-\Delta t/\tau_i}. \tag{18}$$

The probability of transferring to another compartment is then:

$$\sum_{j \neq i} p_{ij} = 1 - e^{-\Delta t/\tau_i},$$

which can be related to the volumetric flows, Q_{ij} between the compartments:

$$p_{ij} = \frac{Q_{ij}}{\sum_{j \neq i}^{N+1} Q_{ij}}(1 - p_{ii})$$

Returning to the work of Harris et al. They present four models in their paper, namely continuous-time and discrete versions of two different compartment configurations. This discussion will focus on their discrete 'four-zone' model, which is also the dominant model in their paper.

A scheme of the four-zone model is shown in Figure 8.

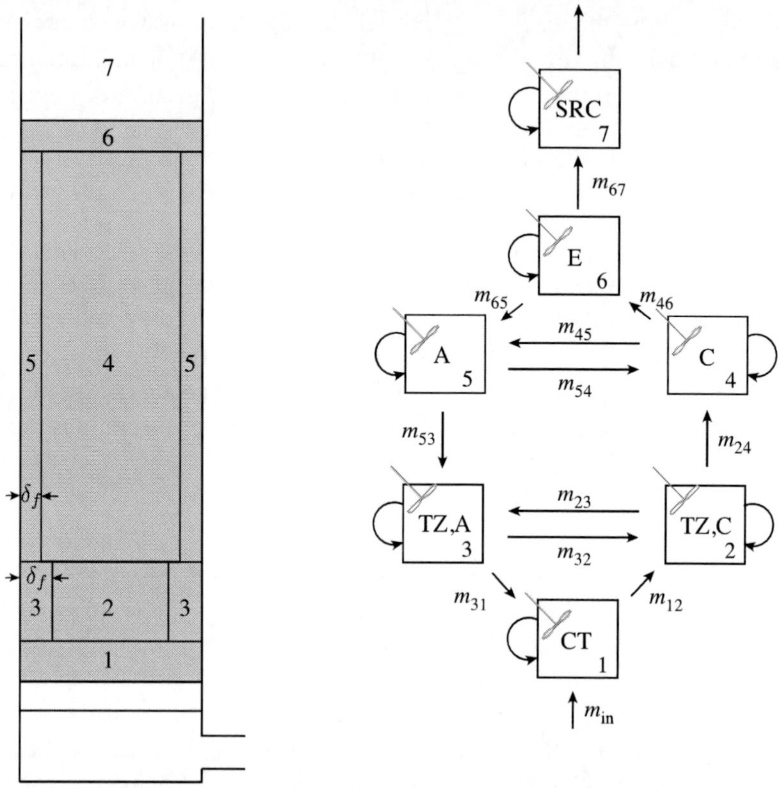

Figure 8: Scheme of the model of Harris et al. [12]. The CSTR compartments have been labelled as in their paper

The compartment labelled CT is a 'churn-turbulent' (turbulent fluidization), dense-phase region at the bottom of the bed. Its height is estimated from the pressure profile over the reactor. The compartments TZ,C and TZ,A are the core and annular regions of the transition region between dense-phase and dilute regions indicated in Figure 7 B. Compartments C and A are the core and annular regions of the dilute phase region, this is the main part of the reactor, and E is an exit region, with SRC the absorbing compartment (or cell or state).

The position of one solid particle is denoted by a stochastic variable, X_n, with state space $\{x_1, x_2, \ldots, x_7\}$, the x_i representing the compartments. X_n is allowed to go through a Markovian process. The Markov chain is, as discussed before, specified by its transition probabilities $p_{ij}^{(n)} \equiv P(X_{n+1} = x_j | X_n = x_i)$. The process is assumed to be homogeneous in time, so that the time index of the p_{ij} can be dropped.

As mentioned, an important aspect to realize is that in the fluidized bed system described the volume fraction taken up by solids, ϵ_s, is not homogeneous and therefore varies between the compartments, so that the material in a sense is *compressible*compressible. It is therefore necessary to consider mass flows between the compartments, m_{ij}, rather than volume flows. The equivalent to the compartment volume in the model of Fan et al. is then for this present case the total mass in the compartment, $M_i = \rho_p \epsilon_{s,i} V_i$, where ρ_p is the particle density.

The mean residence time in compartment i, τ_i, is the total mass in the compartment divided by the total mass flow through it: $\tau_i = M_i / \sum_{j \neq i}^{N+1} m_{ij}$, where $N+1$ is the total number of compartments, including the external absorbing one. This gives rise to the following transfer probabilities:

$$p_{ii} = e^{-\Delta t / \tau_i}, \tag{19}$$

and:

$$p_{ij} = \frac{m_{ij}}{\sum_{j \neq i}^{N+1} m_{ij}} (1 - p_{ii}). \tag{20}$$

To avoid problems arising from the flows through the compartments exceeding the compartment volumes during one time step and that the probabilities of two or more transfers within one time-step become non-negligible a restriction is, as in other cases, imposed on the length of the time step such that for all x_i in the state space of X_n, $\sum_{j \neq i}^{N+1} m_{ij} \Delta t \leq M_i$.

Remark on the relation to the p_{ij} of Harris et al. The quantification of the transfer probability matrix follows, as mentioned, the method of Fan et al. [13] and deviates from the description of Harris et al., whose transfer probabilities are $p_{ij} = m_{ij} / (\sum_{j \neq i}^{N+1} m_{ij} + M_i)$ and $p_{ii} = M_i / (\sum_{j \neq i}^{N+1} m_{ij} + M_i)$, the latter of these is not stated explicitly, but follows from $\sum_j^{N+1} p_{ij} = 1$. These transfer probabilities are not dimensionally consistent, and do not make use of the analytical expression for the mean residence time in CSTRs, Equation (18), ensuring that $p_{ii} \to 0$ for $\Delta t \to \infty$.

There are six internal compartments in the bed, and, to determine the p_{ij}, 11 inter-compartmental flows need to be found, together with the solids inventories, M_i, in each compartment. Harris et al. estimated the solids volume fractions, ϵ_s, and the heights of all the compartments from empirical information. Determining the thickness, δ_f of the annular regions from Equation (17) must be done iteratively, since δ_f depends on the average solids volume fraction, $\bar{\epsilon}_s$, which again depends on δ_f. Knowing the heights and thickness, and therefore the volumes of the regions, yields the solids inventories, M_i. The solids fluxes between the annular and core regions, are also determined empirically, or rather varied in reasonable ranges, from which then the material flows m_{23}, m_{32}, m_{45} and m_{54} could be found knowing the interfacial areas between the regions. The annular $\to$core fluxes are mostly taken as 0.6 of the core$\to$annular fluxes. The value of k_m is, as mentioned, estimated empirically. Having determined these parameters, the m_{ij} are found from steady-state mass balances over the regions. For region i:

$$\sum_{j=1}^{N+1} m_{ij} = \sum_{j=1}^{N+1} m_{ji}$$

The probability distribution for the position of one particle represents the behaviour of a pulse of marked particles in the process. The probability of the particle being in the absorbing state therefore represents the fraction of marked particles having exited the process at any given time, and therefore the cumulative residence time distribution, F, the time-derivative of this is the exit-age distribution, E.

The example calculations of E and F shown below use the following parameters. The compartment heights are, for region 1: 1.25 m, 2 and 3: 0.625 m, 4 and 5: 2.5 m (varies as shown below), 6: 1.25 m. Solids volume fraction, (ϵ_s), are, for region 1: 0.25, 2 and 3: 0.12, 4: 0.01, 5: 0.0375 and 6: 0.1. The relative values for the thicknesses of the annular regions, δ_f/D_v came to 0.132 and 0.060 for the transition and the dilute regions, respectively. k_m was taken as 0.11 (smooth exit), and the solids fluxes between the annular and core regions were taken as equal for the transition and dilute regions with the annular →core fluxes equal to 0.6 of the core→annular fluxes.

Figure 9 shows the effect of the heights of the dilute region and its annular region on the particle residence time distribution in the reactor. The effect is qualitatively the same as that shown in Figure 13 of Harris et al., namely that an increasing core/annular flux leads to a wider residence time distribution. Evaluating the p_{ij} from Equations (19) and (20) clearly makes the effect slightly less strong. These results are also quantitatively similar to those of Harris et al. in spite of the difference in methodology mentioned above. Harris et al found a second peak developing in their E-curves at high core-annular fluxes. Such a feature is not present in Figure 9.

Figure 10 shows the effect of varying the reflux ratio, k_m, on the E and F curves, the effect is seen not to be very significant even though the reflux varies over a wide range.

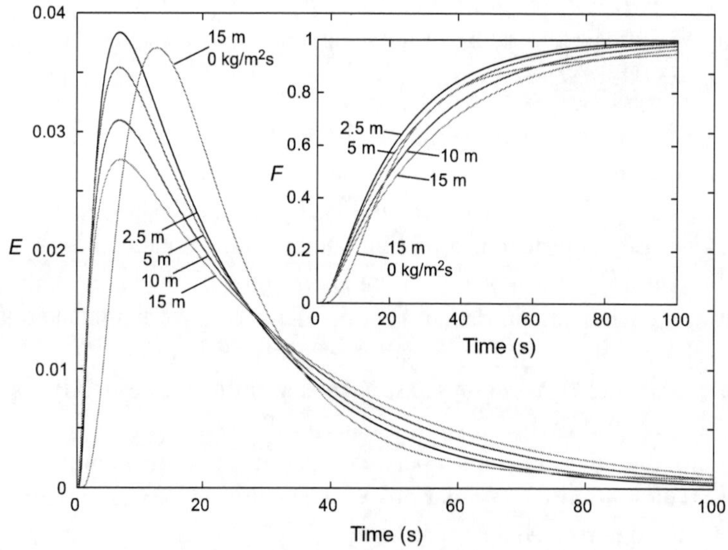

Figure 9: Effect of dilute region height (4 and 5) on the particle RTD according to the model of Harris et al. [12], calculating the p_{ij} using Equations (19) and (20), the core-annulus flux is 5 kg/m^{2}s except for the curve where it is given as 0 kg/m^{2}s

The model of Harris et al. is an example of how the particle flow and residence time in a complex system can be modeled with a simple but powerful and flexible stochastic model, and the results worked out with matrix operations that can be performed in an instant when programmed in a compiled language, such as Fortran.

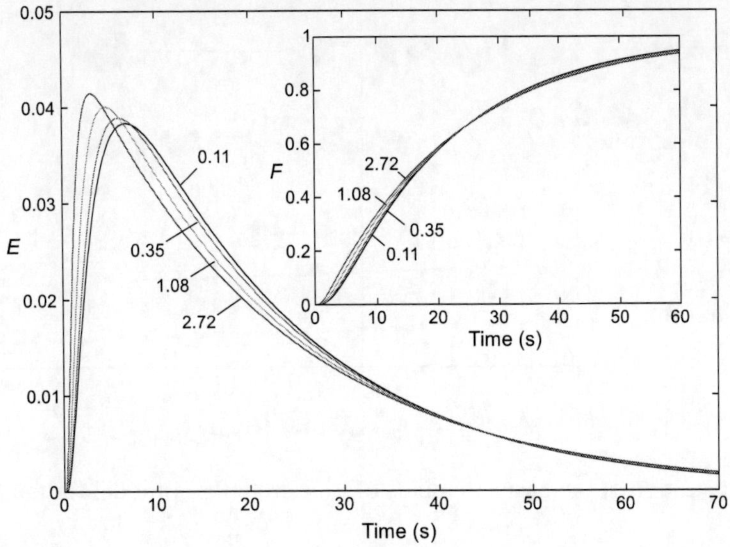

Figure 10: Effect of the reflux ratio, k_m, on the particle RTD according to the model of Harris et al. [12], calculating the p_{ij} using Equations (19) and (20)

4 Population Balance Modeling

An important class of processes in particle technology is processes wherein the size of solid particles changes, either spontaneously or by manipulation.

By far the most-used modeling technique for this type of process is Population Balance Modeling, or PBM [20, 21], a technique whereby balance equations are written for the particle size (or, in further developments of the technique, for other properties), and these equations are solved to yield the particle size distribution as a function of time.

The objective of this section is to show the relation of PBM methods to stochastic modeling by Markov chains. Berthiaux [22] worked out such a relation for a grinding process [3]. Before moving on to the work of Berthiaux a brief general description of PBM modeling of particle size will be given.

Figure 11 illustrates the basic elements in a discrete PBM model for particle size. Small particles may nucleate in a process, for example a crystallization process, suddenly appearing in a given small size class. Particles may also change size class by growth due to deposition of material on their surface. Another reason for particles to change size class is by agglomeration causing the constituent particles to disappear from their respective size classes and the agglomerate to emerge in a higher class. Finally particles may be broken up, disappearing from their size class and the breakage fragments appearing in lower size classes.

The two latter processes are sometimes referred to as 'birth' and 'death' processes, but this is a different use of these terms than in the stochastic literature, as mentioned before.

[3] A general discussion of stochastic PBM processes is given by Ramkrishna [21]. Ramkrishna's discussion, however, is not focused on Markovian modeling.

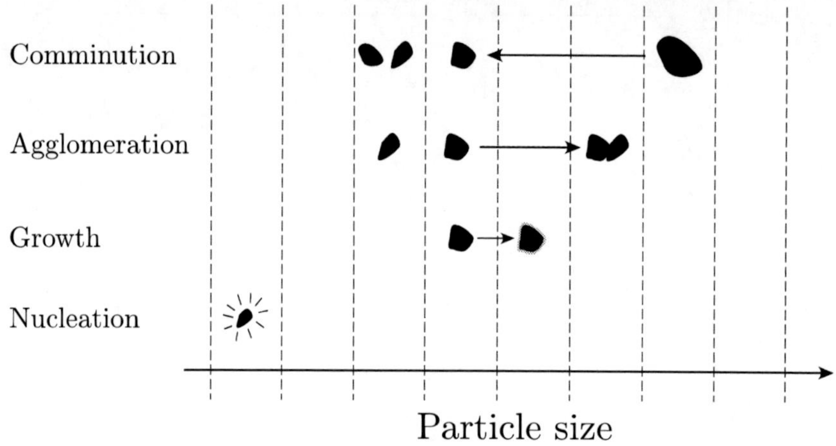

Figure 11: Illustration of the basic elements of a discrete PBM model for the particle size.

Although PBM modeling was first applied to modeling the size distribution of particles it has since been extended and in the most general case the particle properties may include not only their size, but also, e.g. their porosity and their moisture content, such that the particle properties may be described by a whole vector of properties, often called the 'internal property vector'. A further extension of the method involves including the spatial position of the particle in the process in its 'properties', such that the three spatial coordinates constitute another property vector, often called the 'external property vector'.

4.1 Analysis of a Grinding Process

Comminution, or grinding, which is particle size reduction by shredding or crushing, is a ubiquitous process, for instance in the energy, cement and mining industries.

A central issue in the modeling of comminution processes is the size distribution of the fragments produced by a breakage event. For example, when a stone is crushed either by compression or by impact, it tends to break in the pattern indicated in Figure 12. Most of the stone is broken up in relatively coarse particles, while a cone of fine particles is produced in the vicinity of the point of impact. As a result, the particle size distribution of a powder produced in a mill acting by compression or impact tends to be bimodal. Studying the size distribution of a given powder can thus yield information about how it may have been produced.

Berthiaux [22] analyses a grinding (or comminution) process by Markov chains, and links this analysis to traditional PBM analysis.

4.1.1 Traditional Modeling by PBM

The PBM analysis of Berthiaux is based on a discrete PBM analysis in which property space is discretized. The only property considered is the particle size expressed as the particle diameter. The amount of particles in each of the N size classes is expressed as the mass fraction, m_i, of particles in that class.

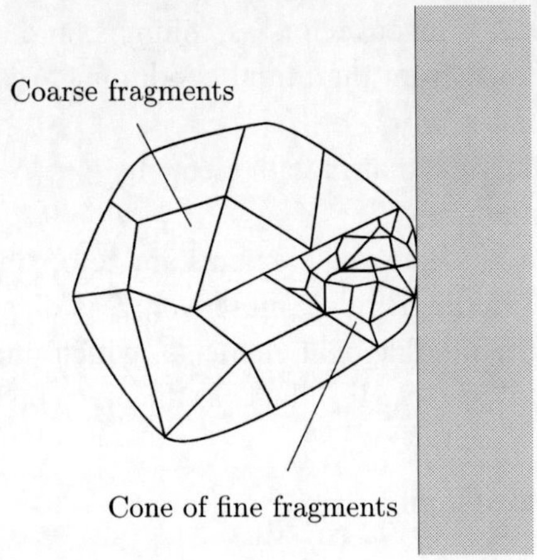

Figure 12: Sketch showing a typical break-up pattern as a result of crushing by impact or compression

In line with traditional PBM analysis two breakage matrices are defined as follows (see also Berthiaux and Dodds [23]):

- The 'breakage function', $\mathbf{B}$, an $N \times N$ matrix with elements b_{ij} that represent the mass fraction of those particles in size class j that break during a short (possibly infinitesimal) time interval that enter size class i. In other words: b_{ij} is the probability of material moving to class i conditional upon it exiting class j. b_{ij} is thus the distribution over the size classes of that material that exits class j[4].

- The 'selection function', a column vector of length N with elements S_i, which represent the mass fraction of particles in size class i that *per second* moves to another size class during a short (possibly infinitesimal) time interval.

With these definitions, the change with time of the mass fraction of material in size class i, m_i becomes:

$$\frac{dm_i}{dt} = -S_i m_i + \sum_{j=1}^{i} b_{ij} S_j m_j.$$

This equation can be written in matrix form:

$$\frac{d\mathbf{M}}{dt} = (\mathbf{B} - \mathbf{I})\mathbf{S}\mathbf{M} = \mathbf{A}\mathbf{M} \tag{21}$$

where $\mathbf{M}$ is the column vector of length n with elements m_i, $\mathbf{I}$ is the identity matrix and $\mathbf{S}$ is a diagonal $N \times N$ matrix containing the elements S_i. The right-hand-side of Equation (21) defines the matrix $\mathbf{A}$.

Equation (21) is a homogeneous system of first order differential equations and can be solved by the standard technique involving matrix diagonalization (see, for example, Kreyszig [24]):

$$\mathbf{M}(t) = \mathbf{\Xi}\mathbf{E}_D(t)\mathbf{\Xi}^{-1}\mathbf{M}(0) = \mathbf{V}(t)\mathbf{M}(0) \tag{22}$$

[4]Note that for the matrix $\mathbf{B}$ the indices i and j are reversed relative to the normal practice in this chapter

where $\boldsymbol{\Xi}$ is the matrix with $\mathbf{A}$'s eigenvectors as columns, and the factor $\boldsymbol{\Xi}^{-1}\mathbf{M}(0)$ is the integration constants calculated from the initial condition. Again, the right-hand-side of Equation (22) defines the matrix $\mathbf{V}(t)$.

The matrix $\mathbf{E}_D$ is the diagonal matrix with elements $\exp(\lambda_i t)$, where λ_i are the eigenvalues of $\mathbf{A}$.

$\mathbf{B}$ is lower triangular and its diagonal elements are zero, since material only moves to smaller size classes due to grinding. Since, moreover, $\mathbf{S}$ is diagonal, Equation (21) shows $\mathbf{A}$ to be lower triangular with the diagonal elements, which due to its lower triangularity are also its eigenvalues, equal to $-S_i$. $\mathbf{E}_D$ thus has the elements $\exp(-S_i t)$.

4.1.2 Modeling by Markov Chains

Berthiaux defines the discrete state space of a Markov chain as the set of possible size classes (N in total, where the index N refers to the smallest size class) for the particles. The Markov chain is discrete also in time, and is defined by its transition matrix [5] $\mathbf{P}$ describing the change in the state of material due to grinding during one time interval of length Δt.

The elements of the transition matrix $\mathbf{P}$, p_{ij}, are defined, as before, as the probability of material being in size class j in time-step $n+1$, conditional on the material having been in class i in time-step n: $p_{ij}^{(n,n+1)} = P(X_{n+1} = j | X_n = i)$. $\mathbf{P}$ is upper triangular, since the grinding can only lead to transfer to a lower size class.

$$\mathbf{P} = \begin{pmatrix} p_{11} & p_{12} & \cdots & \cdots & p_{1n} \\ 0 & p_{22} & & & \vdots \\ \vdots & \ddots & \ddots & & \vdots \\ \vdots & 0 & \ddots & \ddots & \vdots \\ 0 & \cdots & \cdots & 0 & p_{NN} \end{pmatrix}.$$

The elements, p_i, of the probability vector, $\mathbf{p}^{(n)}$, give the probability distribution over the states of, for example, some material marked at the beginning of the grinding process and sum to unity. The index n denotes the time interval.

As before, the evolution of the probability vector can be evaluated by matrix multiplication:

$$\mathbf{p}^{(1)} = \mathbf{p}^{(0)}\mathbf{P}$$

and:

$$\mathbf{p}^{(n)} = \mathbf{p}^{(0)}\mathbf{P}^n. \tag{23}$$

Finding $\mathbf{P}^n$ can be done by diagonalization as described e.g. in Kreyszig [24]. $\mathbf{P}$ can be written:

$$\mathbf{P} = \boldsymbol{\Psi}\mathbf{D}\boldsymbol{\Psi}^{-1} \tag{24}$$

where $\boldsymbol{\Psi}$ is the matrix with columns equal to the right eigenvectors, $\mathbf{u}$, of $\mathbf{P}$, i.e.:

$$\mathbf{P}\mathbf{u} = \lambda\mathbf{u}. \tag{25}$$

[5] The notation of Berthiaux is changed to be consistent with the rest of this chapter

The n-th power of the transition matrix then becomes:

$$\mathbf{P}^n = (\mathbf{\Psi}\mathbf{D}\mathbf{\Psi}^{-1})(\mathbf{\Psi}\mathbf{D}\mathbf{\Psi}^{-1})\cdots(\mathbf{\Psi}\mathbf{D}\mathbf{\Psi}^{-1}) = \mathbf{\Psi}\mathbf{D}^n\mathbf{\Psi}^{-1}$$

so that:

$$\mathbf{p}^{(n)} = \mathbf{p}^{(0)}\mathbf{\Psi}\mathbf{D}^n\mathbf{\Psi}^{-1}. \tag{26}$$

Remark Please note that the developments in Equations (24–26) are done using the right eigenvectors for $\mathbf{P}$. This is normal practice in general, and also in the paper of Berthiaux. Normal practice for transition matrices, however, is to use left eigenvectors, such that Equation (25) would be replaced by $\mathbf{uP} = \lambda\mathbf{u}$ and the matrices $\mathbf{\Psi}^{-1}$ and $\mathbf{\Psi}$ are reversed in Equation (24) and would contain the left eigenvectors of $\mathbf{P}$ as their rows. The reason for this practice for transition matrices is that this is consistent with the equation: $\mathbf{p}^{(n)} = \mathbf{p}^{(n-1)}\mathbf{P}$ where the transition matrix is left-multiplied by the probability vector to find the new probability vector.

4.1.3 Comparison of the Two Models

The elements of both of the vectors $\mathbf{p}$ and $\mathbf{M}(t)$ represent the fraction of marked material in a given size class at time $t = n\Delta t$. Comparing Equations (22) and (23) shows that these two vectors are equal if:

$$\mathbf{P}^n = (\mathbf{V}(t = n\Delta t))^{\mathrm{T}} \tag{27}$$

where $\cdot^{\mathrm{T}}$ denotes the transpose.

Berthiaux continues to compare the approaches in more detail by comparing the breakage functions in PBM to the transfer probability matrix of the Markov-chain model.

To do this, he assumes all the marked material to be concentrated in size class j and time t, and denotes its mass fraction by $m_j^*(t)$. He denotes the mass fraction of marked material in class i after one time step of length Δt as $m_i^*(t + \Delta t)$. The elements of the transfer probability matrix can then be written:

$$p_{ji} = \frac{m_i^*(t + \Delta t)}{m_j^*(t)}. \tag{28}$$

If the time-step, Δt, is small enough the breakage functions can be estimated from first-order approximations (refer to the definitions of b_{ij} and S_i in Section 4.1.1):

$$b_{ij} = \frac{m_i^*(t + \Delta t)}{m_j^*(t) - m_j^*(t + \Delta t)} \tag{29}$$

and:

$$S_j = \frac{m_j^*(t) - m_j^*(t + \Delta t)}{m_j^*(t)\Delta t} = \frac{1}{\Delta t}\left(1 - \frac{m_j^*(t + \Delta t)}{m_j^*(t)}\right). \tag{30}$$

Comparing Equations (28), (29) and (30) gives Berthiaux the following relationships between the breakage functions and the transfer probability matrix:

$$b_{ij} = \frac{p_{ji}}{1 - p_{jj}} \tag{31}$$

and

$$S_j = \frac{1}{\Delta t}(1 - p_{jj}). \tag{32}$$

Solving Equations (31) and (32) for p_{ji} and p_{jj} gives:

$$p_{ji} = b_{ij} S_j \Delta t \tag{33}$$

and

$$p_{jj} = 1 - S_j \Delta t, \tag{34}$$

which can be used to fill in the transfer probability matrix from knowledge of the particle breakage functions. Equations (33) and (34) are in matrix form:

$$\mathbf{P} = \mathbf{I} - \Delta t \mathbf{S} + \Delta t \mathbf{S} \mathbf{B}^{\mathrm{T}} = \mathbf{I} + \Delta t \mathbf{A}^{\mathrm{T}} \tag{35}$$

Diagonalizing $\mathbf{A}$, and taking into account, as stated above, that the eigenvalues of $\mathbf{A}$ are $-S_i$ gives:

$$\mathbf{P} = \mathbf{I} - \Delta t (\mathbf{\Xi}^{-1})^{\mathrm{T}} \mathbf{S} \mathbf{\Xi}^{\mathrm{T}}. \tag{36}$$

Equation (34) shows that $p_{jj} = 1 - \Delta t S_j$, so that $\mathbf{D} = \mathbf{I} - \Delta t \mathbf{S}$ and Equation (24) can be written:

$$\mathbf{P} = \mathbf{I} - \Delta t \mathbf{\Psi} \mathbf{S} \mathbf{\Psi}^{-1} \tag{37}$$

Comparing Equations (36) and (37) gives:

$$\mathbf{\Psi} = (\mathbf{\Xi}^{-1})^{\mathrm{T}}$$

The above developments, and the entire concept of modeling the grinding or comminution process with a stationary, or time-homogeneous, Markov chain model discretized in time, depends on:

a) The breakage functions being independent of time.

b) The time step, Δt, being small enough for the first-order estimates of b_{ij} and S_j in Equations (29) and (30) to be good enough.

Ad a): This may not be the case in practice, which is discussed extensively in the research literature on comminution. However, variation in these functions in time does, as mentioned, in itself not preclude modeling by Markov chains.

Ad b): It was stated in Equation (27) in the beginning of this section that if the two models gave the same result, $\mathbf{P}^n$ would be equal to $\mathbf{V}(n\Delta t)^{\mathrm{T}}$ and if the two are sufficiently close, the model results will be sufficiently close. Comparing the diagonal elements of the two matrices written in terms of the S_i, this means that $1 - \Delta t S_i$ has to be sufficiently close to $\exp -\Delta t S_i$.

Berthiaux provides a discussion of the advantages of the Markov chain model in comparison with the standard PBM model. One major advantage is that the direct experimental determination of the transfer probability matrix [6] is much easier than having to determine

[6] In the paper discussed in this section, Berthiaux determined the transfer probability matrix from the breakage matrices to show the link between PBM and Markov chain modeling, but normally one would determine the elements of the transfer probability matrix directly from experiment.

the breakage matrices experimentally. Berthiaux uses an example to show that the time-discretization for the Markov chain model can be rather rough for an acceptable agreement with the time-continuous PBM model, something, which allows for very economical evaluation of the stochastic model.

4.2 An Example

For this example the particle size domain of interest is again split into N classes with N the smallest size. The lower limit of class i is denoted by $(l_i)_{0 \leq i \leq N}$, where l_0 is the upper limit of class 1. The class limits are logarithmically spaced, so that:

$$l_{N-(i-1)} = \exp\left[\ln(l_1) + \left(\frac{\ln(l_0) - \ln(l_N)}{N}\right)(i-1)\right]$$

The 'class mark' (the size representing each class) is taken as the logarithmic average, or geometric mean size of the class:

$$\ln(x_i) = \frac{\ln(l_{i-1}) + \ln(l_i)}{2}$$

Relations often used for the selection function and the breakage functions in PBM modeling of grinding processes are [25–27]:

$$S_j = A\left(\frac{x_j}{x_0}\right)^{\alpha} \tag{38}$$

and:

$$B'_{ij} = \phi\left(\frac{l_{i-1}}{l_j}\right)^{\gamma} + (1-\phi)\left(\frac{l_{i-1}}{l_j}\right)^{\beta} \qquad i > j \tag{39}$$

where x_j is the particle size of the j'th size class and x_0 some arbitrarily chosen standard particle size. l_i is, as mentioned, the lower limit of the i'th size class[7].

The B'_{ij} should be normalized to $B_{ij} = B'_{ij}/B'_{j+1,j}$, and the B_{ij} are then the components of the cumulative breakage function, i.e. the fraction of material exiting class j entering class i or finer size classes: $B_{ij} = \sum_i^N b_{ij}$ $i > j$ with $B_{j+1,j} = 1$. The parameters A, α, ϕ, γ and β are normally determined empirically.

The values of the parameters for Equations (38) and (39) are relatively uniform in the literature. For this example they are set to [27]: $A = 1.35\,\mathrm{s}^{-1}$, $\alpha = 0.4$, $\phi = 0.4$, $\gamma = 0.6$ and $\beta = 4.4$. Calculating the components of $\mathbf{B}$, b_{ij}, from:

$$b_{ij} = B_{ij} - B_{i+1,j} \quad i > j$$

then gives the following selection vector and breakage matrix if $N = 7$, $l_N = 10\,\mu\mathrm{m}$ and $l_0 = 1000\,\mu\mathrm{m}$:

[7]Please note that the indexing here follows that of Berthiaux and differs from that of many others, e.g. Hoşten and Avşar [26], who has the highest size class indicated by the largest index.

$$\mathbf{S} = \begin{pmatrix} 0.8242 \\ 0.3072 \\ 0.1145 \\ 0.0427 \\ 0.0159 \\ 0.0059 \\ 0 \end{pmatrix}$$

$$\mathbf{B} = \begin{pmatrix} 0 & 0 & 0 & 0 & 0 & 0 & 0 \\ 0.7394 & 0 & 0 & 0 & 0 & 0 & 0 \\ 0.1190 & 0.7526 & 0 & 0 & 0 & 0 & 0 \\ 0.0592 & 0.1212 & 0.7731 & 0 & 0 & 0 & 0 \\ 0.0388 & 0.0603 & 0.1245 & 0.8057 & 0 & 0 & 0 \\ 0.0261 & 0.0395 & 0.0619 & 0.1297 & 0.8613 & 0 & 0 \\ 0.0176 & 0.0265 & 0.0405 & 0.0646 & 0.1387 & 1.0 & 1.0 \end{pmatrix}.$$

Using Equations (33) and (34) to calculate the matrix $\mathbf{P}$ gives:

$$\mathbf{P} = \begin{pmatrix} 0.99176 & 0.00609 & 0.00098 & 0.00049 & 0.00032 & 0.00021 & 0.00014 \\ 0 & 0.99693 & 0.00231 & 0.00037 & 0.00019 & 0.00012 & 0.00008 \\ 0 & 0 & 0.99885 & 0.00089 & 0.00014 & 0.00007 & 0.00005 \\ 0 & 0 & 0 & 0.99957 & 0.00034 & 0.00006 & 0.00003 \\ 0 & 0 & 0 & 0 & 0.99984 & 0.00014 & 0.00002 \\ 0 & 0 & 0 & 0 & 0 & 0.99994 & 0.00006 \\ 0 & 0 & 0 & 0 & 0 & 0 & 1.00000 \end{pmatrix}. \tag{40}$$

The initial probability vector is the initial size distribution of the powder expressed as mass fraction of powder within each size interval. If the powder initially is log-normally distributed, the probability density function is:

$$f(x) = \frac{1}{x\sigma\sqrt{2\pi}} \exp\left(-\frac{(\ln x - \mu)^2}{2\sigma^2}\right) \tag{41}$$

where x is the particle size and σ and μ the standard deviation and the mean of the distribution of the natural logarithm of the particle size.

The components of the initial probability vector are then approximately:

$$p_i' = f(x_i)(l_i - l_{i+1}),$$

remembering that the N'th class if the finest particle size. These values are only approximate, since they do not exactly sum to 1, so that a normalization is necessary to yield the components of the initial probability vector: $p_i = p_i'/\sum_1^N p_i'$. Setting $\mu = \ln(200)$ and

$\sigma = 0.6$ results in:

$$\mathbf{p} = \begin{pmatrix} 0.04497 \\ 0.25593 \\ 0.43773 \\ 0.22498 \\ 0.03475 \\ 0.00161 \\ 0.00002 \end{pmatrix} .$$

The evolution of the particle size distribution can now be found by applying Equation (2) iteratively. Figure 13 shows the development of the cumulative particle size distribution with time in a grinding process using the parameters given above and a time step, Δt of 0.01 s. It can be seen from the form of $\mathbf{P}$ in Equation (40) that the finest size class acts

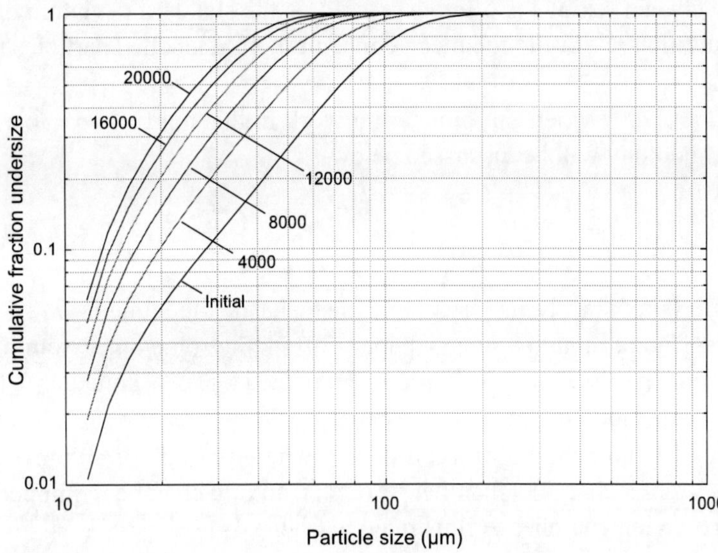

Figure 13: The cumulative mass particle size distribution as a function of time given as number of time steps of 0.01 seconds duration using the parameters stated in the text but with $N = 30$

as an absorbing state. It is normal to plot the particle size as a cumulative distribution, avoiding the problem of how to present the increasing mass in the finest size class in a discrete distribution.

Also Chakraborty and Ramkrishna [28] have modeled grinding processes using Markov chains, emphasizing the advantage of avoiding lengthy computations while still modeling the process in detail.

5 Particle Mixing

Powder mixing is another field where stochastic modeling based on Markov chains has been used.

The method here is to:

- divide the space occupied by the powder in some mixing process, such as a rotary drum mixer, in a set of discrete regions constituting the state space of the Markov process,
- quantify the transfer probability matrix between the regions and compute the time-evolution of the concentration of some labelled particle fraction in each of the regions and finally
- determine the quality of mixing as the variation in the fraction of labelled particles between the spatial regions.

The variation between regions is often expressed as the standard deviation of the labelled particle concentration relative to the mean concentration of labelled particles in the powder as a whole.

Chen and Fan [29] carried out pioneering work in this field. Two papers building on their basic methodology will be discussed below.

5.1 Modeling of a Static Mixer

The group of Berthiaux has been very active in stochastic modeling in powder technology recently and has, in addition to a review paper [30] giving some other examples of the use of stochastic modeling in powder technology, published a paper, Ponomarev et al. [31], on the use of stochastic modeling in powder mixing.

This mixer of Ponomarev et al. is a device wherein a binary mixture of powders is allowed to flow from a charging chamber through a mixing chamber equipped with mixing elements to a receiving chamber as illustrated in Figure 14.

The binary mixture, which initially is completely segregated such that one type of particles occupy the top 5 sections of the charging chamber and the other type the bottom 5, is thus allowed to flow through the mixing chamber and is collected below. The entire vessel is then turned upside-down for a new pass. The model describes the movement of the two types of particles through the mixing section and the filling process of the collection chamber.

In the model the mixing chamber is divided vertically into a number of spatial cells or compartments, which, in accordance with the general modeling concept of Fan et al. [13]—and as described in Section 3.3—each constitutes a CSTR. This is sketched to the right in Figure 14.

The motion of a single particle is again modeled by a Markovian process (X_n), where $(X_n)_{n\geq 0}$ gives the axial location of the particle at time-step n.

The state space for the Markov chain consists of the compartments of the mixer plus the receiving chamber, the receiving chamber constituting an absorbing state.

As before, the time step is kept so short that the transition probabilities of a particle to all compartments except the two that are neighboring the current compartment are negligible. The transition probability matrix for a particle thus forms a sparse matrix, as in a birth-death

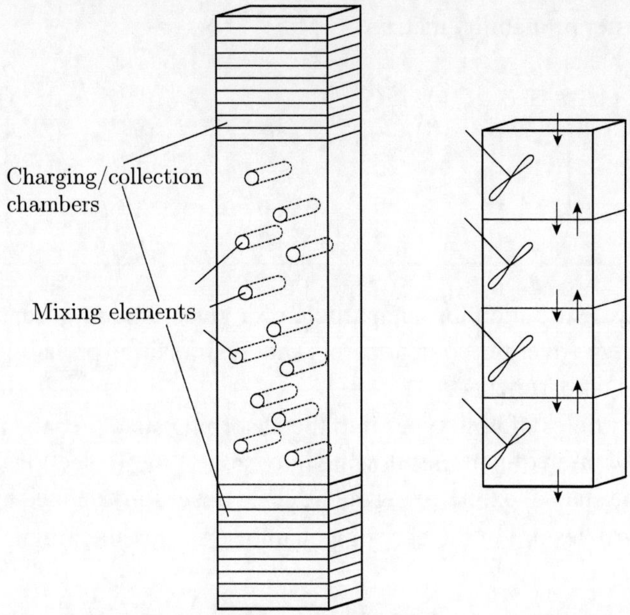

Figure 14: The mixer of Ponomarev et al. After one pass of the binary powder mixture, the whole mixer is turned upside-down for a new pass. The charging/collection chambers are divided into 10 horizontal slices, or sections, as indicated, to investigate the quality of mixing of the powders in the axial direction. To the right the model of the mixing chamber.

Markov chain, with an absorbing state[8]:

$$\mathbf{P} = \begin{pmatrix} \alpha_1 & \beta_1 & 0 & 0 & 0 \\ \delta_2 & \alpha_2 & \beta_2 & 0 & 0 \\ 0 & \delta_3 & \alpha_3 & \beta_3 & 0 \\ 0 & 0 & \delta_4 & \alpha_4 & \beta_4 \\ 0 & 0 & 0 & 0 & 1 \end{pmatrix}$$

where the conditional transfer probabilities are: δ, transferring to the cell above, α, staying in the same cell and β transferring to the cell below.

Since the particles are falling through the mixing section in the gravity field and only can move up by bouncing off the mixing elements, Ponomarev et al. set $\delta = 0$ as a first approximation.

They further set the probability of moving from cell i to the cell below equal to the product of a transfer probability, u, without any mixing elements present in the compartment and a 'retardation factor', ζ_i, due to the presence of mixing elements in the i'th compartment.

[8]Please note that this matrix has been transposed relative to the one given by Ponomarev et al., to be consistent with the rest of this chapter in that p_{ij} is the conditional transition probability from state i to state j and in that the probability vector at time n is obtained by premultiplying the transfer probability matrix with the probability vector at time $n-1$.

This gives the transfer probability matrix:

$$\mathbf{P} = \begin{pmatrix} 1-u\zeta_1 & u\zeta_1 & 0 & 0 & 0 \\ 0 & 1-u\zeta_2 & u\zeta_2 & 0 & 0 \\ 0 & 0 & 1-u\zeta_3 & u\zeta_3 & 0 \\ 0 & 0 & 0 & 1-u\zeta_4 & u\zeta_4 \\ 0 & 0 & 0 & 0 & 1 \end{pmatrix}$$

The probability distribution for one particle of a given type represents the distribution of particles of that type over the compartments and the transition probabilities represent the flows between the compartments.

The flows of particles of both types into the absorbing state, i.e. the receiver, is monitored, and the distribution of particles flowing into each of the 10 sections of the receiver is recorded, giving the state of axial mixing on the scale of scrutiny defined by the 10 sections.

Initially the particles are in the charging chamber, so that the initial probability distribution is:

$$\mathbf{p} = \begin{pmatrix} p_1 \\ p_2 \\ p_3 \\ p_4 \\ p_5 \end{pmatrix} = \begin{pmatrix} 0 \\ 0 \\ 0 \\ 0 \\ 0 \end{pmatrix}$$

where p_1 is the top compartment of the mixing chamber, and p_5 is the absorbing state, i.e. the receiving chamber.

The initial distribution of the two types of particles in the charging chamber is completely segregated, such that one type of particles, type A, fills the bottom 5 sections of the chamber, and the other, B, the top 5 sections. The contents of one whole section is charged to the top compartment in the mixing chamber each time step. Then the development of the probability distribution in time for each of the two particle types becomes:

$$\mathbf{p}^{(n+1)} = (\mathbf{p}^{(n)} + \mathbf{p_f}^{(n)})\mathbf{P}$$

with all the elements of $\mathbf{p_f}^{(n)}$ being zero except for the first, which is:

$$p_{fA,1}^{(n)} = \begin{cases} 1 \text{ for } 1 \leq n \leq 5 \\ 0 \text{ otherwise} \end{cases}$$

for material A and

$$p_{fB,1}^{(n)} = \begin{cases} 1 \text{ for } 6 \leq n \leq 10 \\ 0 \text{ otherwise} \end{cases}$$

for material B.

For the subsequent passes, where the mixer is turned upside-down repeatedly, the $\mathbf{p_f}^{(n)}$ are given by the material distributions in the collector chambers, which become the charging chambers for the next pass.

Figure 15 shows[9] the flows of the two types of particles into the receiver at each time step during the first pass of the material for $u_A = u_B = 0.7$ and $\zeta = 1.0$ for all the mixing compartments.

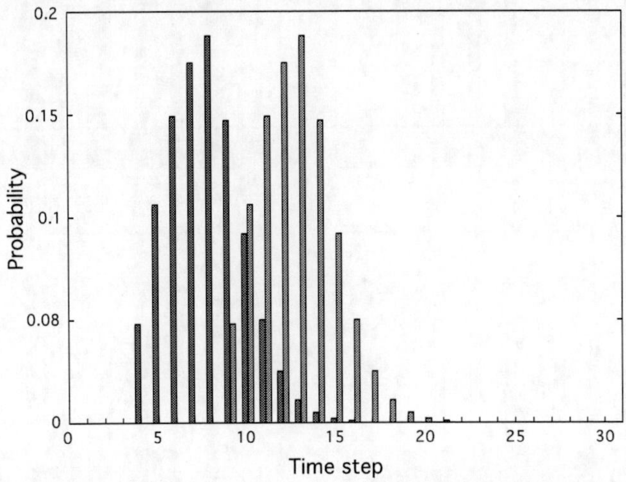

Figure 15: The flow of particles into the receiver at the first pass

Figure 16 show the mixing process as the distribution in the receiver after 1, 3, 5 and 7 passes using the same parameters. Only every other pass is shown to avoid confusion from the powder being turned upside-down every other pass. Section 10 is at the top of the receiver.

If the values of u differ for the two powders, to reflect a tendency for segregation due to a difference in physical properties, such as density and/or size, complete mixing is never achieved. A steady distribution over the sections of the receiver, which is repeated for every pass, is reached also in this situation, but it is not a uniform distribution, as illustrated in Figure 17.

The effect of the retardation factor, ζ, modeling the action of the mixing elements in the mixing chamber, is shown in Figure 18.

5.2 DEM Simulations as an Input to the Markov Model

Doucet et al. [32] used DEM simulations to compute the transition probability matrix for the mixing process in a rotating drum. The process is illustrated in Figure 19.

DEM, or 'Discrete Element Modeling' is an often-used method to study particle mixing processes whereby each particle is followed in a Lagrangian manner as it moves under influence of the forces acting on it, often considered to be the force of gravity and collisional/contact forces with other particles and the wall of the container.

As Doucet et al. point out, DEM simulations are very computationally intensive, and by using a few DEM iterations to compute the Markovian transition probability matrix for

[9]Please note that in this chapter the total amount of each powder is taken as unity so that the amount of powder in each compartment reflects the probability of finding a single particle of that powder in that compartment. In the original paper of Ponomarev et al. the total amount of each powder is taken as equal to five. This is a purely formal difference, and does not influence the results of the model

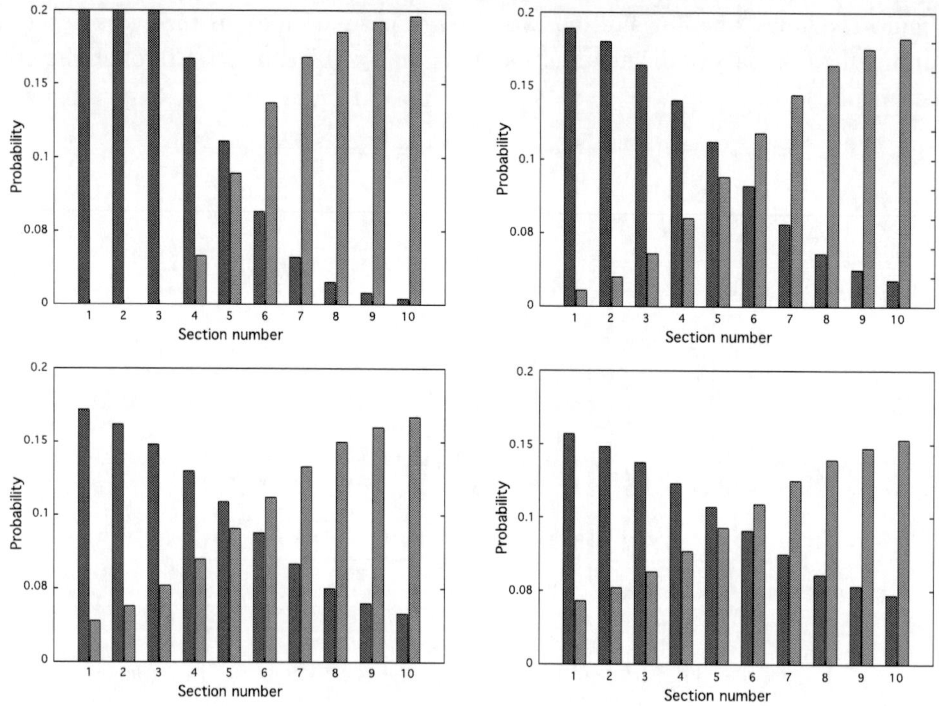

Figure 16: The mixing process illustrated as the distribution of powders in the receiver after 1, 3, 5 and 7 passes (top left to bottom right)

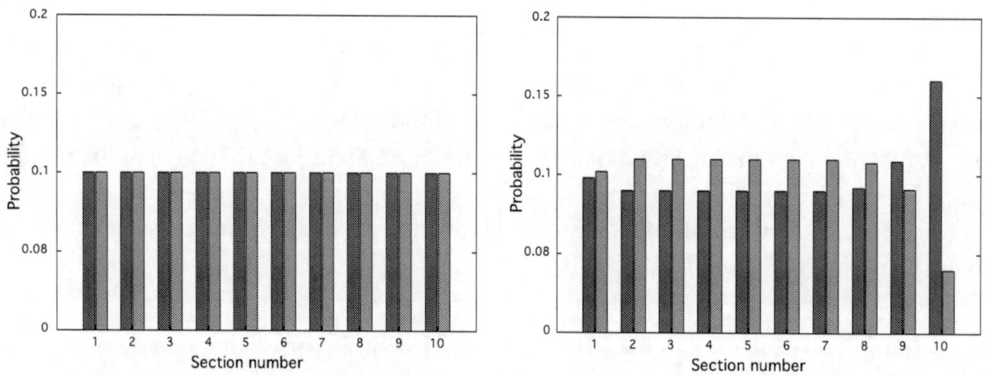

Figure 17: The steady-state distribution (after 77 passes). Left: with both powders having the same u (0.7) and right: for the powders having $u_A = 0.6$ and $u_B = 0.8$.

a given process can yield an elegant and economical model for the process. The assumption is made that the Markov process is stationary, so that the transfer probability matrix does not vary in time.

As before, the motion of a single particle is modeled by a stochastic, Markovian process (X_n), where $(X_n)_{n \geq 0}$ gives the spatial location of the particle at time-step n.

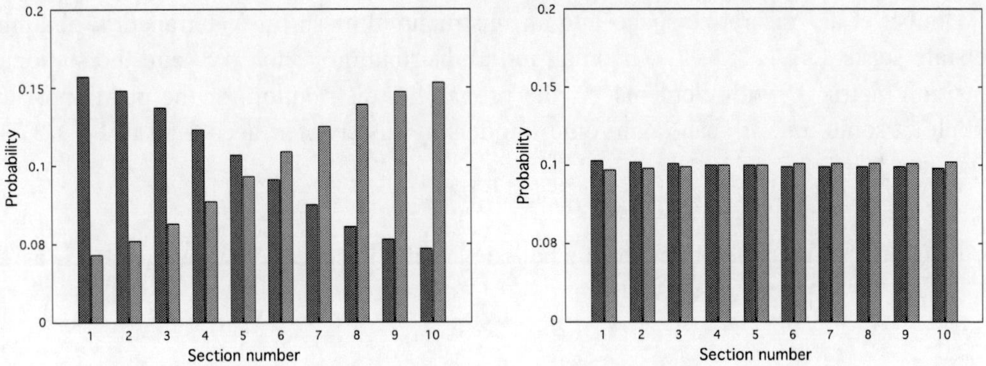

Figure 18: Effect of the retardation factor, ζ, modeling the action of the mixing elements. Left: $\zeta = 1.0$, right: $\zeta = 0.5$, both cases is after seven passes.

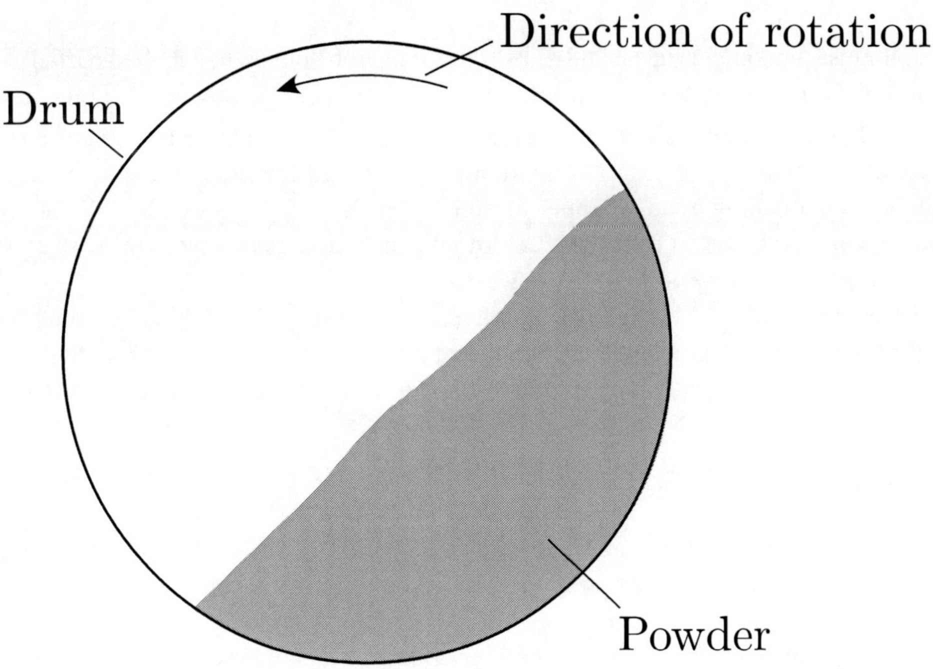

Figure 19: Sketch of powder mixing in a rotating drum mixer

Doucet et al. discretize space into an unstructured mesh of N tetrahedra, obtaining the state space $\{x_1, \ldots, x_N\}$. Given an initial distribution vector $\mathbf{p}^{(0)}$ and the stationary transition matrix $\mathbf{P}$ with elements p_{ij} the probability distribution for the position of the particle at some time n_0 can, as in other models in this chapter, according to Eq. (3), be calculated from

$$\mathbf{p}^{(n)} = \mathbf{p}^{(0)}\mathbf{P}^n. \tag{42}$$

Doucet et al. count the number of labelled particles in state x_i at time $t_n = n\Delta t$ as:

$$\varphi_i^{(n)} = \sum_{p=1}^{N_p} \phi_{p,i}^{(n)}$$

where $\phi_{p,i}^{(n)}$ is an indicator function which takes on the value 1 if particle p is in state i at time t_n, and otherwise is zero. N_p is the number of labelled particles in the system.

Having arranged some initial distribution of labelled particles over the spatial regions, Doucet et al. allow the DEM simulation to proceed for a chosen time interval Δt, N_{LT} times, keeping track of the labelled particles after each time interval and then compute their estimate for the transfer probability matrix from:

$$p_{ij} = \frac{1}{N_{LT}} \sum_{n=1}^{N_{LT}} \frac{1}{\varphi_i^{(n-1)}} \sum_{p=1}^{N_p} \phi_{p,j}^{(n)} \phi_{p,i}^{(n-1)}.$$

The rightmost sum on the right-hand-side of this equation counts the labelled particles that transferred from state x_i to x_j from time $n-1$ to time n. This is divided by the number of labelled particles that were in x_i at time $n-1$, $\varphi_i^{(n-1)}$, giving the fraction of particles that were in x_i that moved to x_j.[10] Summing this over all the time intervals in the DEM simulation and dividing by the number of time intervals [11], N_{LT}, gives the estimate of the probability p_{ij} as the mean of the fraction of labelled particles that were in x_i that moved to x_j during the 'learning period': $N_{LT}\Delta t$.

The authors dedicate the rest of the paper to comparing DEM results with Markov model results, and find, among other things, that the time intervals, Δt, during the learning period have to be sufficiently long for the Markov model to reflect the mixing correctly.

6 Concluding Remarks

The models described here are for quite complex systems in particle technology, in fact for some of the systems formulation of models based on differential conservation equations is not possible at all. Yet the models are, without exception, simple to formulate and are also very fast to evaluate in simple Fortran or C programs or in a matrix oriented package such as MATLAB or, if the matrices are not too large, Mathematica.

[10] In the original paper is written $\varphi_i^{(n)}$ rather than $\varphi_i^{(n-1)}$, but the present formulation ensures that if there is one particle in x_i at time $n-1$ and it moves to be in x_j in time n, then this fraction is 1 rather than ∞. This change is not likely to make any difference in practice to the final results

[11] Note that the time interval, Δt, simulated for the Markovian process is much longer than the time intervals for the DEM simulation itself

References

[1] J. F. Davidson and D. Harrison. *Fluidization*. Academic Press, New York, 1971. chapter 5.

[2] J. M. Matsen, S. Hovmand, and J. F. Davidson. Expansion of fluidized beds in slug flow. *Chem. Eng. Sci.*, 24:1743–1754, 1969.

[3] J. C. Schouten, P. J. M. Valkenburg, and C. M. van den Bleek. Segregation in a slugging fbc large-particle system. *Powder Technol.*, 54:85–98, 1988.

[4] J. C. Abanades and S. Atarés. Investigation of solid mixing in a deep fluidized bed of coarse particles by image analysis. In *Proceedings of the World Conference on Particle Technology 3*, number 211, July 1998.

[5] J. Baeyens and D. Geldart. An investigation into slugging fluidized beds. *Chem. Eng. Sci.*, 29:255–265, 1974.

[6] P. N. Rowe and B. A. Partridge. Particle movement caused by bubbles in a fluidised bed. In *Interaction Between Fluids & Particles*, pages 135–142. Instn Chem. Engrs, London, 1962.

[7] L. G. Gibilaro and P. N. Rowe. A model for a segragating gas fluidised bed. *Chem. Eng. Sci.*, 29:1403–1412, 1974.

[8] H. G. Dehling, H. W. Stuut, and A. C. Hoffmann. Stochastic models for transport in a fluidized bed. *SIAM J. Appl. Math.*, 60:337–358, 1999.

[9] J. F. Davidson, R. Clift, and D. Harrison. *Fluidization*. Academic Press, New York, second edition, 1985.

[10] H. G. Dehling, T. Gottschalk, and A. C. Hoffmann. *Stochastic modelling in process technology*. Elsevier, Amsterdam, London, 2007.

[11] H. G. Dehling, C. Dechsiri, T. Gottschalk, P. C. Wright, and A. C. Hoffmann. A stochastic model for mixing and segregation in slugging fluidized beds. *Powder Technology*, 171:118–125, 2007.

[12] A. T. Harris, R. B. Thorpe, and J. F. Davidson. Stochastic modelling of the particle residence time distribution in circulating fluidised bed risers. *Chem. Eng. Sci.*, 57:4779–4796, 2002.

[13] L. T. Fan, J. R. Too, and R. Nassar. Stochastic simulation of residence time distribution curves. *Chem. Eng. Sci.*, 40:1743–1749, 1985.

[14] W. C. Yang. A model for a dynamics of a circulating fluidised bed loop. In P. Basu and J. F. Large, editors, *Circulating fluidised bed technology*, pages 181–191. Pergamon Press, 1988.

[15] F. Wei and J. Zhu. Effect of the flow direction on axial solid dispersion in gas-solid cocurrent upflow and downflow systems. *Chem. Eng. J.*, 16:109–113, 1993.

[16] A. T. Harris, J. F. Davidson, and R. B. Thorpe. Particle residence time distributions in circulating fluidised beds. *Chem. Eng. Sci.*, 58:2181–2202, 2003.

[17] H. T. Bi, J. Zhou, S.-Z. Qin, and J. R. Grace. Annular wall layer thickness in circulating fluidized bed risers. *Can. J. Chem. Eng.*, 74:811–814, 1996.

[18] E. H. van der Meer, R. B. Thorpe, and J. F. Davidson. Flow patterns in the square cross-section riser of a circulating fluidised bed and the effect of riser exit design. *Chem. Eng. Sci.*, 55:4079–4099, 2000.

[19] L. G. Gibilaro, H. W. Kropholler, and D. J. Spikins. Solutions of a mixing model due to van de vusse by a simple probability method. *Chem. Eng. Sci.*, 22:517–523, 1967.

[20] A. D. Randolph and M. A. Larson. *Theory of Particulate Processes*. Academic Press, 1988.

[21] D. Ramkrishna. *Population Balances, Theory and Applications to Particulate Systems in Engineering*. Academic Press, 2000.

[22] H. Berthiaux. Analysis of grinding processes by markov chains. *Chem. Eng. Sci.*, 55:4117–4127, 2000.

[23] H. Berthiaux and J. Dodds. A new estimation technique for the determination of breakage and selection parameters in batch grinding. *Powder Technol.*, 94:173–179, 1997.

[24] E. Kreyszig. *Advanced Engineering Mathematics*. John Wiley and Sons, sixth edition, 2006.

[25] L. G. Austin. A discussion of equations for the analysis of batch grinding data. *Powder Technology*, 106:71–77, 1999.

[26] Ç. Hoşten and Ç. Avşar. Variation of back-calculated breakage rate parameters in Bond-mill grinding. *Scandinavian Journal of Metallurgy*, 33:286–293, 2004.

[27] Hoon Lee, Cho Heechan, and Jihoe Kwon. Using the discrete element method to analyze the breakage rate in a centrifugal/vibration mill. *Powder Technology*, 198:364–372, 2010.

[28] J. Chakraborty and D. Ramkrishna. Identification of Markov matrices of milling models. *Ind. Eng. Chem. Res*, 48:9763–9771, 2009.

[29] S. J. Chen and L. T. Fan. The mixing of solid particles in a motionless mixer: a stochastic approach. *AIChE Journal*, 18:984–989, 1972.

[30] H. Berthiaux and V. Mizonov. Applications of Markov chains in particulate process engineering: A review. *Canadian Journal of Chemical Engineering*, 82:1143–1168, 2004.

[31] D. Ponomarev, V. Mizonov, C. Gatumel, H. Berthiaux, and E. Barantseva. Markov-chain modelling and experimental investigation of powder-mixing kinetics in static revolving mixers. *Chemical Engineering and Processing*, 48:828–836, 2009.

[32] J. Doucet, N. Hudson, F. Bertrand, and J. Chaouki. DEM-Markov combination. Computing the transition probability matrix from DEM models. *Computers and Chemical Engineering*, 32:1334–1341, 2008.

In: Powder Engineering, Technology and Applications
Editor: Jason M. Barker, pp.257-309
ISBN: 978-1-61761-216-9

Chapter 10

EVALUATION OF MECHANICAL ENERGY APPLIED TO POWDERS IN DRY PROCESSES AND ITS APPLICATION FOR DESIGN AND PREPARATION OF FUNCTIONAL PARTICULATE MATERIALS*

Tomohiro Iwasaki†
Department of Chemical Engineering, Osaka Prefecture University,
Sakai, Osaka, Japan

ABSTRACT

Dry mechanical treatments of particulate materials such as mixing and grinding are frequently carried out in various industries such as foods, pharmaceuticals, cosmetics, electronics, etc., because of its simplicity and versatility. In the mechanical treatments, the physical properties and characteristics of powders treated depend strongly on both amount and history of the mechanical energy applied to the powders during treatment. For controlling their physical properties and characteristics, full understanding and precise control of the net applied mechanical energy are necessary. The author has proposed a simple method for measuring the applied energy and evaluating the energy characteristics of powder processing machines. In the proposed method, the stress-sensitive materials, the calcium carbonate powder containing the brittle agglomerates of fine pigment (iron oxide) particles and the copper powder with a spherical particle shape, are properly used as a standard material depending on the amount of the external force exerting on the powder to be treated. In the former, the applied energy can be estimated by analyzing the disintegration phenomenon of the agglomerates in the mixing process, and this method is available for powder mixers conducting relatively mild mixing such as a vessel-rotating type and an agitating type of powder mixers. On the other hand, the latter can estimate the applied energy from the plastic deformation of copper particles and is employed for grinding and compounding machines in which relatively larger

* A version of this chapter was also published Chemical Engineering Research Trends, edited by Leon P. Berton, Nova Science Publishers. It was submitted for appropriate modifications in an effort to encourage wider dissemination of research.

† E-mail: iwasaki@chemeng.osakafu-u.ac.jp

mechanical energy applies to the powders. Based on the applied energy per unit time thus estimated, various types of powder mixers and mills could be characterized on a uniform scale regardless of their mechanical structure and mixing and grinding mechanisms. Subsequently, the author studied on the mechanical treatments for powder design of functional materials as examples of utilization of the applied energy, in which the surface of easily deformable particles with a spherical shape were mechanically modified without changing their original shape: e.g., the formation of polymer film on the surface of lead-free solder alloy particles for oxidation resistance. The author has proved through the application of the evaluation results of the applied energy to practical treatments that the operating conditions of treatment processes can be optimized based on the physical properties of raw materials to be treated and that advanced powder design can be achieved in simple dry mechanical processes.

INTRODUCTION

Particulate materials are widely used as raw materials, intermediate products and end products in various industrial processes because they have many useful characteristics differing from those of gaseous, liquid and solid materials. Recently, industrial products have been required to have more advanced, multiple functions and higher performance, leading to the increase in the request for further improvement of characteristics of particulate materials in initial, intermediate and/or final stages of manufacturing processes. In order to satisfy such request, to add new function(s) suitable for the usage purpose to the particulate materials, that is, functionalization of particulate materials, has been frequently performed.

In the functionalization of particulate materials, the powder design, i.e., to determine elaborately the constituents and suitable features of the functional particulate material, plays an important role. To add multiple functions to the particulate materials, powder designs using multi components are effective in many cases because the functions of a single material are limited. Various methods using electrical, magnetic, physical and chemical interactions have been developed to introduce different components into the base materials. Particle compounding (hybridizing) methods are very useful since various combinations of raw materials are possible. Especially, mechanical compounding in dry (gas or solid phases) processes has been widely employed in various industrial fields because of the following advantages: energy saving, contamination control, simplification of the manufacturing process and simple operational procedure and handling.

In mechanical compounding methods, strong external forces applied to the particles induce their physical and/or chemical changes: for example, formation of chemical bonds between the particle surfaces due to the mechanical forces, which is called a mechanochemical reaction. Since it depends on the intensity of the external forces (the amount of the mechanical energy) whether the composite particles exhibit the desired functions or not, it is important to select compounding methods suited for the combination of raw materials and their chemical composition and physical properties.

The mechanical energy applied to the particles during treatment is principally dominated by the following three factors.

(i) Equipment structure: the factors determining the type of mechanical energy, i.e., its direction (shear, attrition, compression, bending, etc.) and duration (instantaneous, continuous, intermittent, periodical, etc.).
(ii) Operating conditions: the factors determining the amount of the mechanical energy, e.g., the charge amount of the particles, the feeding rate of the particles into a high energy zone in the equipment, the components and pressure of the atmosphere gas, etc.
(iii) Starting materials: the factors determining the receiving of the mechanical energy, that is, the external factors (size, shape, surface friction, hardness, etc.) and internal factors (softening and melting points, transition point, crystal structure, etc.) of the particles.

In order to conduct advanced powder designs, a grasp of the mechanical energy and its control with high accuracy on the compounding processes is necessary. Thus, the applied mechanical energy must be evaluated considering these factors. Furthermore, the selection of equipment and the optimization of its conditions should be done based on the powder design. However, those have been often carried out based on the engineer's experience and intuitions with trial and error because the methods to determine the optimum equipment structure and operating conditions have not been developed yet. Under this situation, it is difficult to prepare the functional particulate material with better functions than those by the conventional techniques. Accordingly, in order to solve such problem, it is necessary to develop an industrially usable method for evaluating the net applied mechanical energy and utilizing it to the powder design.

The evaluation methods of the applied mechanical energy are mainly classified into two categories: direct measurements and indirect ones. In indirect measurements, the applied energy is qualitatively (relatively) evaluated through changes in the size, shape and color of the processed particles using a scanning electron micrograph, an image analyzer and a colorimeter. For example, in relatively low stress fields, a thermosensitive pigment powder (microcapsules) and pigment granules (agglomerates) can be used as tracer materials, i.e., stress-sensitive materials, to colorimetrically measure the flow patterns and to characterize the equipment [1, 2]. Satoh et al. classified various types of powder mixers into three groups based on the mixing performance by using a mixture of ferric oxide agglomerates and calcium carbonate powder as the stress-sensitive material [3]. Murakami et al. added short and straight annealed copper wires into a high viscosity fluid and estimated the shear rate distributions in the mixer vessel using the degree of bend in the deformed wires [4]. Measurements of the power requirement and the torque acting on the rotating shaft of an agitator or a vessel during the operation are included in this category.

Direct measurements to quantitatively evaluate the applied energy have been actively attempted by installing various kinds of sensors to the equipment. Yoshino et al. measured the compression and friction forces applied to the particles in the Angmill using several strain gauges attached to the horizontal stay of the inner piece [5]. The advance of measurement technique accompanied by increases in the accuracy of sensors has accelerated to control the applied mechanical energy. However, the evaluation of the net applied energy considering the utilization for the practical applications has lacked in experimental approaches.

The author has proposed two types of simple methods for estimating experimentally the net applied energy using stress-sensitive materials, in which the applied energy is estimated based on the disintegration (comminution) of the agglomerates of fine pigment (iron oxide) particles in the powder bed of calcium carbonate [6] and the plastic deformation of spherical soft metal particles [7–9]. The former is applied to the characterization of vessel-rotating and stirring type of powder mixers in which relatively low stress fields are formed in the powder bed; the later for grinding and compounding machines in which higher compressive and shear stresses apply to the particles. In the following sections, the details of these methods are described and some examples of application to a practical powder design are presented.

EVALUATION OF APPLIED MECHANICAL ENERGY

Energy Estimation in Vessel-Rotating Type and Agitating Type of Powder Mixers

In this estimation method, soft dye granules (brittle agglomerates of ferric oxide (Fe_2O_3), Toda Color 140ED supplied by Toda Kogyo Corp.) were used as a stress-sensitive material, of which the degree of disintegration, i.e., the agglomerate size, changes depending on the amount of the mechanical energy applied to the Fe_2O_3 agglomerates [3]. First, in order to determine the correlation between the agglomerate size and the applied energy, a high-speed agitation mixer (see Figure 1) having a simple mechanism of stress transmission from the stirring impellers to the powder was used and the disintegration phenomenon of agglomerates in a bulk powder bed of calcium carbonate ($CaCO_3$, Super #1500 supplied by Maruo Calcium Co., Ltd.) was theoretically analyzed. The disintegration process was simulated by means of a simplified model of comminution of brittle materials [10–12].

comerates and bulk $CaCO_3$ powder, two kinds of different processes, disintegration and agglomeration of the Fe_2O_3 agglomerates, proceed simultaneously. During the disintegration process, the agglomerates dispersed uniformly in the bulk powder bed are broken down gradually, and then become smaller lumps [13]. On the other hand, during the agglomeration process, re-agglomeration occurs among smaller lumps generated by disintegration. In the following, the model equations based on the population balance in each phenomenon are described by assuming that the powder mixture forms a uniform flow without local residence during the mixing operation and that the agglomerates are homogeneously dispersed in the bulk powder bed before the mixing operation is started. The disintegration and agglomeration of $CaCO_3$ agglomerates were ignored.

Temporal change in the volume fraction of the agglomerates of diameter D_{px} with the progress of disintegration is described by a modified comminution model [12] as follows (see Figure 2):

$$\frac{\partial^2 D(D_{px},t)}{\partial D_{px}\partial t} = \int_{D_{px}}^{D_{pxm}} \frac{\partial D(D_{px},t)}{\partial D_{py}} S(D_{py}) \frac{\partial B(D_{px},D_{py})}{\partial D_{py}} \mathrm{d}D_{py} - S(D_{px}) \frac{\partial D(D_{px},t)}{\partial D_{px}} \quad (1)$$

where $D(D_{px}, t)$ indicates the volume fraction of the agglomerates less than diameter D_{px} at the mixing time t. $S(D_{py})$ and $B(D_{px}, D_{py})$ are the selection function and the distribution function of the breakage, respectively. The selection function represents the probability that the agglomerates of diameter D_{py} are disintegrated per unit time. The distribution function indicates the volume fraction of agglomerates smaller than D_{px} produced by disintegration of the agglomerates of D_{py}. The formula expressing the disintegration rate of agglomerates is given by rewriting Eq. (1) as follows:

$$\left\{\frac{\mathrm{d}v_i(t)}{\mathrm{d}t}\right\}_{\mathrm{Breakage}} = \sum_{\substack{j=1 \\ i>j}}^{i-1} S_j\, b_{ij}\, v_j(t) - S_i\, v_i(t) \tag{2}$$

Where

$$v_i(t) = \{\partial D(D_{\mathrm{px}}, t) / \partial D_{\mathrm{px}}\} \mathrm{d}D_{\mathrm{px}} \tag{3}$$

$$v_j(t) = \{\partial D(D_{\mathrm{py}}, t) / \partial D_{\mathrm{py}}\} \mathrm{d}D_{\mathrm{py}} \tag{4}$$

$$b_{ij} = \{\partial B(D_{\mathrm{px}}, D_{\mathrm{py}}) / \partial D_{\mathrm{py}}\} \mathrm{d}D_{\mathrm{py}} \tag{5}$$

$$S_i = S(D_{\mathrm{px}}) \tag{6}$$

The term of agglomeration rate in the population balance model is expressed as follows:

$$\left\{\frac{\mathrm{d}v_i(t)}{\mathrm{d}t}\right\}_{\mathrm{Agglomeration}} = C\left\{-v_i(t) + \frac{1}{8} v_k(t)\right\} \tag{7}$$

where C is the rate constant of agglomeration. The first and second terms in the right side of Eq. (7) respectively mean the disappearance due to the agglomeration between the agglomerates of diameter D_{py} and the generation from eight smaller lumps of diameter $D_{py}/2$. Using Eqs. (2) and (7), the population balance equation of the agglomerates is expressed as:

$$\frac{\mathrm{d}v_i(t)}{\mathrm{d}t} = \sum_{\substack{j=1 \\ i>j}}^{i-1} S_j\, b_{ij}\, v_j(t) - S_i\, v_i(t) + C\left\{-v_i(t) + \frac{1}{8} v_k(t)\right\} \tag{8}$$

The Fe_2O_3 agglomerates are disintegrated on collision with the impeller blades and the vessel wall and in the shear stress field generated by the local velocity difference of powder flow. The degree of disintegration depends strongly on the physical properties (size, density, adhesive property, etc.) of the agglomerates [14] and the operating conditions of the powder mixer. Thus, it is very difficult to express theoretically the external force applied to the agglomerates based on both the flow pattern and the stress distribution in the powder bed. In this analysis, assuming that the agglomerates are disintegrated mostly in the stress field around the impeller blades, the disintegration phenomenon is simplified to express the selection and distribution functions.

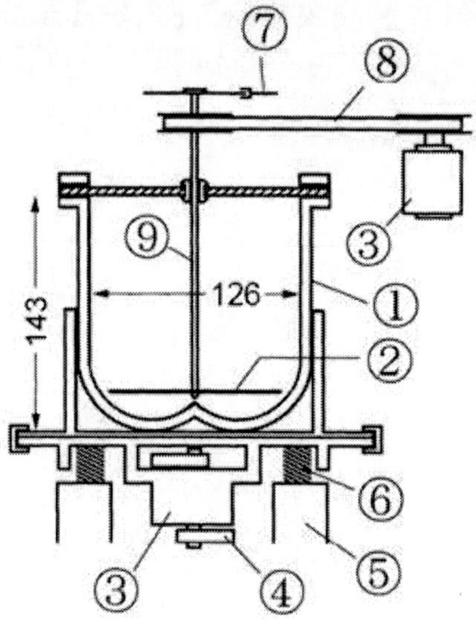

Figure 1. Schematic diagram of agitation type of powder mixer (unit: mm); 1. vessel, 2. impeller, 3. motor, 4. weight, 5. base, 6. spring, 7. tachometer, 8. belt, 9. shaft

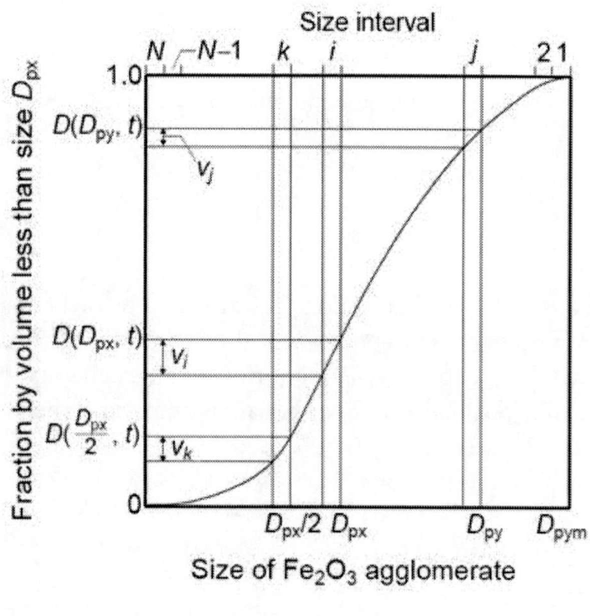

Figure 2. Size distribution of agglomerates and nomenclature

It is assumed that the selection function $S(D_{py})$ is proportional to the following factors: (i) the volume ratio of the region where the agglomerates of D_{py} are selected per unit time (= the breakage region, see Figure 3) to the powder bed, (ii) the probability that the agglomerates of D_{py} are taken into the breakage region moved by the rotation of the impeller (= the ratio of the inertial force of an agglomerate to the fluid resistance force applied to the agglomerate), (iii) the number n of impeller blades, (iv) the frequency of impact force applied to the agglomerates (= the function of impeller speed N), and (v) the powder pressure in the powder bed.

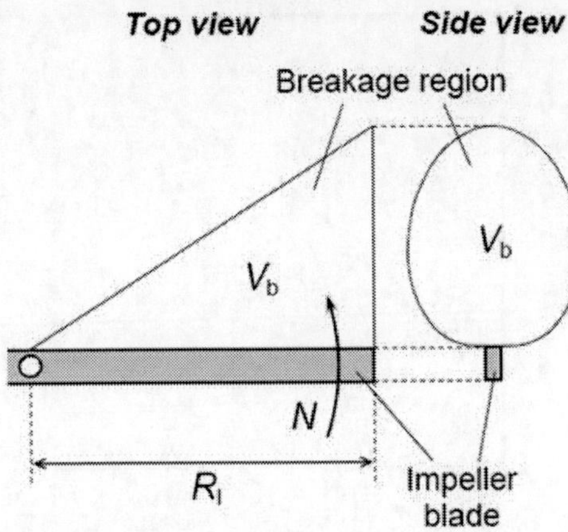

Figure 3. Schematic illustration of breakage region formed in front of impeller blade

$$S(D_{py}) = k_1 \cdot \frac{V_b}{\pi R_V^2 H} \cdot \frac{D_{py}^3 \rho_A}{D_{py}^2} \cdot n \cdot N^c \cdot \rho_B g H \cdot \frac{2R_I}{H} \cdot \frac{2R_I}{2R_V} \tag{9}$$

where V_b indicates the volume of the breakage region, R_I and R_V respectively the radii of the impeller and the cylindrical mixer vessel, H the height of the powder bed, ρ_A and ρ_B respectively the bulk densities of the agglomerate and the bulk powder bed, k_1 and c the constants, and g the gravity acceleration. To take the mixer size into consideration, the parameters $2R_I/H$ and $2R_I/(2R_V)$ are introduced to $S(D_{py})$.

When adhesive fine particles form the agglomerates, larger-sized agglomerates have a tendency to contain larger voids (i.e., crack) inside themselves. Thus, assuming that the intra-void fraction ε_i in an agglomerate increases with the diameter D_{py}, ε_i is given as a function of D_{py}:

$$\varepsilon_i = 1 - A(D_{py}/D_{pym})^{-B} \tag{10}$$

where D_{pym} is the maximum diameter of agglomerate (= 2.8 mm) and A and B are the constants. Therefore, the apparent density ρ_A of the agglomerate of D_{py} including the intra-void is expressed as follows:

$$\rho_A = \rho_{F0}\,(1 - \varepsilon_i) = A\,\rho_{F0}(D_{py}/D_{pym})^{-B} \tag{11}$$

where ρ_{F0} indicates the true density of the Fe_2O_3 particle (= 5200 kg·m^{-3}). As shown in Figure 4, A and B were determined to be 0.13 and 0.21, respectively, so that two conditions were satisfied as follows: (i) the Fe_2O_3 agglomerate of maximum diameter D_{pym} has the same loose bulk density (including the void) as the Fe_2O_3 powder bed (= 658 kg·m^{-3}) and (ii) the intra-void fraction inside the agglomerate formed by a regular tetrahedron structure consisting of four primary Fe_2O_3 particles with a size of 0.6 μm is equal to that of the Fe_2O_3 powder bed at the state of rhombohedral regular packing (= 0.259). The volume V_b of the breakage region was determined as follows. Hayashi et al. [15] expressed the propagation of stress in solids using the formula of Boussinesq's equation as illustrated in Figure 5.

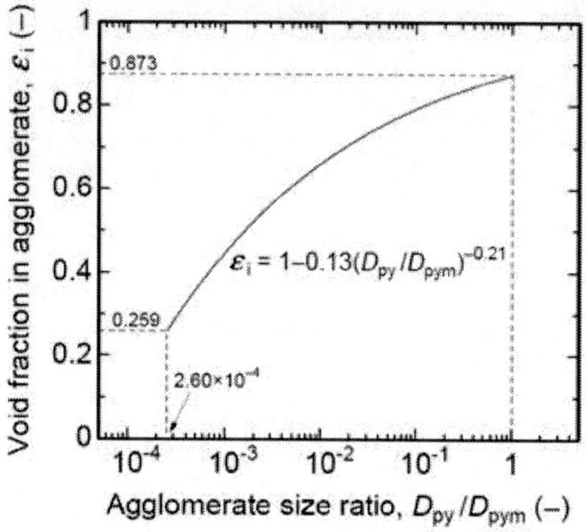

Figure 4. Relationship between intra-void fraction ε_i and agglomerate diameter ratio D_{py}/D_{pym}

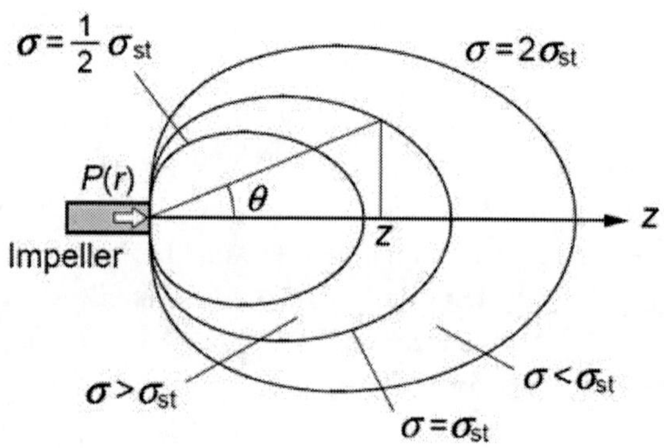

Figure 5. Model of breakage region (side view)

In this analysis, the breakage region is represented by using this model to determine the selection function $S(D_{py})$. The region where the stress σ is propagated is given as a function of z and θ indicating a position in the powder bed.

$$\sigma = \frac{3P(r)}{2\pi z^2}\cos^3\theta \tag{12}$$

where $P(r)$ is the impact force that the impeller applies to the powder bed at a radius r on the impeller blade and proportional to the momentum of the impeller blade, i.e., the peripheral velocity.

$$P(r) = k_2 N r \tag{13}$$

k_2 is the constant. In the region where σ is larger than the fracture strength σ_{st} of the agglomerate, the agglomerates included in the region are disintegrated. Here, σ_{st} of brittle materials at unconfined compression is related to the tensile strength σ_{tm} of the material [16, 17].

$$\sigma_{\mathrm{st}} = \sigma_{\mathrm{tm}} / 0.7 \tag{14}$$

σ_{tm} of the agglomerate is approximately expressed by Rumpf's equation:

$$\sigma_{\mathrm{tm}} = \frac{1-\varepsilon_{\mathrm{i}}}{\pi} k \frac{F_{\mathrm{t}}}{d^2} \tag{15}$$

where d is the mean diameter of the primary Fe_2O_3 particles, and k is the coordination number and can be calculated by using the empirical formula proposed by Ridgeway and Tarbuck [18].

$$k = 13.8\left(1 - \sqrt{1.21\varepsilon_{\mathrm{i}} - 0.297}\right) \tag{16}$$

F_{t} in Eq. (15) is the breaking tensile force acting on a single contact point between primary Fe_2O_3 particles and was measured with a separable cell type tensile tester (Cohetester, model CL-II, Hosokawa Micron Corp.) for the Fe_2O_3 powder bed [19] according to the following procedure. The Fe_2O_3 agglomerates were filled in the separable cell consisting of a movable cell and a fixed cell, and the void fraction was adjusted by applying the normal stress of 4.9–12.7 kPa for 10 min. After removing the normal stress, the movable cell was pulled at a constant rate of 0.017 $\mathrm{mm \cdot s^{-1}}$. The tensile force at the breaking point was measured. Figure 6 shows the measured relationship between the maximum tensile stress and the void fraction of the Fe_2O_3 powder bed. By applying Eq. (15) to the experimental data, the breaking tensile force F_{t} was determined to be 4.55 nN. By using Eqs. (10), (14)–(16), σ_{st} is given as a function of the agglomerate diameter D_{py}.

$$\sigma_{\mathrm{st}} = \frac{6.28AF_{\mathrm{t}}}{d^2}(D_{\mathrm{py}}/D_{\mathrm{pym}})^{-B}\left[1 - \sqrt{0.913 - 1.21A(D_{\mathrm{py}}/D_{\mathrm{pym}})^{-B}}\right] \tag{17}$$

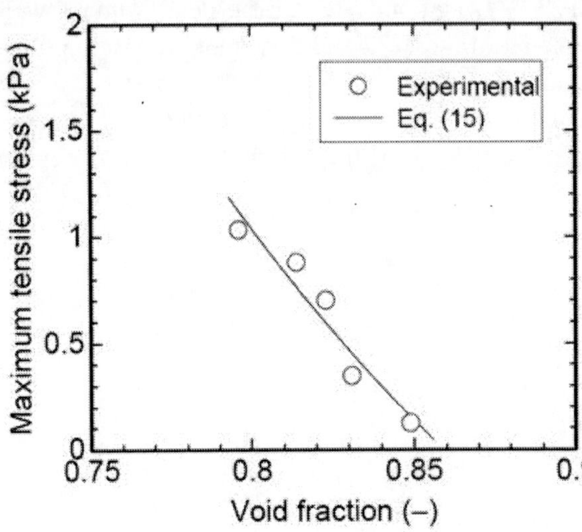

Figure 6. Relationship between maximum tensile stress and void fraction of Fe_2O_3 powder bed

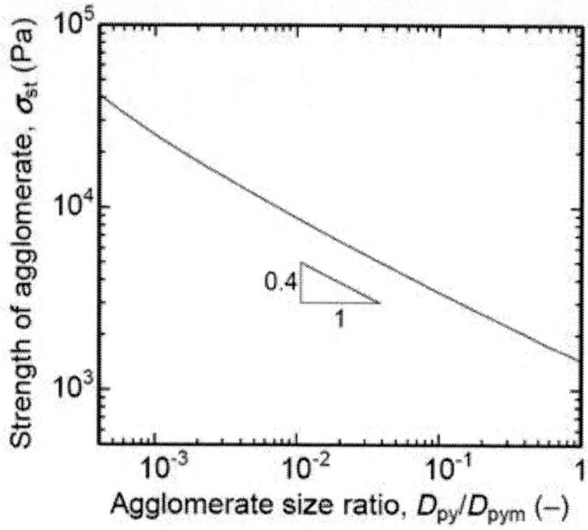

Figure 7. Change in estimated fracture strength σ_{st} with agglomerate size ratio D_{py}/D_{pym}

This equation indicates that the fracture strength of the Fe_2O_3 agglomerate decreases with increasing in the particle size similar to brittle materials such as quartz and lime. Figure 7 shows the relationship between the fracture strength σ_{st} of the agglomerate calculated by using Eq. (17) and the agglomerate size D_{py}. When D_{py} = 1 to 1000 μm, σ_{st} was in the range of 1 to 100 kPa, similarly to vulnerable agglomerates [14], and could be almost approximated by a simple formula:

$$\sigma_{st} = k'(D_{py}/D_{pym})^{-b} \tag{18}$$

where k' and b are the constants depending on the physical properties of the material. Eq. (18) can be generally applied for brittle materials, and b is in the range of 0.2 to 0.9 [20]. For the Fe_2O_3 agglomerate, the value of b was approximately 0.4, as shown in Figure 7; this means that the fracture of the Fe_2O_3 agglomerate can be dealt with as well as that of a brittle material. The fracture model of the Fe_2O_3 agglomerate proposed in this work also expressed well the actual phenomenon.

The volume V_b of the region where the agglomerates with a diameter of more than D_{py} are disintegrated, that is, the volume of the region of $\sigma \geq \sigma_{st}$, is expressed by the following equation.

$$V_b = \int_0^{R_I} \int_0^{z_{max}} 2z \tan\theta \, dz \, dr \tag{19}$$

By using Eqs. (12), (13) and (18), V_b is expressed as follows.

$$V_b = \frac{3k_2 N R_I^2}{k'\pi}(D_{py}/D_{pym})^b \tag{20}$$

Consequently, the selection function $S(D_{py})$ is obtained from Eqs. (9), (11) and (20).

$$S(D_{\mathrm{py}}) = K_{\mathrm{A}}(D_{\mathrm{py}}/D_{\mathrm{pym}})^{-B+b+1} \tag{21}$$

where

$$K_{\mathrm{A}} = \frac{6Ak_1k_2 g\,\rho_{\mathrm{F0}}\rho_{\mathrm{B}} D_{\mathrm{pym}}}{k'\pi^2}\frac{R_{\mathrm{I}}^{\ 4}}{R_{\mathrm{V}}^{\ 3}H} nN^{\,c+1} \tag{22}$$

The distribution function $B(D_{\mathrm{px}}, D_{\mathrm{py}})$ is approximated by a logarithm normal distribution depending only on the ratio $D_{\mathrm{px}}/D_{\mathrm{py}}$ of the agglomerate diameters before and after the disintegration.

$$B(D_{\mathrm{px}}, D_{\mathrm{py}}) = \frac{1}{\sqrt{2\pi}\log\sigma_{\mathrm{Bg}}} \times \int_0^{D_{\mathrm{px}}/D_{\mathrm{py}}} \exp\left\{-\frac{[\log(D_{\mathrm{px}}/D_{\mathrm{py}}) - \log(D_{\mathrm{B50}}/D_{\mathrm{py}})]^2}{2(\log\sigma_{\mathrm{Bg}})^2}\right\} \mathrm{d}\left\{\ln\left(\frac{D_{\mathrm{px}}}{D_{\mathrm{py}}}\right)\right\} \tag{23}$$

where D_{B50} and σ_{Bg} are the logarithm geometric mean diameter and standard deviation of the disintegrated agglomerates, respectively.

The parameters K_{A}, D_{B50} and σ_{Bg} in the selection and distribution functions were determined experimentally according to the mono-sized-fraction method [21, 22]. In the method, the bulk powder in which Fe_2O_3 agglomerates within a size division (lower limit: D_{pxm}) were dispersed was mixed for a proper time t, and then the decrease of the volume fraction $\Delta D(D_{\mathrm{pxm}}, t)$ and the size distribution of the agglomerates were measured. The values of the selection and distribution functions were calculated by using the following equations [21, 22]:

$$S_i = \frac{1}{t}\ln\left\{\frac{1}{\Delta D(D_{\mathrm{pxm}}, t)}\right\} \tag{24}$$

$$B(D_{\mathrm{px}}, D_{\mathrm{pxm}}) = \frac{D(D_{\mathrm{px}}, t)}{1 - \Delta D(D_{\mathrm{pxm}}, t)} \tag{25}$$

and the parameters can be determined using Eqs. (21) and (23). In the experiments for determining these parameters, the Fe_2O_3 agglomerates were divided into six classifications from 106 to 1180 μm with seven standard (Japanese Industrial Standard: JIS) sieves. The additional amount of the sieved agglomerates to the bulk powder was kept constant at 5 mass% regardless of the agglomerate size. The charge volume (including the void) of the powder mixture was kept constant at 33% of the effective volume of the mixer vessel; the height of the powder bed was $H = 0.048$ m. The powder mixture was mixed for $t = 2$ and 6 s at the impeller speed $N = 8.3$ and 16.7 s^{-1}. The mixed powder mixture was sampled and the size distribution of the agglomerates (Heywood diameter, volume basis) was measured with an image analyzer (Luzex-FS, Nireco Corp.) mounted on a metallographic microscope. Figure 8 shows the relationship between the selection function S_i and the agglomerate size (lower limit) D_{pxm} used in the mono-sized-fraction method. S_i depended on the impeller speed N and was independent of the mixing time t. This tendency agreed with Eq. (21) and the experimental data could be approximated by Eq. (21). The parameter K_A in Eq. (21) was determined as follows: $K_A = 0.159\ s^{-1}$ at $N = 8.3\ s^{-1}$ and $K_A = 0.434\ s^{-1}$ at $N = 16.7\ s^{-1}$. The constant c in Eq. (22), which represented the degree of influence of the impeller speed on the impact frequency between the impeller blade and the agglomerates, was determined to be 1.45.

Figure 9 shows the change in the value of the distribution function $B(D_{px}, D_{pxm})$ with D_{px}/D_{pxm} in the case of $N = 16.7\ s^{-1}$ as an example. $B(D_{px}, D_{pxm})$ was independent of both agglomerate size and mixing time, and Eq. (23) could be applied to the experimental results. The parameters D_{px}/D_{py} and σ_{Bg} in Eq. (23) were determined as follows: $D_{px}/D_{py} = 0.3$ and $\sigma_{Bg} = 1.4$. Similar values were obtained in the case of $N = 8.3\ s^{-1}$; this means that $B(D_{px}, D_{pxm})$ is also independent of the impeller speed.

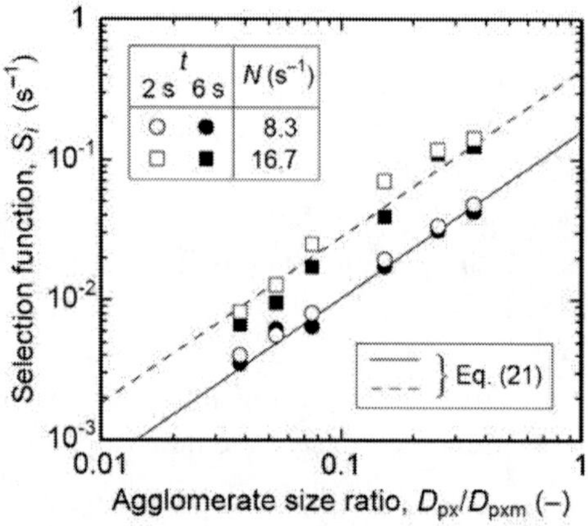

Figure 8. Plots of values of selection function S_i against agglomerate size ratio D_{px}/D_{pxm}

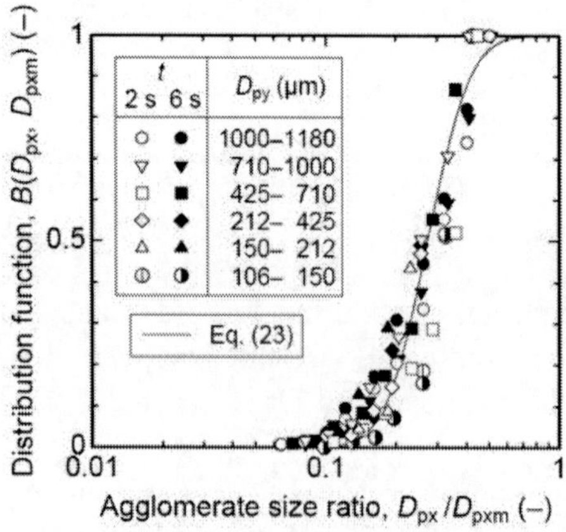

Figure 9. Plots of values of distribution function $B(D_{px}, D_{pxm})$ against agglomerate size ratio D_{px}/D_{pxm} at impeller speed $N = 16.7\ s^{-1}$

In order to investigate the influence of the rate constant C of agglomeration on the temporal change in the agglomerate size, the model calculation was carried out under several values of C, and the mean diameter D_{50} and standard deviation σ_g of the size distribution of the agglomerates were obtained. The mixing experiments were carried out under the same conditions, in which an appropriate amount of the $CaCO_3$ powder and the Fe_2O_3 agglomerates, in the weight proportion 95:5 [3], was put into the vessel. The size distributions measured with the image analyzer were approximated in terms of a log-normal distribution, and the geometric mean diameter D_{50} and the standard deviation σ_g were calculated. Figures 10 and 11 show the temporal changes in D_{50} and σ_g together with the experimental data, respectively. Both D_{50} and σ_g were varied with C, and especially, remarkable differences among the calculation results were observed at a longer mixing time. In the change in D_{50}, a similar tendency was seen regardless of C. On the contrary, the calculation results of σ_g showed obviously a different tendency as C increased; the calculation value of σ_g increased as the mixing time proceeds at larger values of C although the experimental data were almost constant. These results reveal that the agglomeration among smaller lumps hardly occurred, that is, $C = 0$. Accordingly, the calculation results of D_{50} obtained at $C = 0$ are shown in the following.

Figure 12 shows the effect of the impeller speed N on the dispersion of the agglomerates. The calculated values of D_{50} agreed well with the experimental data. The Fe_2O_3 particles in the bulk powder were dispersed rapidly at higher speeds due to increasing in both the impact force and collision frequency of the agglomerates with the impeller blade. To investigate the effects of both factors on the dispersion process, the mixing experiments were carried out under conditions in which the total number of impeller revolutions was common by adjusting both the number n of the impeller blades and the impeller speed N. The results of the calculation and experiment are shown in Figure 13.

The dispersion of the Fe_2O_3 particles proceeded rapidly at the higher speed due to applying a stronger impact force, even though the total number of revolutions was the same.

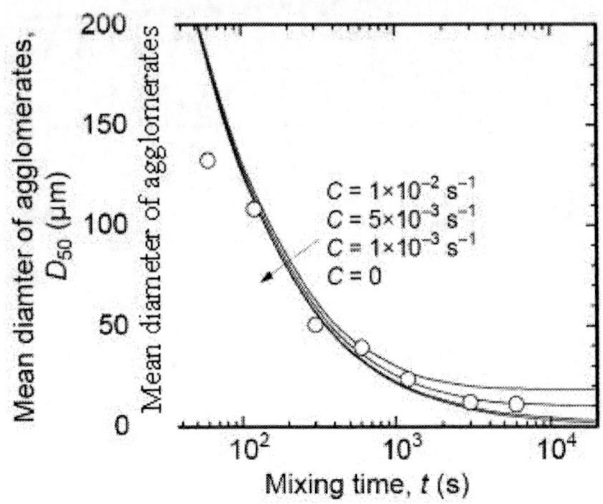

Figure 10. Comparison of calculation results with experimental data of temporal change in mean agglomerate diameter D_{50}.

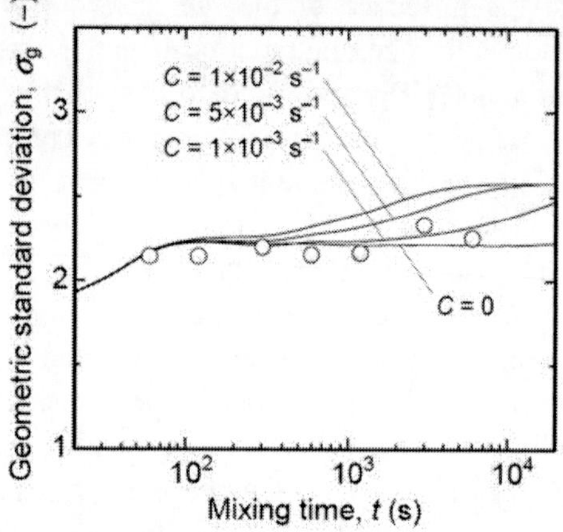

Figure 11. Comparison of calculation results with experimental data of temporal change in geometric standard deviation σ_g of agglomerate size distribution

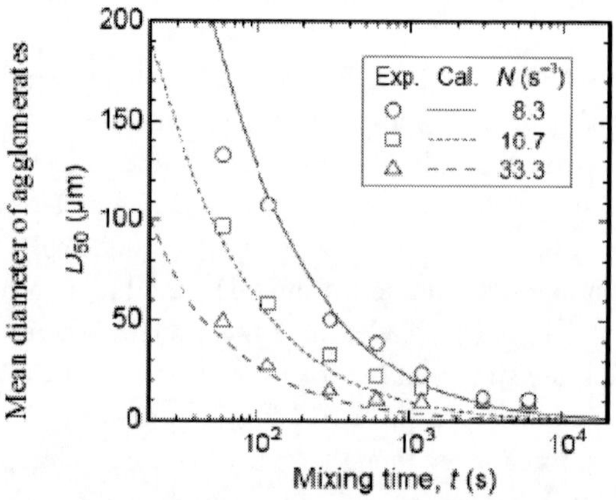

Figure 12. Effect of impeller speed N on size reduction of agglomerates (H = 0.047 m, R_V = 0.065 m, R_I = 0.050 m, n = 4).

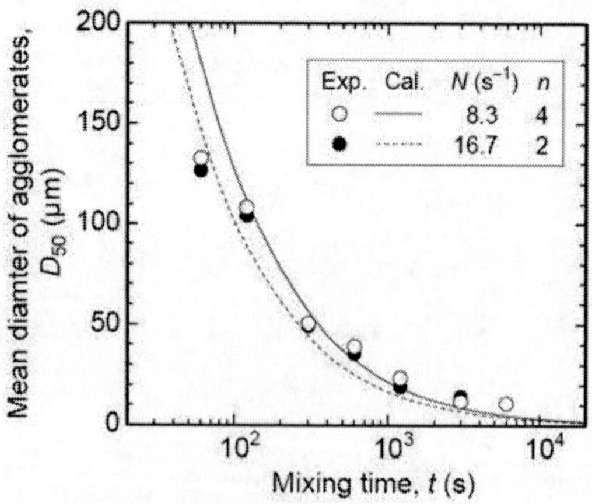

Figure 13. Temporal change in D_{50} under equal total number of impeller revolutions (H = 0.047 m, R_V = 0.065 m, R_I = 0.050 m)

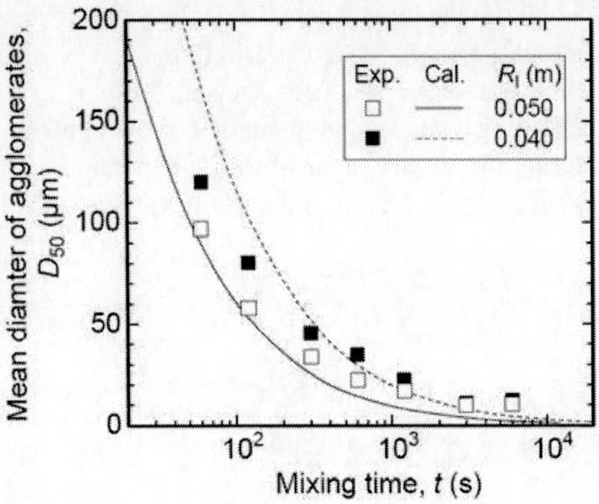

Figure 14. Effect of impeller radius R_I on size reduction of agglomerates ($H = 0.047$ m, $R_V = 0.065$ m, $n = 4$, $N = 16.7$ s^{-1}).

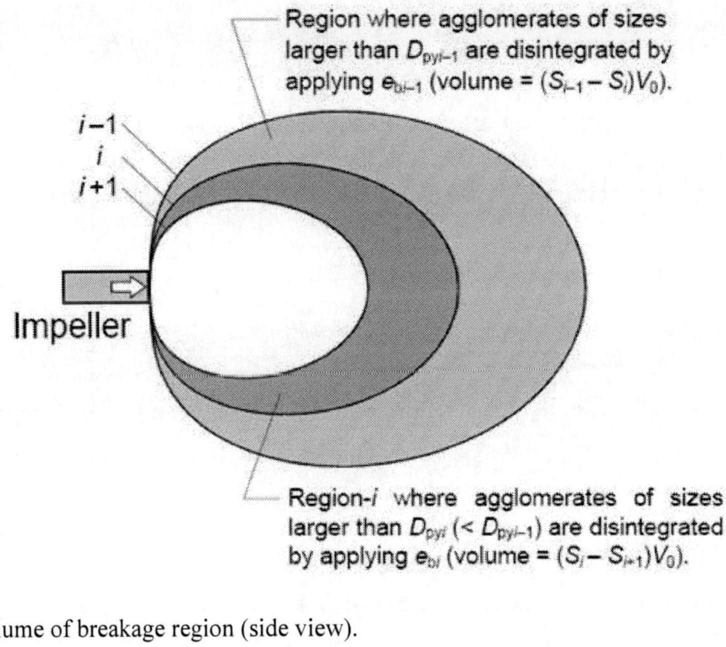

Figure 15. Volume of breakage region (side view).

Figure 14 shows a comparison of the results of the model calculation with those of the experiment using impellers with a different length R_I. The radius ratio of the impellers was 0.8. In this case, the volume of the breakage region V_b for the smaller impeller was 0.64 times as much as the use of the larger. The calculation and experimental results were in close agreement. Consequently, it was proved that the dispersion process of the Fe_2O_3 particles could be described by using the presented model in which the effects of the operating conditions are considered.

Subsequently, the mechanical energy applied to the powder bed was estimated based on the temporal change in the size distribution of Fe_2O_3 agglomerates as follows. The solid volume (except for the void) of the region where the agglomerates with a diameter larger than D_{py} are disintegrated per unit time is expressed as $S_i V_0$ (= $S(D_{py})V_0$), where V_0 is the total solid volume of the particles in the mixer vessel. Thus, the solid volume of the region to which the energy e_{bi} required for breaking a single agglomerate of D_{py} is applied is given as $(S_i – S_{i+1})V_0$ (see Figure 15). Considering the surface areas of the agglomerates and the bulk powder, the disintegration energy E_s per unit true volume and time is expressed as follows:

$$E_s = \frac{1}{V_0}\left[\sum_{i=1}^{N-1} \frac{e_{bi}}{\pi D_{py}^{\,2}} s_v (S_i - S_{i+1}) V_0 + \frac{e_{bN}}{\pi D_{py}^{\,2}} s_v S_N V_0\right]$$
$$= \frac{s_v}{\pi}\left[\sum_{i=1}^{N-1} \frac{e_{bi}}{D_{py}^{\,2}} (S_i - S_{i+1}) + \frac{e_{bN}}{D_{py}^{\,2}} S_N\right] \quad (26)$$

where s_v is the specific surface area of the bulk powder (= 3.9×10^{-6} $m^2 \cdot m^{-3}$). e_{bi} is expressed approximately by the tensile stress σ_t of the Fe_2O_3 powder bed.

$$e_{bi} = \int_0^{\Delta} A_b \, \sigma_t \, d\Delta \quad (27)$$

where A_b and Δ indicate the total area of fractured section and the extension displacement, respectively. Assuming that σ_t is proportional to Δ and that A_b is equivalent approximately to the half of increased surface area of the agglomerates by disintegration [13], e_{bi} is rewritten as follows:

$$e_{bi} = \frac{1}{2} A_b \, \sigma_{tm} \, d \quad (28)$$

where σ_{tm} is the tensile strength of the agglomerate and A_b is given by the following equation:

$$A_b = \frac{\pi}{2}\left(\sum_{i=1}^{N} D_{px}^{\,2} - D_{py}^{\,2}\right) \quad (29)$$

Here, the number of the broken agglomerates at each time was calculated by using Eq. (28). Figure 16 shows the temporal change in the accumulated value of the disintegration energy under the conditions of N = 8.3 s^{-1}, n = 4 and R_I = 0.050 m as a representative example. The cumulative value of the disintegration energy was proportional to the mixing time; the energy that the mixer gave to the powders per unit time was constant. Similar results were obtained under different conditions. These results confirm that the method properly estimates the applied energy. The energy E_s applied per unit time to the powders of solid volume of 1 m^3 was determined by using Eq. (26): E_s = 0.335 $kJ \cdot m^{-3} \cdot s^{-1}$ under the conditions of N = 8.3 s^{-1}, n = 4 and R_I = 0.050 m. By using E_s and the experimental data of D_{50} shown in Figure 12, the relationship between the cumulative value E_v of the applied energy and the mean diameter D_{50} is expressed by the empirical formula as shown in Figure 17.

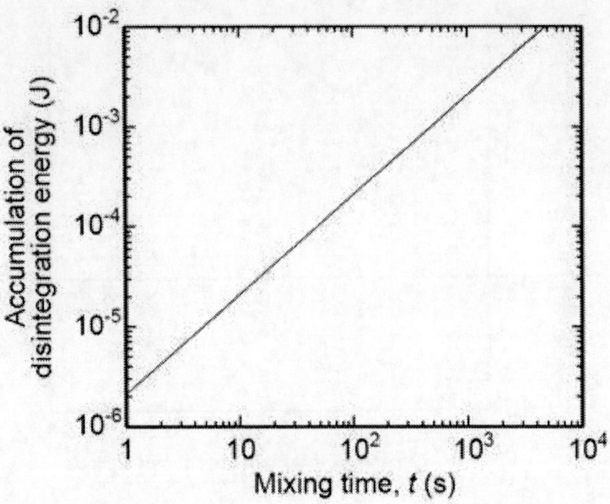

Figure 16. Change in disintegration energy with mixing time ($N = 8.3\ s^{-1}$, $n = 4$, $R_I = 0.050$ m)

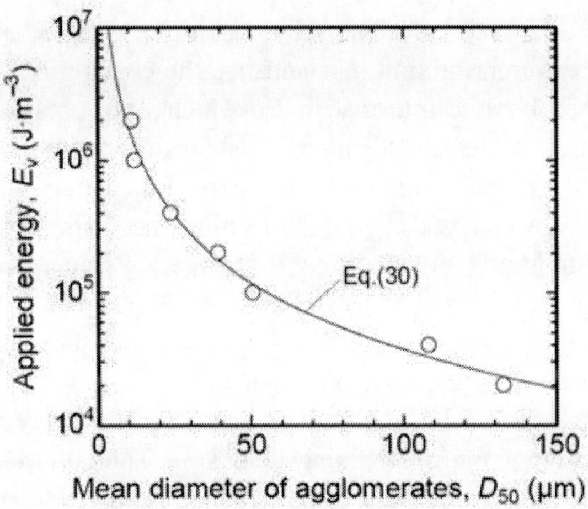

Figure 17. Relationship between applied energy E_v and mean agglomerate diameter D_{50}

$$E_v = 7.46\times10^{-3}\, D_{50}^{-1.67} \qquad (30)$$

D_{50} can be also given as a function of the degree η of attainment to the final stage of dispersion [3], which is determined from the lightness of the color of the Fe_2O_3–$CaCO_3$ powder mixture [23].

$$D_{50} = 2.08\times10^{-4}\,(1–0.997\eta) \qquad (31)$$

The E_v–D_{50} relationship expressed by Eq. (30) can be converted into the E_v–η relationship by using Eq. (31) (see Figure 18).

$$E_v = 1.05\times10^{4}\,(1–0.997\eta)^{-1.67} \qquad (32)$$

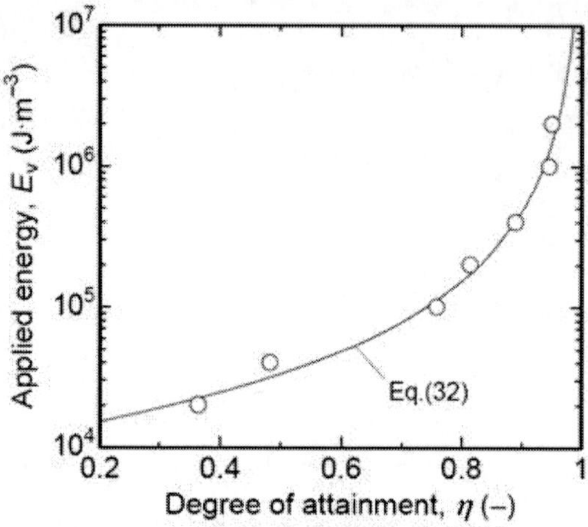

Figure 18. Relationship between applied energy E_v and degree η of attainment to final stage of dispersion

Assuming that Eq. (32) expresses the relationship between E_v and η in other types of powder mixers having different mixing mechanisms, the change in η with the mixing time t for various mixers was converted into that in E_v, which were obtained under representative operating conditions in each mixer [3]. Figure 19 shows the temporal change in E_v, i.e., the energy characteristics of powder mixers. The energy characteristic curves were roughly classified into group A (a rotary vessel type), B (a high-speed stirring (shearing) type) and C (a rotary vessel type with impellers) according to the mixing principle. The dispersion rate of the particles in the mixing process is faster in the order of group B, C and A [3, 24]. In group B, the energy E_v was almost proportional to the mixing time t because the energy was used for disintegration of the Fe_2O_3 agglomerates. On the other hand, in group A, η increased slightly at an early stage of mixing as the time proceeds because the Fe_2O_3 agglomerates were moved in the bulk powder while maintaining their size. Thus, the change in E_v with t was relatively small in the time period; this meant that most of the energy was used not for disintegration but for movement (dispersion) of the Fe_2O_3 agglomerates. At the final stage of mixing, E_v was proportional to t even in group A. Consequently, it is considered that the shearing mixing is dominant in the range in which E_v is proportional to t and that the convective mixing principally occurs in the range in which E_v is constant regardless of t. When the intermediate change of E_v is observed, the mixing process depends on these mixing mechanisms and it can be regarded as the transition period from the convective to the shearing mixing. The mixing mechanism in the powder mixers can be clarified by means of an energy characteristic curve.

Energy Estimation in Media Mills

Media mills have been widely used not only for size reduction of particulate materials but also for synthesis of functional composite particles using mechanical alloying [25] and mechanochemical treatments [26, 27]. In media mills, relatively larger mechanical energy applies to the particulate materials to be treated, and this enables to conduct such milling treatments. Thus, it is difficult to precisely estimate the applied energy in media mills using the disintegration of Fe_2O_3 agglomerates mentioned in the previous section because the disintegration rate of Fe_2O_3 agglomerates is too fast. Therefore, the applied energy in media mills was estimated from the plastic deformation of soft metal particles, gas-atomized spherical copper particles (Cu-At 200–350 supplied by Fukuda Metal Foil and Powder Co., Ltd., Figure 20) [7], used as a stress-sensitive material, of which the median diameter was 69.1 μm and the true and bulk densities were 8930 and 5020 kg·m^{-3}, respectively.

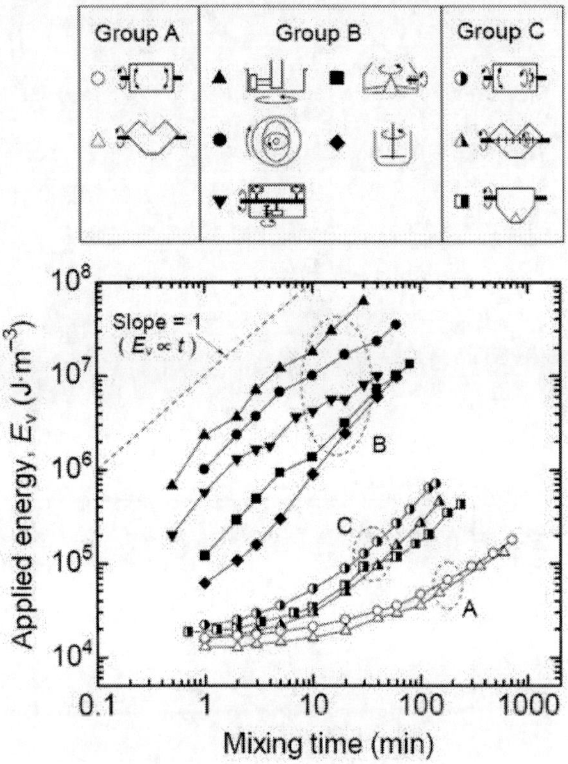

Figure 19. Characterization of various types of powder mixers based on mechanical energy applied to powder (group A: rotary vessel type, B: high-speed stirring (shearing) type, C: rotary vessel type with impeller)

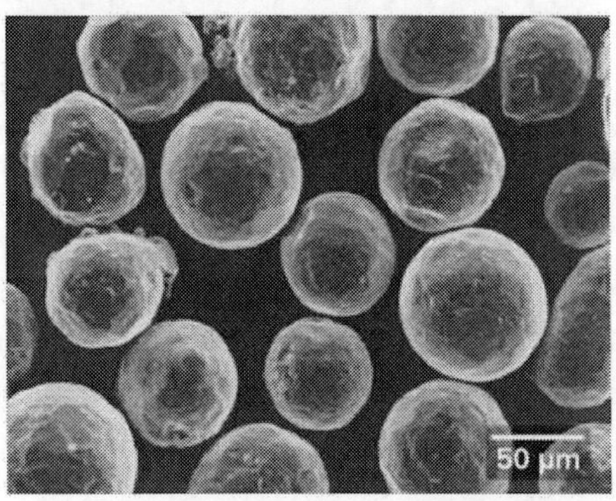

Figure 20. Scanning electron micrograph of gas-atomized copper powder

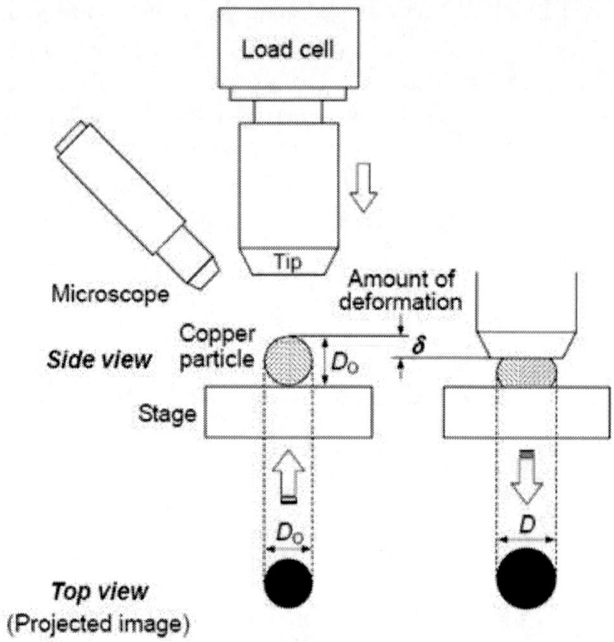

Figure 21. Unconfined compression tester

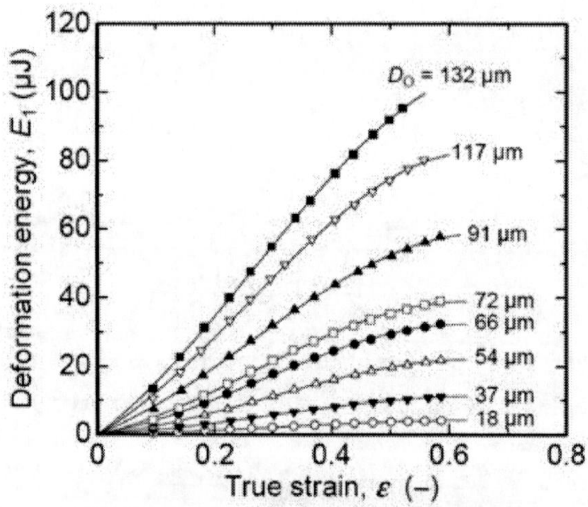

Figure 22. Relationship between deformation energy E_1 and true strain ε of single copper particle obtained from unconfined compression test. The symbols and solid lines indicate the experimental data and approximate curves, respectively

In order to estimate precisely the applied energy, an unconfined compression test for a single copper particle of 18–132 μm was carried out with a strength tester shown in Figure 21, and the correlation between the deformation of a copper particle and the applied energy was investigated. The original diameter D_O of a copper particle to be compressed was measured beforehand with an image analyzer (Luzex-FS, Nireco Corp.) mounted a metallographic microscope, and then it was transferred on a stage in the strength tester. A cylindrical stainless steel tip with a diameter of 2 mm descended at a constant rate of 0.01

$mm{\cdot}s^{-1}$ and compressed the copper particle placed on the stage. The compressive force and the amount of the plastic deformation were measured simultaneously with a load cell and a displacement sensor, respectively. The amount of deformation was defined as the difference between D_O and the height of the deformed particle. The true compressive stress and the true strain were estimated from the compressive force and the amount of deformation thus measured, and the true stress-true strain curve was obtained. The compressive energy per unit volume was calculated by integrating the true compressive stress over the true strain, and the deformation energy E_1 was determined using the compressive energy per unit volume and the particle volume ($= \pi D_O^3/6$). Figure 22 shows the relationship between the deformation energy E_1 of a single copper particle and the true strain. E_1 increased with increasing in the original diameter D_O of copper particles and was arranged as an empirical formula containing the true strain ε.

$$E_1 = \alpha\varepsilon + \beta\varepsilon^2 - \gamma\varepsilon^3 \tag{33}$$

where α, β and γ are the constants depending on the physical properties of copper particles and were expressed as an empirical formula containing D_O as shown in Figure 23.

$$\alpha = 1.51\times10^4\, D_O^{2.08} \tag{34}$$

$$\beta = 87.9\, D_O^{1.40} \tag{35}$$

$$\gamma = 2.31\times10^2\, D_O^{1.46} \tag{36}$$

Figure 24 shows the comparison of calculated values of E_1 using Eqs. (33) to (36) with the experimental data. The calculated values agreed well with the experimental data.

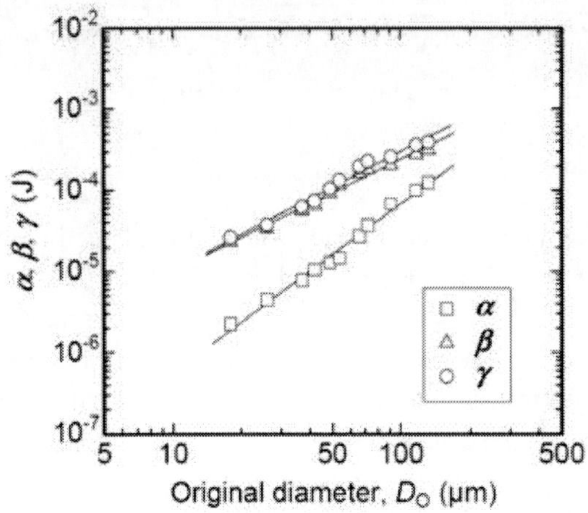

Figure 23. Relationships between constants α, β and γ in Eq. (33) and diameter D_O of original copper particle

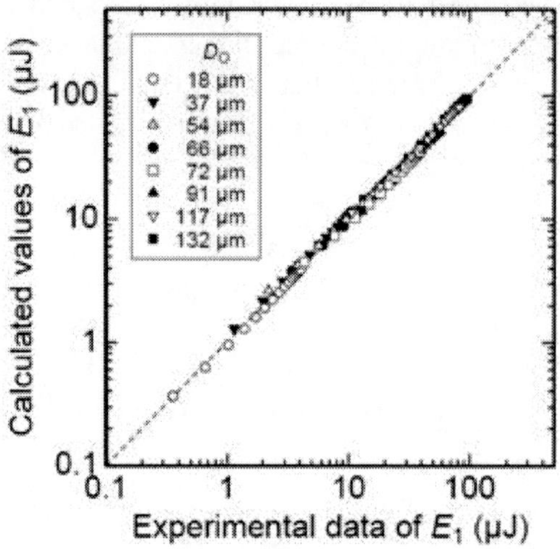

Figure 24. Comparison of calculated values with experimental data of deformation energy E_1

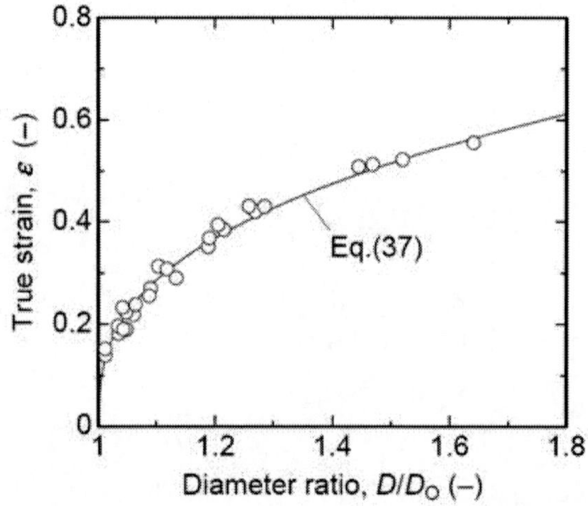

Figure 25. Relationship between true strain ε and diameter ratio D/D_O of deformed particle to original.

Subsequently, the diameter D of circular projected area of the deformed particle with a barrel shape was measured with the image analyzer. Each of these operations was repeated for copper particles having the same diameter at least three times, and the average of the measurements was adopted as the experimental data. The correlation among the true strain, D_O and D was investigated to estimate the true strain in terms of D_O and D. Figure 25 shows the relationship between ε and the diameter ratio D/D_O of the deformed particle to the original. ε was expressed as an empirical formula containing D/D_O.

$$\varepsilon = 0.664\,[(D/D_O)-1]^{0.367} \tag{37}$$

The deformation energy of a single copper particle could be estimated from the diameters of copper particles before and after deformation by using Eqs. (33) to (37). Using this estimation method, several media mills having a different milling mechanism were characterized on a uniform scale based on the applied energy. Here, three types of media mills, a tumbling mill, agitating mill and bead mill, having different media-driving mechanisms, were used as representatives of media mills. A horizontal tumbling ball mill (TBM) with a cylindrical alumina pot of an inside diameter of 76 mm and a capacity of about 400 ml was used. Stainless steel (SUS304) balls of 3.2 mm in diameter were used as milling media, and the bulk volume of the balls containing the void formed among the balls was 40% of the pot capacity.

Figure 26 illustrates the structure and dimensions of a vertical agitating ball mill (VAM). This mill consists of a vessel with an annularly hollowed bottom and an agitator composed of two round bars, made of SUS304. The vessel's effective volume (= vessel capacity – volume of agitator) is about 300 ml. The SUS304 balls of 3.2 mm were used as milling media, and the balls occupied 10% of the vessel's effective volume. A vibrator was attached to the vessel bottom for assisting the movement of balls, and the vibration frequency was kept at 60 Hz.

Figure 27 shows a schematic illustration of a high-speed elliptical-rotor-type powder mixer (HEM) used as a bead mill [7, 28]. It consists of a rotating elliptical barrel-type vessel and an elliptical rotor that rotates inside the vessel at a fairly high speed co-axially in the opposite direction. The ratio of minor to major axis of the elliptical rotor is the same as that of the vessel. The rotor and vessel are made of zirconia and SUS304, respectively. The effective volume (= vessel capacity – volume of rotor) is about 110 ml. The major axis length of the rotor is shorter than the minor axis length of the vessel so that a narrow clearance appears periodically between the interior wall surface and the tip of the rotor. The minimum clearance is 3 mm. Strong compressive and shear stresses are applied repeatedly to the particles in this clearance region. Zirconia beads with a diameter of 1 mm were used as milling media and occupied 20% of the effective volume of the vessel. The rotational speed of the vessel was kept at a constant rate of 30 rpm.

The physical properties of milling media (stainless steel ball and zirconia bead) were summarized in Table 1.

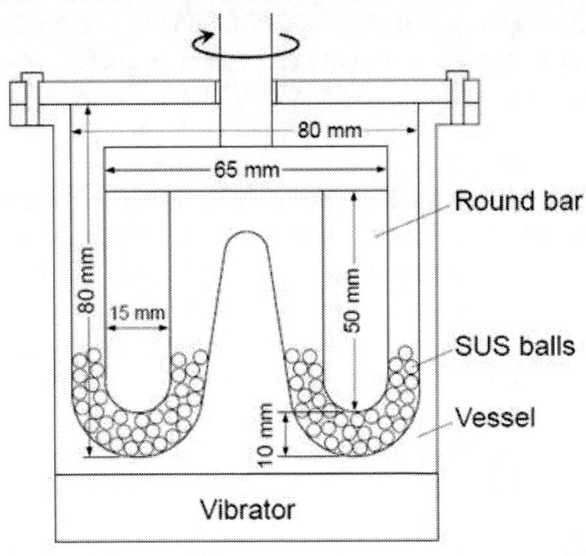

Figure 26. Vertical agitating ball mill (VAM)

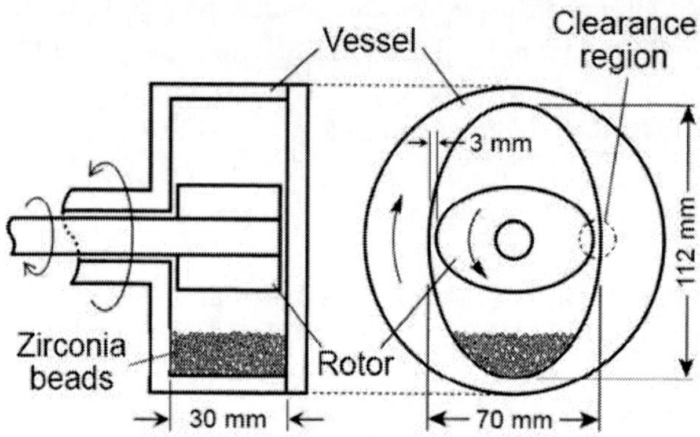

Figure 27. High-speed elliptical-rotor-type powder mixer (HEM) with zirconia beads

Table 1. Physical properties of stainless steel (SUS304) ball and zirconia bead

Material		SUS304	zirconia
Diameter	(mm)	3.2	1.0
True density	(kg·m^{-3})	7930	6050
Packing density	(kg·m^{-3})	5030	3640
Young's modulus	(GPa)	197	210
Poisson's ratio	(–)	0.30	0.31

The amount of copper powder charged into the mills was expressed in terms of the volume ratio V_C/V_v, where V_C is the bulk volume of the charged copper powder and V_v is the total volume of the void formed among media in the vessel, and V_C/V_v was varied from 0.1 to 0.5. The mills were operated at various rotational speeds of the pot, the agitator and the rotor for a proper time. The milling-treated copper particles were sampled and then distributed on a glass plate without overlapping, and their size distributions were measured with the image analyzer. Using the diameters D_O and D of copper particles before and after the treatments, the total mechanical energy applied to the whole copper particles in the mill during the treatments was estimated. The relationships between the applied energy and the operating conditions were obtained using the results of energy estimation.

In the milling treatments of copper particles for estimating the applied energy, it is impossible to measure the strain ε of each copper particle for the whole and to determine their original diameter D_O before deformation. In this estimation method, the median diameters, D_O and D, before and after the milling treatments were used as the representative diameters of the whole copper particles, and the applied energy E_v to the whole copper particles could be obtained as the total of deformation energy E_1 of each copper particle calculated by using Eqs. (33) to (37) and the amount of copper particles charged into the mill.

Figure 28 shows the relationship between the applied energy E_v per unit volume of copper particles and the operating time in TBM, VAM and HEM at $V_C/V_v = 0.2$. The results of energy estimation in HEM, in which both median diameter and width of size distribution of copper powder changed most remarkably among the mills, were confirmed to almost agree with those using the correspondence of the particle diameter intervals in the size distributions before and after the milling treatments, presented in Ref. [7]. This means that this simple estimation method could be applied to the characterization in which the relative evaluation of the applied energy on a uniform scale is principally aimed at. In addition, it was confirmed that the applied energy estimated from the disintegration of Fe_2O_3 agglomerates was almost agreed with that from the deformation of copper particles [6]. As can be seen in Figure 28, the applied energy E_v increased with increasing in the rotational speed and was almost proportional to the operating time in the mills within the rotational speed ranges regardless of the milling mechanism. This result reveals that the applied energy E_s per unit time was constant at a constant rotational speed, and E_s was given as a function of the rotational speed, except for at an early stage of the milling treatments in TBM. E_s in HEM was the largest among these mills because HEM was operated within a high rotational speed range and the number of contact points among the beads where the particles received the compressive and shear forces increased due to using the beads having a relatively small size. E_s in VAM was larger than that in TBM since VAM was operated at higher agitation speeds although the media balls used in VAM were the same as those in TBM. At an early milling stage in TBM, E_s was relatively low, and the deformation of each copper particle was small and varied widely. This resulted in decreasing in a temporal increment of E_v.

Figure 29 shows the influence of the charging volume ratio V_C/V_v on the applied energy E_v for 30 min. E_v decreased slightly with increasing in V_C/V_v; however, the influence of V_C/V_v on E_v was very small comparing with the rotational speed and the operating time as seen in Figure 28.

Figure 30 shows the relationship between E_v and the accumulation of revolutions of the rotating structure corresponding to a pot, a single bar and a rotor tip in TBM, VAM and HEM, respectively, at $V_C/V_v = 0.2$. For VAM and HEM, the accumulation of revolutions was estimated as twice the number of revolutions of the agitator and the rotor, respectively. The relationship in VAM was almost expressed by a single straight line; however, E_v in TBM and HEM increased with increasing in both the accumulation of revolutions and the rotational speed.

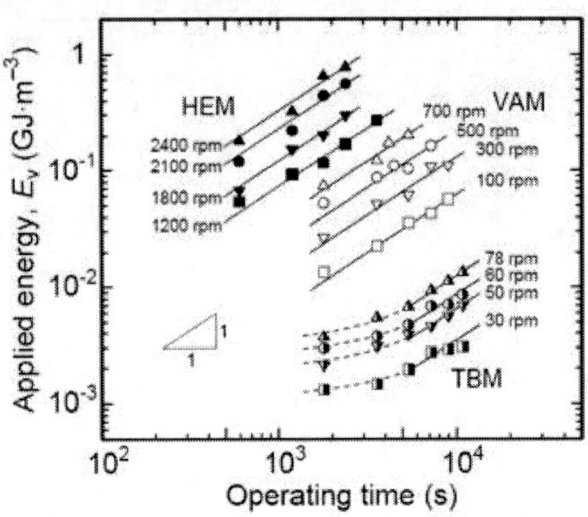

Figure 28. Relationships between applied energy E_v per unit volume and operating time t in TBM, VAM and HEM (charging volume ratio $V_C/V_v = 0.2$)

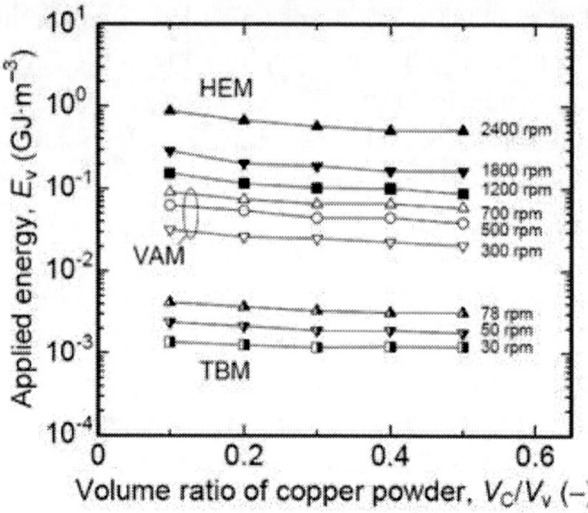

Figure 29. Influence of charging volume ratio V_C/V_v on applied energy E_v (operating time = 30 min)

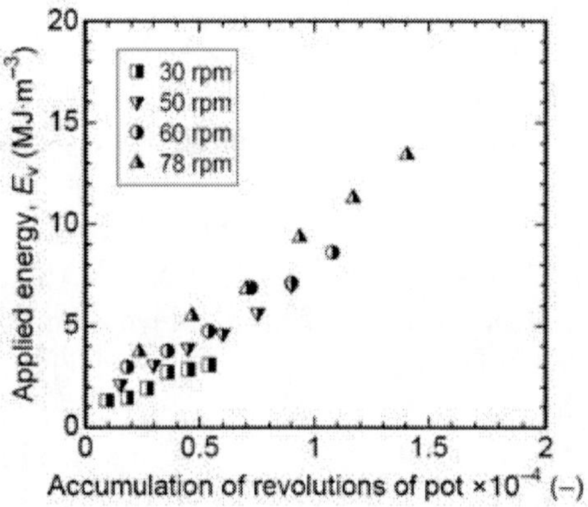

(a)

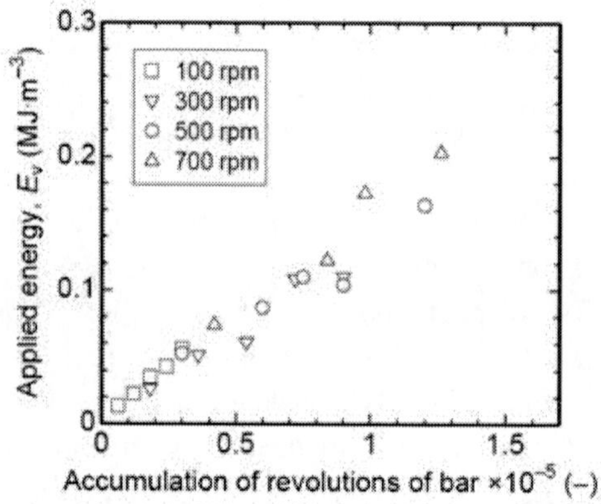

(b)

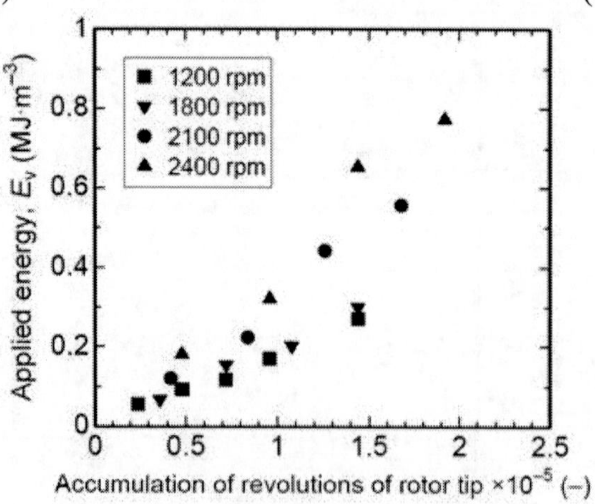

(c)

Figure 30. Relationship between applied energy E_v per unit volume and accumulation of revolutions of rotating structure in (a) TBM, (b) VAM and (c) HEM (charging volume ratio V_C/V_v = 0.2)

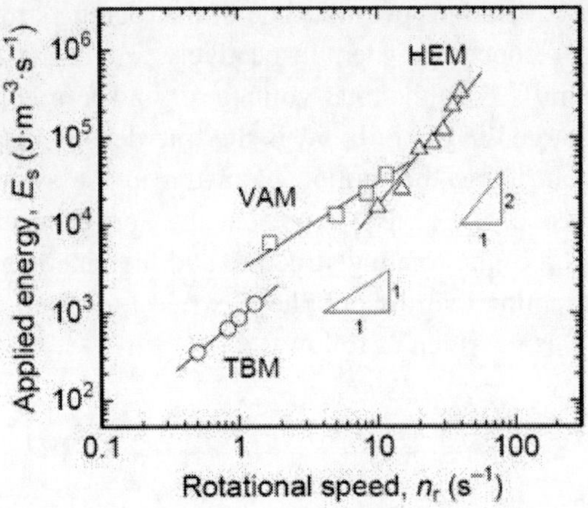

Figure 31. Relationships between applied energy E_s per unit time and rotational speed n_r of rotating structure (charging volume ratio $V_C/V_v = 0.2$)

In VAM, the motion of media balls in the vessel was restricted by a protrusion on the vessel bottom and the contribution of agitator's kinetic energy to the motion of media balls was little. Therefore, the influence of the rotational speed on E_v was small and it was expressed as a function of only the accumulations of revolutions of a single bar. On the other hand, the media in HEM and TBM could move relatively freely and the motion of media was mainly controlled by the rotational speed. Consequently, E_v depended strongly not only on the accumulation of revolutions but also the rotational speed, and the variation of E_v in the relationship corresponded to the degree of contribution of the rotational speed to E_v.

Figure 31 shows the relationship between the applied energy E_s per unit time and the rotational speed n of the rotating structure at $V_C/V_v = 0.2$. E_s in these mills was classified corresponding to the operating range of the rotational speed and the velocity dependence of E_s was remarkably observed in this result. E_s in VAM was almost proportional to n; in HEM E_s was proportional to the square of n. The slope of the log-log plotting of E_s against n in TMB was an intermediate value (about 1.3) between them. The result means that the slope changes within 1 to 2 depending on the motion of media in the mill. It approaches 2 when the media vigorously move receiving a part of the kinetic energy of the rotating structure; on the other hand it becomes to be 1 in the mills where the motion of media is restricted.

The mechanical energy applied to the particles strongly depends not only on the milling mechanism and the rotational speed of the rotating structure but also on the physical properties (e.g., size and density) of the media. In order to characterize various types of media mills on a uniform scale, two dimensionless parameters corresponding to the applied energy and the rotational speed respectively were introduced. The dimensionless parameter Ξ expressing the amount of the applied energy was defined as the ratio of the applied energy in a single revolution of the rotating structure to the kinetic energy of media, in which the applied energy E_s per unit volume of particles was converted into the applied energy per unit volume of media by using several parameters as follows:

$$\Xi = \frac{(E_s/n')(\rho_C'/\rho_C)(V_C/V_v)[\varepsilon_M/(1-\varepsilon_M)]}{\rho_M u_r^2} \tag{38}$$

where n' is the number of revolutions per unit time of the rotating structure, ρ_C' and ρ_C are the bulk and true densities of copper powder, respectively, V_C is the bulk volume of copper powder charged into the mill, V_V is the total volume of void formed among media, ε_M is the void fraction of media charged in the mill, ρ_M is the true density of a medium, and u_r is the representative velocity controlling the motion of media in the system. The numerator and denominator in the right side of Eq. (38) represent the applied energy per unit volume of media in a single revolution of the rotating structure and the kinetic energy of media per unit volume, respectively. Assuming that the peripheral speed ($= 2\pi r_r n_r$) of the rotating structure corresponds to u_r, Eq. (38) is rewritten as follows:

$$\Xi = \frac{(E_s/n')(\rho_C'/\rho_C)(V_C/V_v)[\varepsilon_M/(1-\varepsilon_M)]}{\rho_M(2\pi\, r_r\, n_r)^2} \tag{39}$$

where r_r is the radius of rotation of the rotating structure (e.g., for TBM the inside radius of the pot, for HEM the half length of the major axis of the elliptical rotor). Thus, Ξ means the contribution of the kinetic energy of media to the applied energy, that is, the degree of energy transfer from the media to the particles.

The condition of media motion (i.e., the intensity of collision between media) in the vessel depends not only on the structure and operating conditions of the mill but also the physical properties (e.g., mass and hardness) of the medium. Therefore, the dimensionless parameter Θ' expressing the condition of media motion was introduced, which was defined as the ratio of the inertial force to the viscous force of a single medium; Θ' corresponded to stirring Reynolds number in a stirring system of viscous fluid. In order to calculate the viscous force (viscous dissipation) from the physical properties of the medium, the author introduced the contact force f acting on a single contact point between media at the collision.

$$\Theta' = \frac{(\pi/6)\rho_M d_M{}^3 a_r}{\kappa f} \tag{40}$$

where d_M is the diameter of a medium and a_r is the representative acceleration in the system. κ is the coordination number of a medium in the mill and is estimated by using a simple formula of ε_M [29].

$$\kappa = \pi/\varepsilon_M \tag{41}$$

The contact force f is expressed in terms of the damping coefficient η_d reflecting the viscous dissipation of kinetic energy of media at the inelastic collision, which is wildly used in the discrete numerical model of media motion [30].

$$f = \eta_d\, u_r \tag{42}$$

where u_r is the representative velocity. Tsuji et al. [31] expressed η_d as follows:

$$\eta_d = \lambda' \left[(\pi/6)\rho_M d_M{}^3 k_s\right]^{1/2} \delta^{1/4} \tag{43}$$

where λ' is the constant and δ is the elastic deformation of a medium at the collision. k_s is the stiffness (spring constant) corresponding to the elasticity of a medium and is calculated by the Hertzian contact theory [32].

$$k_s = Y d^{1/2}/[3(1-v^2)] \tag{44}$$

where Y and v are Young's modulus and Poisson's ratio, respectively. δ is approximated by the integration of the stress-strain curve of a medium over δ.

$$\delta = d_M (2e/Y)^{1/2} \tag{45}$$

e indicates the impact energy per unit volume of media and is assumed to be proportional to the kinetic energy of media.

$$e = \varphi \rho_M u_r^2 \tag{46}$$

where φ is the constant.

Substituting Eqs. (41) to (46) into Eq. (40) and approximating a_r and u_r by the centrifugal acceleration ($= r_r(2\pi n_r)^2$) and the peripheral speed of the rotating structure, respectively, Eq. (40) is rewritten as follows:

$$\Theta = \Theta'\{\lambda' [\varphi/(2\pi^2)]^{1/8}\} = \varepsilon_M (1-v^2)^{1/2} (\rho_M/Y)^{3/8} d_M r_r^{-1/4} n_r^{3/4} . \tag{47}$$

Eq. (47) represents that Θ could be determined using the physical properties of the medium and the structure and operating conditions of the mill and varies corresponding to the motion of media.

Figure 32 shows the change in Ξ with Θ in TBM, VAM and HEM. Ξ decreased with increasing in Θ. TBM had relatively large values of Ξ; on the other hand, Ξ in HEM was considerably small. The applied energy per unit time in TBM was smaller than those in VAM and HEM since TBM was operated under lower rotational speeds and the motion of media was relatively mild. This resulted in increasing in the frequency that the particles were caught between media. On the contrary, in HEM the frequency decreased because the media moved vigorously although larger energy applied to the particles per unit time. Consequently, it has been clarified that the energy transfer efficiency depends strongly on the motion of media and decreases remarkably under operating conditions in which the media have larger kinetic energy.

APPLICATION EXAMPLES

Using the evaluation results of the applied energy for media mills, some particulate composite materials were successfully prepared with a media mill in a dry process. In the following sections, practical powder design methods using the applied energy, mechanical coating treatments of soft spherical metal particles with fine ceramic particles [33, 34] and wax powder [8] maintaining the original shape of metal particles, are presented.

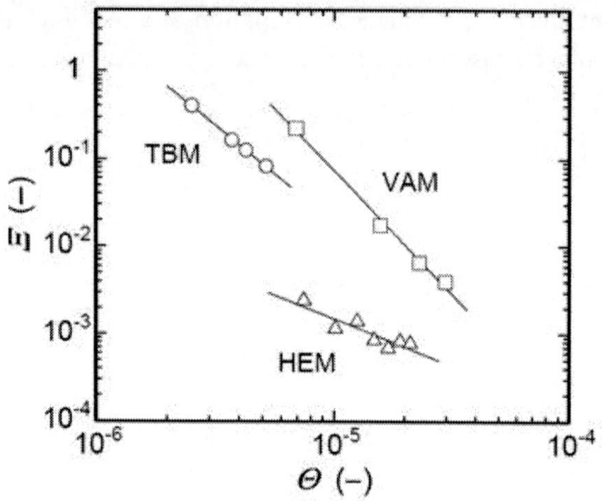

Figure 32. Energy characteristics of TBM, VAM and HEM

Case 1: Particle Coating of Soft Metal Particles with Ceramic Fines

The surface modification of particles is often utilized in various industrial fields so that the core particles are given new functions maintaining its fundamental physical properties. The coating treatment of coarser particles with fine materials is one of the effective methods in surface modification treatments. Especially, dry coating process on which the fine particles (guest) are directly attached onto the surface of core particles (host) by mechanical means without any binders is used most frequently because of the simplicities of process and operation [35–45].

If the particles coated with the fine particles by relatively weak adhesive forces such as electrostatic and liquid bridge forces are used in the processes where relatively strong forces are applied to the particles, such as mixing, pneumatic transport and compression forming, the fine particles are often peeled off from the surface of core particles during the processing because of insufficient adhesive strength. Therefore, the guest particles are desirable to be immobilized on the surface of host particles in such cases, for example, by forming a chemical bond between host and guest particles using a mechanochemical reaction [40, 45–47] or by embedding the guest particles onto the surface of host particles [48]. For the immobilization of guest particles, stronger external forces are necessary to be applied to the particles in the coating treatments. However, if the external forces are simply increased by making the operating conditions severe, the characteristics of host and guest particles are changed by deformation of the particles [49] and coalescence among the particles. The optimum operating conditions of equipment must be determined properly based on the features of the particles designed so that the particles have effective functions.

Therefore, the minimum energy required for the immobilization was estimated from the desired feature of an end product, and the optimum operating conditions in the coating process were determined by the calculation based on the energy. Here, a coating process where submicron-sized ceramic guest particles were immobilized on the spherical copper host particles [39] was adopted as a model compounding process using a bead mill (HEM) [43–45, 50, 51]. The energy requirement for the immobilization in the system corresponds to the energy for embedding of the ceramic particles onto the copper particles. The net mechanical energy applied to the particles was estimated experimentally from the deformation of spherical copper particles mentioned above and analyzed the collision energy of media beads numerically using the discrete element method [28]. Utilizing these results, the preparation of the coated particles having the smoothed surface and maintaining the shape and size distribution of original copper particles was studied.

Gas-atomized spherical copper particles (Cu-At 200–350, supplied by Fukuda Metal Foil and Powder Co., Ltd.) were used as the host particles, which were used for the energy estimation as a stress-sensitive material mentioned in the previous section. As the guest particles, fine particles of α-alumina (AMS-12 and -2, supplied by Sumitomo Chemical Co., Ltd.) and tungsten carbide (WC06N, supplied by Tokyo Tungsten Co., Ltd.) were adopted. Table 2 gives their physical properties. Al_2O_3-1 and Al_2O_3-2 correspond to AMS-12 and AMS-2, respectively. The Vickers hardness of a copper particle given in Table 2 was approximated by that of a 30 mm × 30 mm × 5 mm copper plate, which was measured with a micro-Vickers hardness tester (HMV-2000, Shimadzu Corp.).

The amount of copper particles placed in the vessel is expressed in terms of the volume ratio V_P/V_v, where V_P is the bulk volume of the mixture of copper and ceramic particles and V_v is the total void volume formed among the beads including V_P. Here, $V_P/V_v = 0.2$ was chosen as a representative condition. The content (volume basis) v of ceramic particles in the ceramic–copper composite particles was varied from 0.006 to 0.103. The ceramic particles ideally form a randomly packed mono-layer on the surface of copper particles in the case of $v = v_r$ which depended on the size and density of ceramic particles and was determined as follows: $v_r = 0.012$ for Al_2O_3-1, 0.038 for Al_2O_3-2 and 0.021 for WC [52]. When the coating treatment was carried out under the condition of $v > v_r$, the ceramic particles were placed on the vessel in several batches in order to prevent the adhesion of the ceramic particles to the vessel wall. For example, when the Al_2O_3-1 particles of $v = 0.044$ were compounded, they were divided into three batches corresponding to $v = 0.012$ (= v_r) and one batch corresponding to $v = 0.008$, and then these were successively fed to the vessel. Using the method described in the following, the processing time t and the rotor speed N_R were adjusted depending on a given v and on the shape of composite particles, respectively.

The optimum rotor speed and processing time can be determined based on the three factors as follows:

(i) the mechanical energy required to embed all the ceramic particles in the copper particles,
(ii) the mechanical energy applied to the ceramic–copper powder mixture in a single compression by the rotor, and
(iii) the allowable deformation of copper particles in a single compression.

The specific procedure for determining the operating conditions is described below.

The mechanical energy required to embed the ceramic particles is estimated by expressing approximately the embedding process of a single ceramic particle by means of the process in which a quadrangular pyramid type of diamond indenter intrudes into the copper plate (that is, the measurement of Vickers hardness). From the definition of Vickers hardness H_V, the force f_i required to intrude the indenter is given as:

$$f_i = H_V S_c \tag{48}$$

where S_c indicates the contact surface area of the indenter with the copper plate. On the assumption that the shape of the ceramic particles is spherical and that no deformation or fracture of the ceramic particles occurs during the embedding process, the contact area S_c of a ceramic particle with a copper particle is expressed as a function of the depth d_e to which the ceramic particle was embedded (see Figure 33).

$$S_c = \begin{cases} \pi d_g d_e & (0 \le d_e/d_g \le 0.5) \\ (\pi/2) d_g^{\,2} \text{ (const.)} & (0.5 \le d_e/d_g \le 1) \end{cases} \tag{49}$$

where d_g indicates the median diameter of the ceramic particles. Therefore, the energy e_1 required to embed a single ceramic particle completely (i.e., $d_e/d_g = 1$) is given from Eqs. (48) and (49).

$$e_1 = \int_0^{d_g} f_i \, \mathrm{d}d_g = \frac{3\pi}{8} H_V d_g^{\,3} \tag{50}$$

Table 2. Physical properties of host and guest particles

Material	Median diameter* (μm)	True density ($kg \cdot m^{-3}$)	Bulk density ($kg \cdot m^{-3}$)	Vickers hardness (GPa)
Host				
Cu	69.1	8930	5020	0.935 (plate)
Guest				
Al_2O_3-1	0.4	3900	1060	–
Al_2O_3-2	1.3	3900	710	–
WC	0.7	15600	3640	–

* Heywood diameter, number basis.

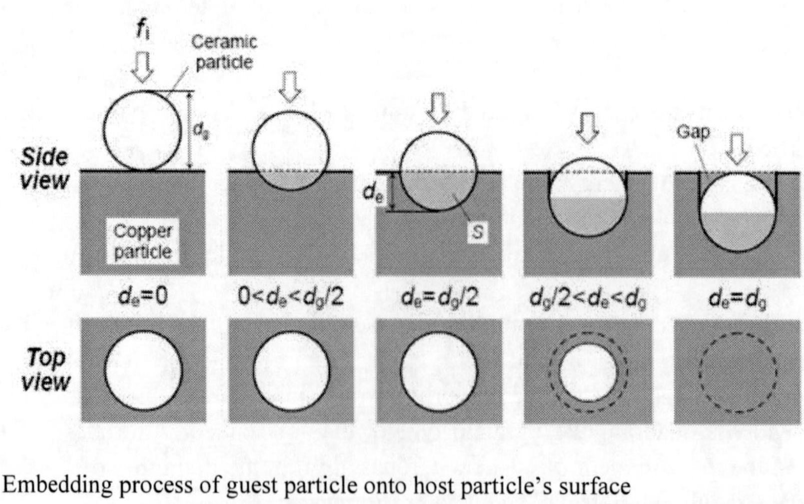

Figure 33. Embedding process of guest particle onto host particle's surface

Thus, the energy e' required to embed all the ceramic particles is expressed as:

$$e' = e_1 \frac{m}{(\pi/6) d_g^{\,3} \rho_f} = \frac{9 H_V m}{4 \rho_f} \quad (51)$$

where m and ρ_f indicate the total mass and true density of the ceramic particles, respectively.

At the initial stage of the coating process, the ceramic particles form agglomerates and are not dispersed uniformly on the surface of copper particles. If the compressive energy is applied to such copper particles, their plastic deformation will be caused. Here, the operating condition on the coating process is determined by considering the allowable deformation of copper particle in a single compression through the beads by the rotor.

Figure 34 illustrates the motion of the particle–bead mixture caused by the revolutions of rotor and vessel. The mixture is lifted up by the rotational motion of the vessel, and then slips down due to the force of gravity. Strong compressive and shear forces are applied to the particle–bead mixture in this region every time a special condition is satisfied: the particle–bead mixture passes through the minimum clearance region and the rotor's major axis is superimposed on the vessel's minor axis.

The time interval in which this condition is satisfied, that is, the time period λ of a single compression by the rotor, is given by

$$\lambda = [2(N_R + N_V)]^{-1} \quad (52)$$

where N_R and N_V are the rotational speeds of rotor and vessel, respectively [28]. The single circulation time t_p of the particle–bead mixture is equal to that of a half turn of the vessel:

$$t_p = (2N_V)^{-1}. \quad (53)$$

All the copper particles pass through the minimum clearance region while t_p elapses. Therefore, the average mass flow rate u of the copper particles fed into the minimum clearance region is expressed by

$$u = M/t_p \quad (54)$$

where M is the total mass of the copper particles in the vessel. Thus, the average number n_c of copper particles fed into the minimum clearance region within a single compression is expressed as follows:

$$n_c = \frac{u \lambda}{(\pi/6) D_h^{\,3} \rho_C} \quad (55)$$

where D_h and ρ_C indicate the original median diameter and true density of the copper particles, respectively. n_c is dependent on the energy E_c applied to the copper particles in a single compression.

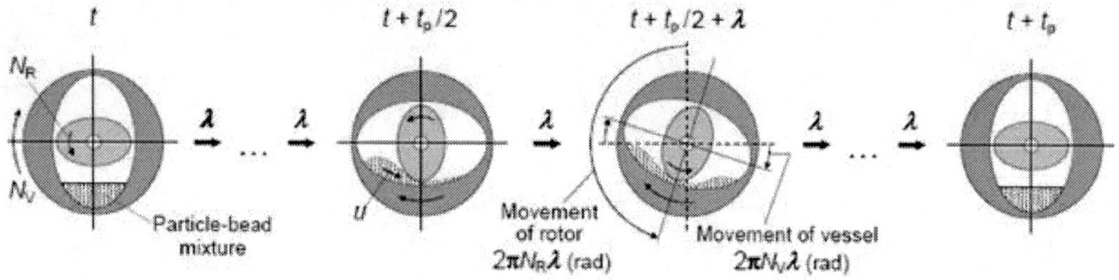

Figure 34. Compression of particle–bead mixture by rotor

$$E_c = E_1 n_c \tag{56}$$

where E_1 is the compressive energy applied to a single copper particle and is given by a simple formula as a function of the deformation δ_h of a copper particle [53, 54].

$$E_1 = \frac{\pi}{8} k_f D_h {\delta_h}^2 \tag{57}$$

where δ_h is defined as the difference between the original diameter D_h of the copper particle and the height of the deformed barrel- or disk-shaped particle [7]. k_f is an experimental coefficient that represents the particle hardness and depends on the physical properties. For the copper particles, k_f is independent of D_h within the range of 13 to 165 μm and its value is 0.722 GPa [54]. E_c is expressed as a function of the rotor speed N_R [7, 28].

$$E_c = \xi E_0 {N_R}^2 \tag{58}$$

where ξ is the degree of energy transmission from the rotor to the copper particles via the beads and is given as a function of the charge volume ratio V_P/V_v of the copper particles. When $V_P/V_v = 0.2$, the value of ξ is determined to be 0.024 [28]. E_0 is a constant determined by discrete element analysis of the bead motion in the vessel: $E_0 = 6.63\times10^{-5}$ kg·m^2 [28]. By rearranging Eqs. (52)–(58), the relationship between the rotor speed N_R and the dimensionless deformation δ_h/D_h of the copper particle is obtained as follows:

$${N_R}^2\left(1+\frac{N_R}{N_V}\right) = \frac{3k_f}{4\xi E_0}\left(\frac{M}{\rho_C}\right)\left(\frac{\delta_h}{D_h}\right)^2 \tag{59}$$

where δ_h/D_h corresponds to the allowable deformation of copper particles in a single compression.

On the other hand, the energy e' given by Eq. (51) is equal to the product of the energy E_c in a single compression and the compression frequency for the operation time t (= t/λ).

$$e' = E_c \cdot (t/\lambda) \tag{60}$$

Using Eqs. (51), (52) and (58)–(60), t is expressed as follows:

$$t = \frac{3H_V(m/\rho_f)}{2k_f N_V(M/\rho_C)}\left(\frac{\delta_h}{D_h}\right)^{-2} \tag{61}$$

Accordingly, by considering the allowable deformation δ_h/D_h (that is, the required sphericity) of the copper particles, the rotor speed and the processing time can be determined depending on the physical properties of ceramic and copper particles.

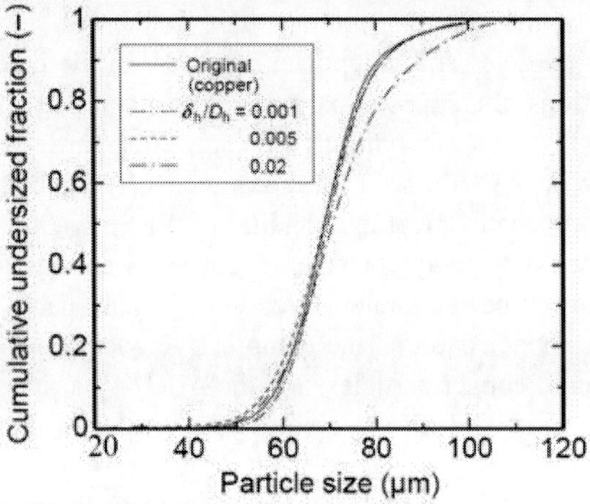

Figure 35. Size distribution curves of copper and coated particles

δ_h/D_h (–)	0.001	0.005	0.02
e' (J)		26.9	
N_R (s^{-1})	5.51	16.4	41.6
t (ks)	46.5	1.86	0.116
Shape 50 μm	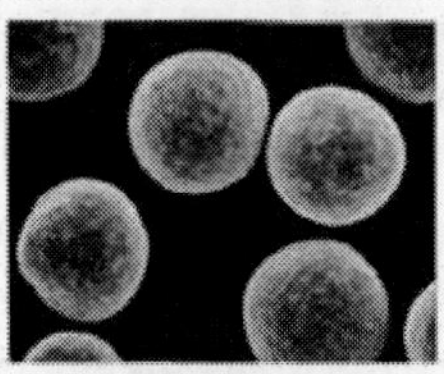	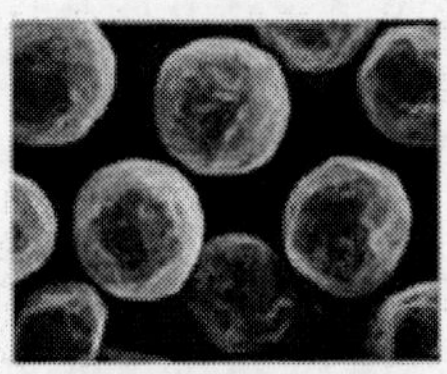	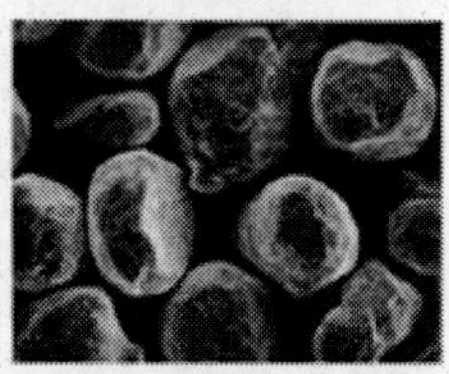
Surface condition 1 μm		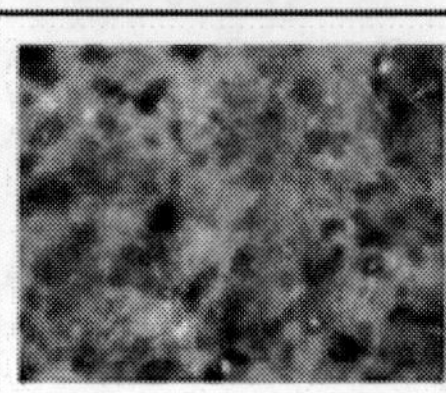	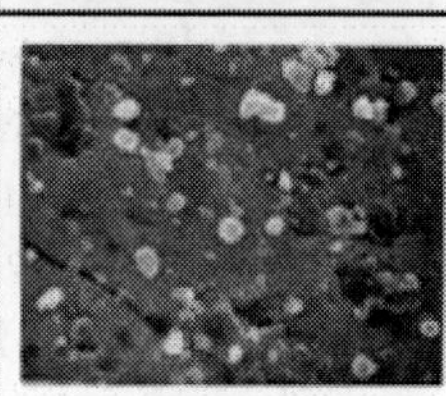

Figure 36. Change in shape and surface condition of coated particles with allowable deformation δ_h/D_h (d_e/d_g = 1.0)

The preparation of the coated particles having a spherical shape and a smooth surface was attempted by embedding the Al_2O_3-1 particles of $v = v_r$ completely onto the surface of copper particles under the operating conditions of several δ_h/D_h. In the treatments, the rotor speed N_R and the operation time t were determined by substituting $d_e/d_g = 1$ and $\delta_h/D_h = 0.001$, 0.005 and 0.02 into Eqs. (59) and (61). $\delta_h/D_h = 0.001$ and 0.02 correspond to the deformation of a copper particle in the elastic and plastic regions, respectively, and $\delta_h/D_h = 0.005$ is included in the intermediate region [55]. The size distribution (Heywood diameter, number basis) of the composite particles was measured with an image analyzer (Luzex-FS, Nireco Corp.) and expressed in terms of a log-normal distribution. The geometric mean diameter D_{50} and standard deviation σ_g of the size distribution were calculated. Figure 35 shows the size distributions of copper particles and composite particles. In the cases of $\delta_h/D_h = 0.001$ and 0.005, the size distributions of composite particles almost agreed with that of original copper particles. The shape and surface conditions of composite particles were observed with a scanning electron microscope (SEM). The appearance of composite particles is shown in Figure 36 together with each operating condition. When $\delta_h/D_h = 0.001$, the composite particles were maintained the spherical shape; its surface was rough because the embedding of Al_2O_3-1 particles had not been completed yet. On the other hand, when $\delta_h/D_h = 0.02$, i.e., in the plastic region of copper particle, the compressive energy applied to the particles was spent to the deformation of copper particles, and the Al_2O_3-1 particles were hardly embedded though larger energy was applied. In the case of $\delta_h/D_h = 0.005$, the shape of the composite particles was almost spherical (polyhedral). The surface of the composite particles was smoothed due to the particle-to-particle and particle-to-wall collisions, and the Al_2O_3-1 particles were hardly protruded from the surface of copper particles. In order to confirm the embedding of Al_2O_3-1 particles, the surface of the composite particles was etched in a nitric acid solution. The Al_2O_3-1 particles were appeared from the inside of coated particles; these results proved that the Al_2O_3-1 particles were layered uniformly on the surface of copper particle and that the composite particles maintaining the size distribution and having the desired feature (i.e., spherical shape and smooth surface) is obtained under the operating condition of $\delta_h/D_h = 0.005$. Thus, for maintaining the shape of soft host particle throughout the coating process, the allowable deformation δ_h/D_h of host particle must be properly determined by considering its mechanical properties such as hardness.

For Al_2O_3-1, Al_2O_3-2 and WC particles, the optimum rotor speed N_R and processing time t at $\delta_h/D_h = 0.005$ were calculated under various contents v as shown in Figures 37 and 38, respectively. Figure 39 shows the SEM photographs of composite particles containing the Al_2O_3-2 and WC particles of $v = v_r$ together with the mechanical energy e' required to completely embed all the ceramic particles, the optimum operating conditions determined using Figures 37 and 38. The composite particles were almost spherical regardless of the kind of ceramic particle. Figure 40 shows the electron probe micro analyzer (EPMA) mapping images of sections of composite particles. The ceramic particles were confirmed to exist near the surface of the composite particles.

Figure 41 shows the changes in the geometric mean diameter D_{50} and standard deviation σ_g of the size distribution of composite particles with the content v of ceramic particles. Both D_{50} and σ_g were kept constant independently of v (i.e., the processing time) and the kind of ceramic particle. Consequently, the composite particles with the desired shape and surface conditions could be prepared by means of this method regardless of the physical properties of the ceramic particles.

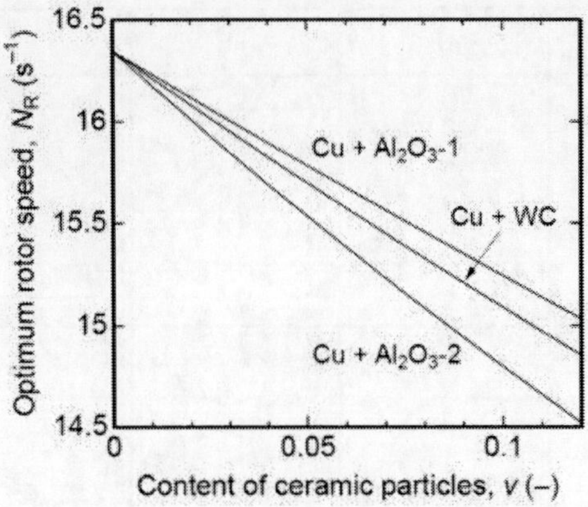

Figure 37. Relationship between optimum rotor speed N_R and content v of ceramic particles (δ_h/D_h = 0.005)

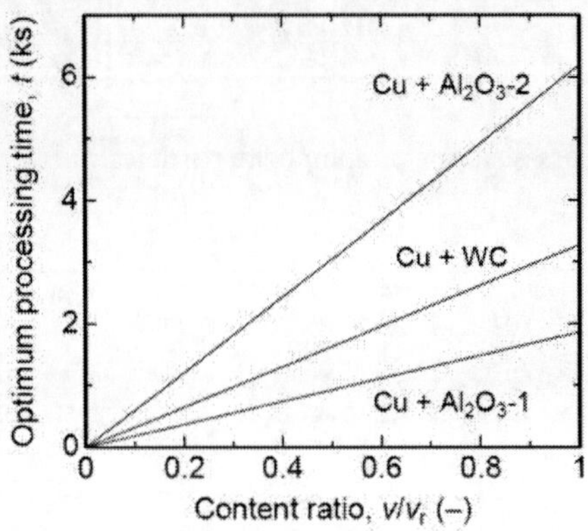

Figure 38. Relationship between optimum processing time t and content ratio v/v_r (δ_h/D_h = 0.005)

Guest particle	Al_2O_3-2	WC
e' (J)	79.0	44.7
N_R (s^{-1})	15.7	16.1
t (ks)	6.19	3.28
Shape 50 μm	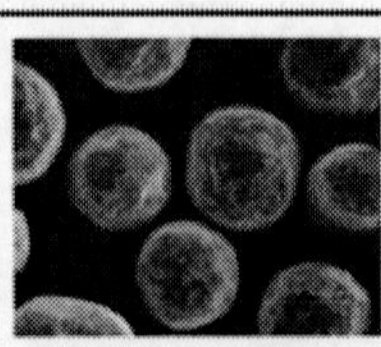	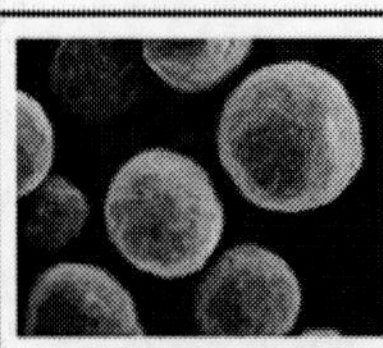
Surface condition 1 μm	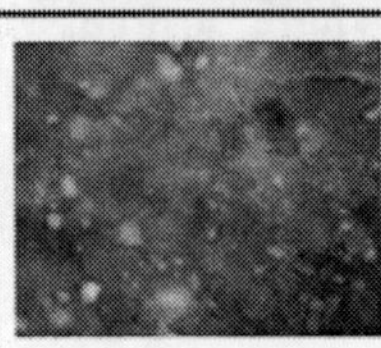	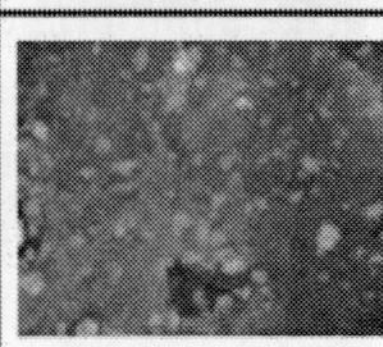

Figure 39. Shape and surface conditions of composite particles at $v = v_r$

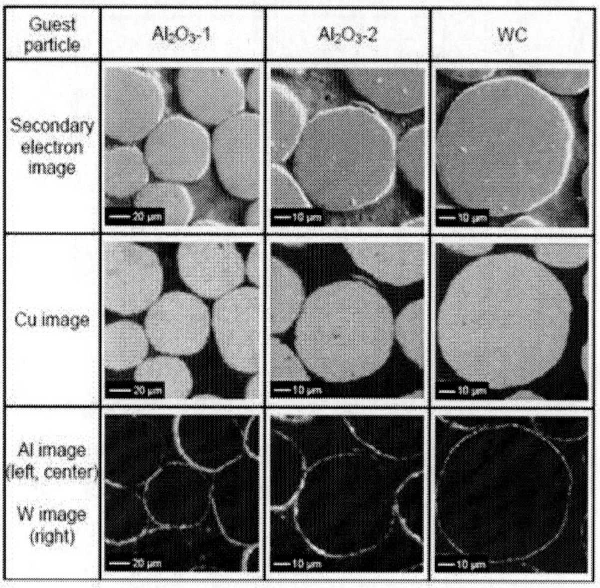

Figure 40. EPMA images of cross-sections of composite particles

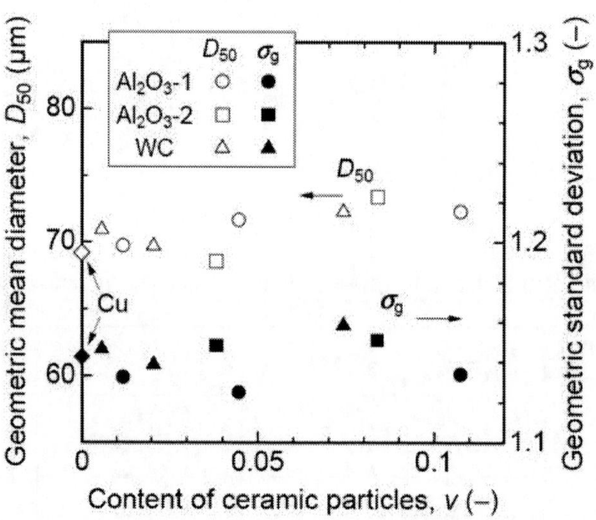

Figure 41. Changes in geometric mean diameter D_{50} and standard deviation σ_g of size distribution of composite particles with content v of ceramic particles

In order to investigate the hardness of the composite particles, the composite particles were deformed in HEM for an appropriate length of time. In this case, the rotor speed was set to 40 s^{-1} so that the composite particles were surely deformed. The size distribution of deformed composite particles was measured with the image analyzer. The deformation of each composite particle was estimated from the change in the size distribution before and after the treatment [7]. On the assumption that the mechanical energy applied to the composite particles was the same as that to the copper particles without any ceramic particle under the same operating conditions, the mechanical energy applied to each composite particle can be given by Eq. (58). The coefficient k_f of composite particles was determined using Eq. (57). Figure 42 shows the change in the coefficient k_f of the composite particles containing Al_2O_3-1 of $v = 0.012$ ($= v_r$) with the operation time t_d in the deformation treatment. k_f was constant within the range of t_d = 30 to 60 min, although the total mechanical energy applied to the composite particles increased with t_d because the composite particles were deformed, depending on the applied energy. In the range of $t_d \geq 60$ min, k_f increased with t_d because work-hardening made the composite particles difficult to deform. This means that the deformation in this time range was not proportional to the applied energy. Thus, the constant value of k_f obtained in the range of relatively small deformation represents accurately the hardness of composite particles.

The copper particles coated with ceramic particles of various volumes were used for the deformation treatment for 60 min, and k_f of the composite particles was estimated. As shown in Figure 43, k_f increased linearly with v in each composite particle. The slope of the linear-relationship was greater in the coating of finer particles. To cause plastic deformation of composite particles, the dislocation line in the composite particles must be moved [56]. This becomes more difficult when finer particles are used. In addition, finer ceramic particles were immobilized more compactly on the surface of copper particles, and the distance between the ceramic particles decreased even though v was the same. Thus, movement of the dislocation line was considerably prevented when finer particles were used.

In the linear-relationships shown in Figure 43, k_f in the case of $v \rightarrow 0$ corresponds to that of copper particles (= 0.722 GPa). However, the value was 1.59 GPa in the experimental results regardless of the kind of ceramic particles. Immobilization of the ceramic particles changed the surface conditions of the copper particles, and the coefficient of the surface friction decreased. This led to decreasing in the energy transmission from the rotor to the particles and resulted in apparently increasing in k_f. The ratio of k_f at $v = 0$ (that is, 0.722/1.59 = 0.46) may represent decreasing in the degree ξ of energy transmission; ξ was decreased from 0.024 to 0.011 by immobilization of the ceramic particles. Using this value of ξ, k_f was recalculated, as shown in Figure 44. By using the corrected k_f, k_f', the effect of the addition of ceramic particles on the hardness of composite particles could be clarified.

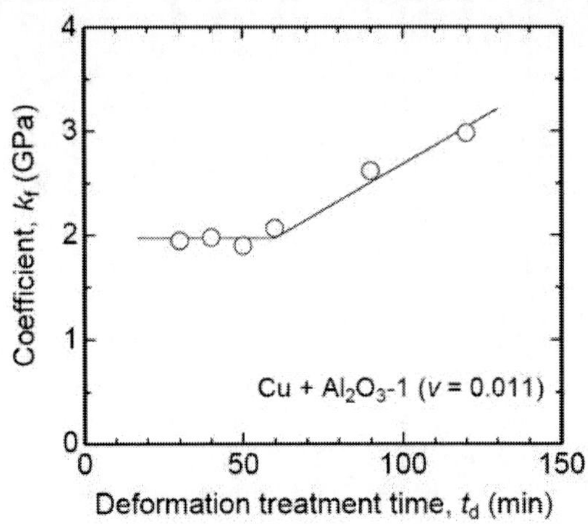

Figure 42. Change in coefficient k_f of Al_2O_3-1–Cu composite particles ($v = 0.012$) with deformation treatment time t_d

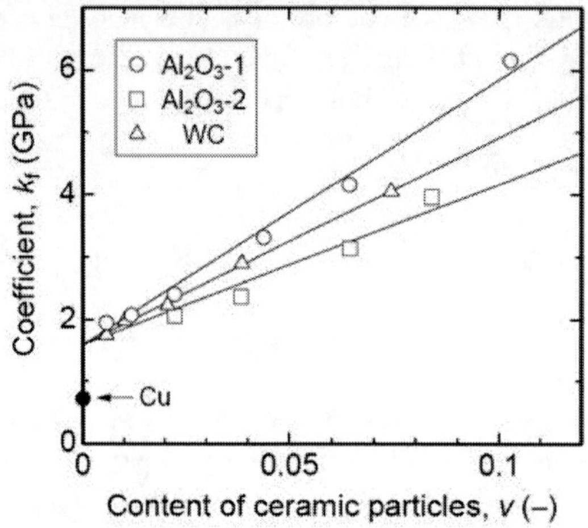

Figure 43. Change in coefficient k_f with content v of ceramic particles

Generally, the hardness of particle-dispersion-strengthened materials is proportional to the volume of the dispersing particles and inversely proportional to the distance between the dispersing particles in the matrix [56]. Therefore, the data of k_f' shown in Figure 44 were arranged in terms of the ratio v/κ_g of the content v to the distance κ_g as a parameter. In the calculation of v/κ_g, when v was larger than the concentration in which the ceramic particles coated the copper particles in the hexagonal closest packing state (for example, $v \geq 0.022$ for Al_2O_3-1), it was assumed that κ_g was equal to the diameter d_g of the ceramic particles. As shown in Figure 45, the relationship between k_f' and v/κ_g was expressed by a single straight line regardless of the kind of ceramic particles. This result reveals that the change in the hardness of composite particles with the amount of ceramic particles could be evaluated similar to that of particle-dispersion-strengthened materials.

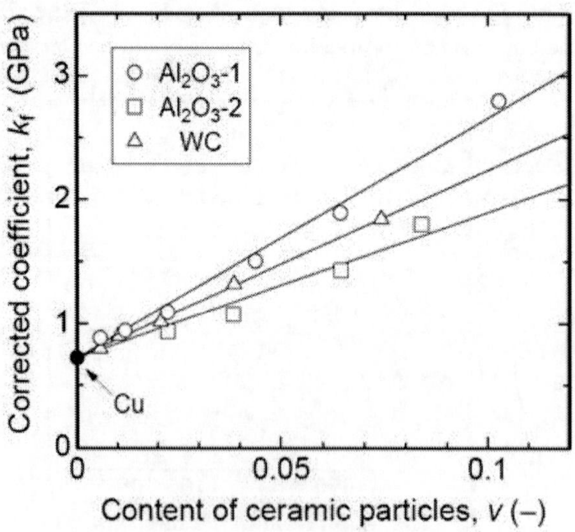

Figure 44. Change in corrected coefficient k_f' with content v

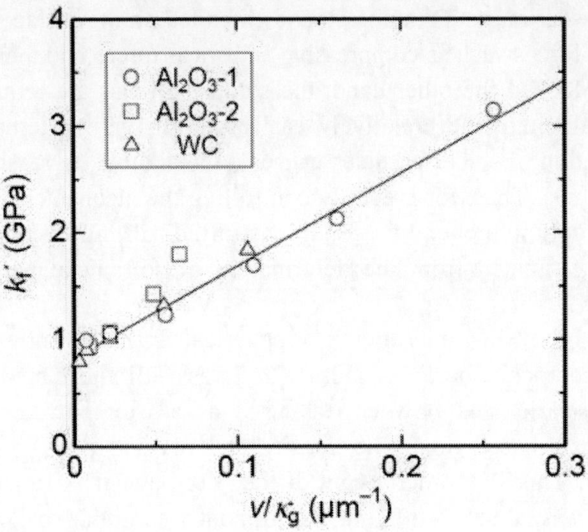

Figure 45. Plots of corrected coefficient k_f' against parameter v/κ_g

Case 2: Encapsulation of Lead-Free Solder Alloy Particles by Coating with Wax Powder

In recent years, a variety of lead-free solders has been developed as substitutes of Sn/Pb eutectic solders and some lead-free solder has already been practically used in precision electronic devices, for example, note PCs and cellar phones by Japanese manufacturers [57]. Lead-free solders are required to have excellent properties such as a suitable melting point and good solder joints and, in addition, it is desirable that the characteristics of lead-free solders are similar to those of conventional Sn/Pb solders to use present processing equipment, conditions and fluxes optimized for Sn/Pb solders. Sn/Zn/Bi solder alloys have been studied actively on since they have a similar melting point to Sn/Pb solder alloys and excellent mechanical properties such as strength, expansion, fatigue and creep properties comparing with Sn/Pb solder alloys and their price is close to that of Sn/Pb solder alloys [57–59]. However, the industrial usage of Sn/Zn/Bi solder alloys is very difficult because they have a fatal defect, the poor corrosion resistance, which shortens the storage time of solder powder and solder pastes due to the marked oxidation of alloys and rapid chemical reaction between alloys and fluxes [59, 60]. The poor wettability of Sn/Zn/Bi alloys is also obstacle to practical applications [61–63]. To solve such problems, the oxidation resistance and reactivity of alloys and the stability of solder pastes were improved by modifying the surface of alloy particles with silane compounds in a wet process, doping minute amounts of elements such as Al and Mg into the alloys, and adjusting the flux components [64–67]. The treatment processes employed in such methods are complicated and simple processes have been required to reduce the treatment costs.

Therefore, in order to improve both the oxidation resistance and the wettability of Sn/Zn/Bi alloys and to achieve a long-term storage and high stability of the solder pastes, the solder particles were encapsulated by coating with a wax powder. A dry mechanical process using a tumbling ball mill was employed to simplify the treatment process. In the treatment process, the wax powder is ground, adhered and extended on the surface of alloy particles by exerting mechanical forces such as compressive and shear forces and then the thin film (shell) of wax is formed [68]. On the other hand, the solder particles are required to maintain the original sphericity although they relatively easily deform with external forces exerted on because the deformation of solder particles causes a remarkable decreasing in the flowability of the solder pastes [69]. Thus, for precisely controlling the mechanical energy applied to the solder particles, the optimum operating conditions of the ball mill were determined based on the applied energy estimated from the deformation of soft metal particles, as mentioned above.

Figure 46 shows the SEM micrograph of spherical Sn/Zn/Bi alloy powder used in this work, which has a composition of Sn-8mass%Zn-3mass%Bi and a median diameter of 40.5 μm. A 12-hydroxystearic acid powder was used as a wax powder, which was one of components of a rosin-based flux. The density and melting point were 1000 kg/m^3 and 76.4°C, respectively. The SEM micrograph of the wax powder is shown in Figure 47. The particle size range was about 2–20 μm. A horizontal tumbling ball mill (TBM) with a cylindrical alumina pot of an inside diameter of 76 mm and a capacity of about 400 ml was used in the wax-coating treatments. Stainless balls of 3.2 mm in diameter were used as ball media and occupied 40% of the pot capacity. The optimum operating conditions in the treatments were determined based on the applied energy in which the alloy particles with an allowable deformation were obtained, as presented in the previous section. The wax-coating treatments were carried out with the ball mill in an argon atmosphere under the optimum operating conditions thus determined. The content of the wax powder was varied in the range of 0.3–1.2mass% of the alloy particles. The shape of wax-coated alloy particles was observed with the SEM and the size distribution was measured with the image analyzer.

Before the wax-coating treatments, the correlations between the applied energy and the operating conditions were investigated. The applied energy E to the whole alloy particles could be obtained as the total of deformation energy E_1 of each alloy particle calculated by using Eqs. (33)–(37), and given as the product of E_1 and the number n_a of alloy particles to be treated.

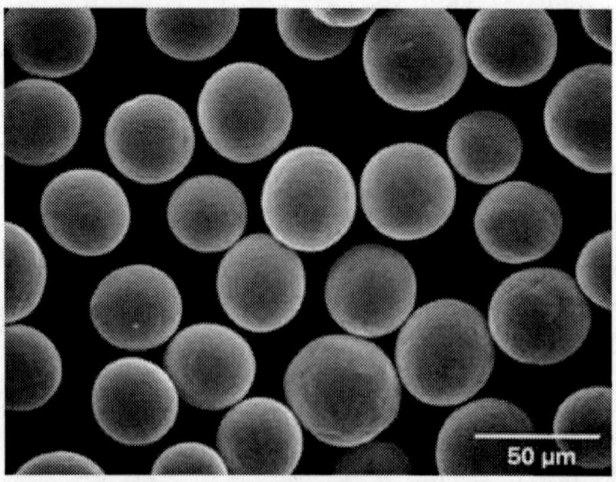

Figure 46. Sn-8mass%Zn-3mass%Bi alloy solder powder

Figure 47. Wax (12-hydroxystearic acid) powder used to coat Sn/Zn/Bi solder powder

$$E = E_1\, n_a \tag{62}$$

where n_a is expressed as a function of the surface area ratio S of the alloy particles to the media balls.

$$n_a = A_s S/(\pi D_O^2) \tag{63}$$

where A_s indicates the total surface area of media balls placed in the mill pot, and D_O is the original median diameter of alloy particles. The ball-milling treatments of the alloy particles without adding the wax powder were carried out to determine the optimum operating conditions in the wax-coating treatments. The operating conditions of TBM in the treatments were varied as follows: the rotational speed of the pot n_r = 0.5–1.3 s^{-1}, the treatment time t = 1.8–7.2 ks and the amount of the alloy particles to be treated in a batch, which was expressed by the surface area ratio of the alloy particles to the media, S = 4.4–13.6. Following to the procedure presented in the previous section, the applied energy E was estimated under various operating conditions of the rotational speed n_r of the pot, the treatment time t and the surface area ratio S, as shown in Figure 48. The relationships between them were expressed by empirical formulas.

$$E = 1.5\times10^2\, n_r + 1.4\times10^2\, n_r^2 \text{ at } t = 7.2 \text{ ks and } S = 6.3 \tag{64}$$

$$E = 58\, t \text{ at } n_r = 1.3 \text{ s}^{-1} \text{ and } S = 6.3 \tag{65}$$

$$E = 3.8\times10^2 + 8.4\, S \text{ at } n_r = 1.3 \text{ s}^{-1} \text{ and } t = 7.2 \text{ ks} \tag{66}$$

The results show the facts that (i) the kinetic energy of rotational motion of the pot contributes to the applied energy, (ii) the applied energy per unit time is constant, and (iii) the energy transfer to the alloy particles increases slightly with increasing in the amount of alloy particles.

The mechanical energy suitable for wax-coating was estimated by taking into consideration an allowable deformation of alloy particles in the treatments. A slight deformation of the alloy particles was allowed in the wax-coating treatments, and the allowable true strain was introduced to determine the optimum operating conditions of the ball mill, in which the solder pastes containing the wax-coated alloy particles had a similar flowability to that containing only the original alloy particles. The viscosity of a solder paste containing 89.5mass% uncoated, deformed alloy particles having the true strain less than 0.1 almost agreed with that of a paste containing the original alloy particles regardless of the shear rate [69]. Accordingly, the allowable true strain in the wax-coating treatments was determined to be 0.1 and the mechanical energy E_1 applied to each alloy particle in the wax-coating treatments could be estimated to 1.8 μJ for the alloy particle of the median diameter (40.5 μm). As shown in Figure 48, the operating conditions in which the true strain was 0.1 were obtained as follows: (n_r, t, S) = (1.3 s^{-1}, 7.2 ks, 6.3), (1.3 s^{-1}, 7.4 ks, 6.3) and (1.3 s^{-1}, 7.2 ks, 6.4).

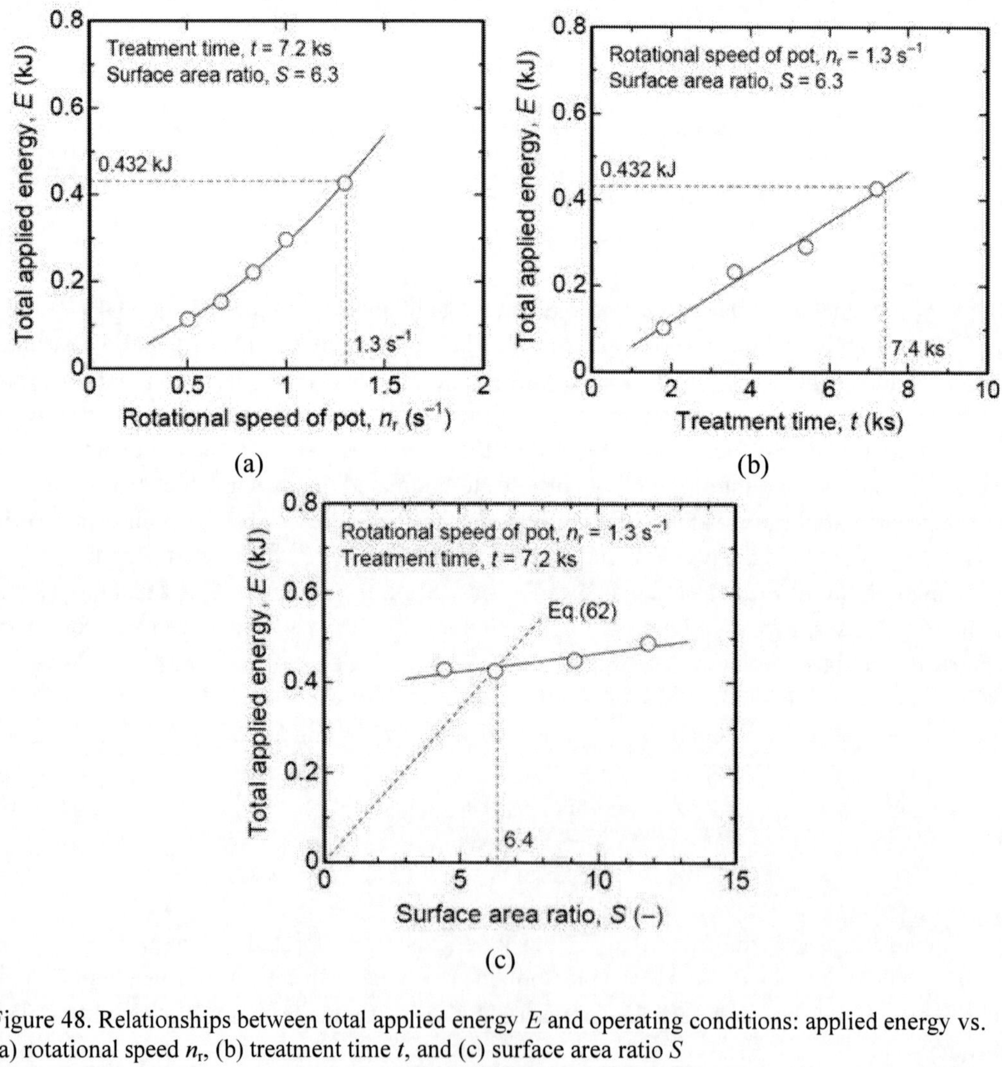

Figure 48. Relationships between total applied energy E and operating conditions: applied energy vs. (a) rotational speed n_r, (b) treatment time t, and (c) surface area ratio S

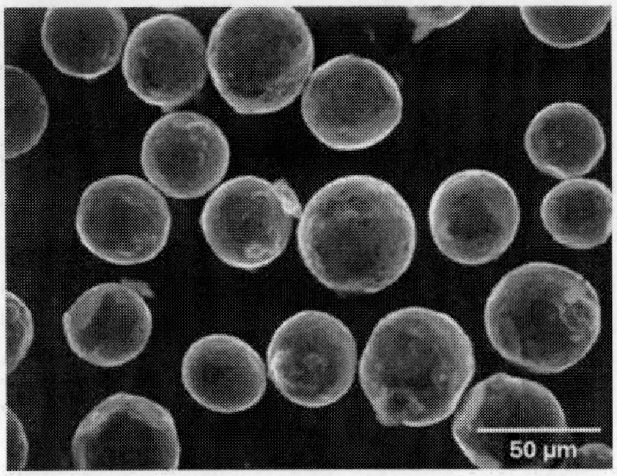

Figure 49. 0.9mass%-wax-coated alloy particles treated under optimum operating conditions

Table 3. Size distributions of original and wax-coated alloy particles

Wax Content (mass%)	Median diameter* (μm)	Geometric standard deviation (–)
0 (original)	40.5	1.15
0.3	41.7	1.13
0.6	41.8	1.15
0.9	40.5	1.14
1.2	41.5	1.14

* Heywood diameter, area basis

Comparing with these conditions, slightly moderate conditions, that is, a lower rotational speed, a shorter treatment time and a smaller surface area ratio, were corresponded to a representative of optimum operating conditions. Therefore, the wax-coating treatment was carried out under the conditions of $n_r = 0.83\ s^{-1}$, $t = 7.2$ ks and $S = 6.3$ as an example. Figure 49 shows the SEM photograph of the 0.9mass%-wax-coated alloy particles prepared under the operating conditions. The alloy particles maintained the spherical shape after the wax-coating treatment. The similar results were obtained in the treatments of 0.3 to 1.2mass%-wax-coating. The median diameter and geometric standard deviation of the size distribution were presented in Table 3. The median diameter of wax-coated alloy particles was almost the same as that of original alloy particles. The wax powder was ground and extended on the surface of alloy particles by acting compressive and shear forces on them, and then the uniform wax film was formed on the particle surface. On the other hand, it was confirmed that the most of alloy particles were considerably deformed to be barrel- and disk-shape under severe conditions comparing with the treatments under the optimum operating conditions. It was proved that the proposed method is useful to determine the optimum operating conditions in the wax-coating treatment based on the mechanical energy applied to the alloy particles during the treatments.

The wettability of molten solder of wax-coated alloy particles was evaluated. The procedure of the wettability test followed the Japanese Industrial Standard (JIS) Z 3284. After the wax-coated alloy particles were stored at 10°C in argon and at room temperature in air for a proper time, the solder pastes containing the alloy particles of 89.5mass% were prepared by hand-mixing of the alloy particles and a rosin-based flux for 5 min with a spoon, and then immediately the solder pastes were printed on a copper plate cleaned with ethanol using a

200-μm-thick stencil with an aperture diameter of 6.5 mm. The solder paste-printed copper plates were heated on a heating plate in air as follows: 150°C for 90 s (the preheating period) and then 230°C for 20 s. After cooling, the projected area of molten solder was measured with the image analyzer and the wettability was determined, which was defined as the spreading area ratio of the projected area measured to the printed area of the solder paste. The average of 20 measurements for each sample was adopted as the experimental data. Figure 50 shows the wettability of wax-coated alloy particles together with that of the mixture of the wax powder and the alloy particles. The wettability of original alloy particles was improved by just adding the wax. Especially, the improvement of the wettability was remarkable in the wax-coated alloy particles even when the wax content was 0.3mass%, and it was almost constant in the range of 0.3 to 0.9mass%. The wettability of the wax–alloy mixture, however, slightly increased with increasing in the wax content although the spreading area of melted flux greatly increased. In the wax-coated alloy particles, the wax contributed efficiently to the spreading of molten solder because each particle was homogeneously coated with the wax. On the other hand, the spreading of molten solder of the wax–alloy mixture decreased because of insufficient dispersion of the wax in the solder paste. In the 1.2mass%-wax-coated ones, there was the excess wax powder which could not adhere to the surface of alloy particles. This reduced the wettability, resulting in the similar wettability to that of the wax–alloy mixture regardless of the storage conditions. The influence of wax-coating on the wettability was small within the range of the wax content in which all the wax powder added adhered on the alloy particles.

Furthermore, the wettability of wax-coated alloy particles was maintained after 17 days even when they were stored at room temperature in air. The results represent that the wax film formed on the particle surface has isolated the external atmosphere and inhibited the formation of oxide layers on the surface of alloy particles since the wetting properties depend strongly on the conditions of the oxide layers. The encapsulation treatment was effective to control the oxidation of alloy particles and improved the wettability.

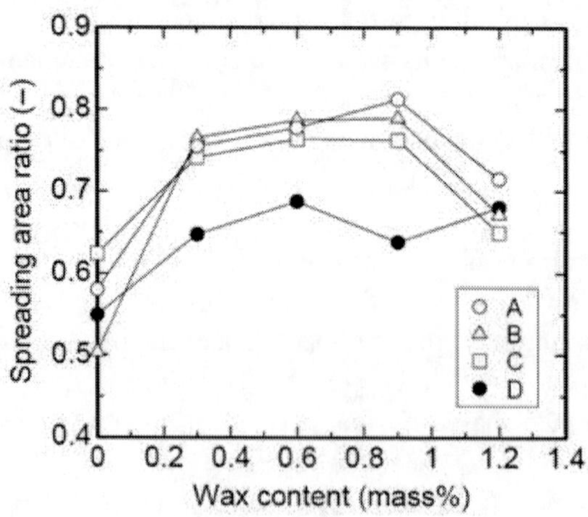

Figure 50. Effect of storage conditions of solder powders on spreading area of molten solder in wettability test carried out immediately after wax-coating treatment of alloy particles (A), after storing wax-coated alloy particles for 17 days at 10°C in argon (B), after storing wax-coated alloy particles for 17 days at room temperature in air (C), and immediately after simple mixing of alloy particles and wax powder (D).

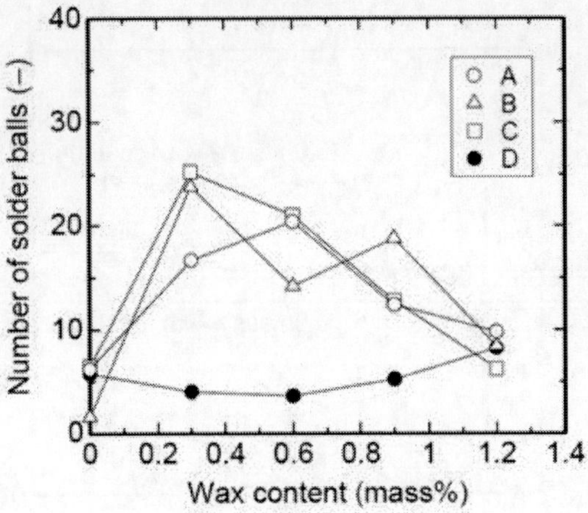

Figure 51. Effect of storage conditions of solder powders on number of solder bolls in solder balling test. Symbols in this figure are the same as those in Figure 50.

The solder balling test was carried out followed by the JIS Z 3284. The solder paste was printed on an ethanol-cleaned standard alumina substrate with the stencil used in the wettability test immediately after the solder paste was prepared. The alumina substrate was heated on a heating plate in air under the same profile as in the wettability test and observed with an optical stereoscopic microscope after cooling. The number of solder balls remaining around the coalesced solder was counted. The average of 9 measurements for each sample was adopted as the experimental data. Figure 51 shows the effect of wax-coating on the coalescence properties of molten solder. The number of solder balls for the 0.3 to 0.9mass%-wax-coated alloy particles was larger than that for the wax–alloy mixture regardless of the storage conditions. The wax adhered strongly to each alloy particle by forming the metallic soaps in the wax-coated alloy particles, resulting in preventing the coalescence of molten solder of the wax-coated ones. This tendency was remarkable at lower wax contents and the number of solder balls was the highest when the wax content was 0.3mass%. The spreading area of melted flux increased with increasing in the wax content; this assisted partially the coalescence of molten solder and the number of solder balls decreased with increasing the wax content. Therefore, their coalescence properties would be improved by adjusting the components of the flux and/or by soldering in a nitrogen atmosphere.

When the wax content was 1.2mass%, the coalescence properties were almost the same as that of the wax–alloy mixture due to presenting the excess wax powder among the wax-coated ones. The influence of the storage conditions on the coalescence properties of the wax-coated ones was small similar to the results of the wettability test. It has been proved that the coalescence properties are unchanged even though the wax-coated alloy particles are stored at room temperature in air.

Figure 52 shows the temporal variation of the soldering properties of the original (without adding the wax) and 0.9mass%-wax-coated alloy particles stored at room temperature in air. The soldering properties of wax-coated alloy particles had been maintained under a severe storage condition since when the wax-coating treatments were finished. Consequently, the encapsulation of readily oxidizable lead-free solder powder by wax-coating was effective in the improvement of their oxidation resistance and soldering properties.

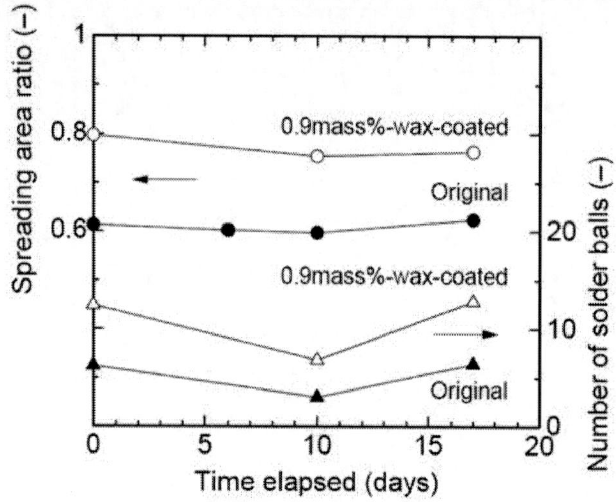

Figure 52. Influence of storage time on soldering properties of solder pastes containing original and 0.9mass%-wax-coated alloy particles stored at room temperature in air. Circular and triangular symbols indicate the results of the wettability test and the solder balling test, respectively

Conclusion

The methods for estimating the mechanical energy applied to the particulate materials in mechanical treatment processes have been developed, and it has been demonstrated that the evaluation results of the applied energy can be applied to practical powder design and preparation of functional particulate materials. For advanced powder designs using mechanical means, it is particularly important to understand the energy characteristics of processing equipment used in treatment processes and to estimate the energy requirements for the treatment. The evaluation methods of mechanical energy presented in this chapter could be widely used for various types of equipment because the evaluation methods do not require that the equipment to be evaluated satisfy special conditions. The author hopes that this study can contribute to the development and progress of technology in the dry mechanical syntheses of functional composite particles, to the clarification of dry compounding mechanisms and to the achievement of microscopic powder designs.

Nomenclature

A	constant in Eq. (10), dimensionless
A_b	area of fractured section of agglomerates, m^2
A_s	total surface area of media, m^2
a_r	representative acceleration, $m \cdot s^{-2}$
B	constant in Eq. (10), dimensionless
$B(D_{px}, D_{py})$	distribution function of breakage accumulation, dimensionless
b_{ij}	distribution function of breakage frequency defined by Eq. (5), dimensionless
b	constant in Eq. (18), dimensionless

C	agglomeration rate constant, s^{-1}
c	constant in Eq. (9), dimensionless
D	diameter of a deformed copper particle, m
D_{50}	geometric mean diameter, m
$D(D_{pxm}, t)$	volume fraction of agglomerates having a diameter less than D_{pxm} at time t, dimensionless
D_{B50}	geometric mean diameter, m
D_h	median diameter of host particles, m
D_O	original diameter of a particle, m
D_{px}, D_{py}	diameters of an Fe_2O_3 agglomerate, m
D_{pxm}	minimum diameter of sieved agglomerates, m
D_{pym}	maximum diameter of agglomerates, m
d	diameter of a primary Fe_2O_3 particle, m
d_e	depth to which a guest particle was embedded, m
d_g	median diameter of guest particles, m
d_M	diameter of a medium, m
E	energy applied to whole particles, J
E_1	deformation energy applied to a single particle, J
E_c	energy applied to particles in a single compression, J
E_s	applied energy per unit volume and time, $J{\cdot}m^{-3}{\cdot}s^{-1}$
E_v	energy applied to powder per unit volume, $J{\cdot}m^{-3}$
E_0	constant in Eq. (58), $kg{\cdot}m^2$
e	impact energy applied to a medium per unit volume, $J{\cdot}m^{-3}$
e'	embedding energy, J
e_1	energy required to embed a single guest particle, J
e_{bi}	disintegration energy of a single agglomerate defined by Eq. (27), J
F_t	breaking tensile force acting on a single contact point, N
f	contact force caused by viscous dissipation, N
f_i	force required to intrude a single guest particle into a host particle, N
g	acceleration of gravity, $m{\cdot}s^{-2}$
H	height of a powder bed, m
H_V	Vickers hardness, Pa
K_A	breakage rate constant, s^{-1}
k	coordination number of a particle, dimensionless
k'	constant in Eq. (18), Pa
k_1	constant in Eq. (9), $m^3{\cdot}s^{1+c}{\cdot}kg^{-2}$
k_2	constant in Eq. (13), $kg{\cdot}s^{-1}$
k_f	coefficient in Eq. (57), Pa
k_f'	corrected k_f, Pa
k_s	stiffness of a medium, $kg{\cdot}m^{-1/2}{\cdot}s^{-2}$
M	total mass of host particles, kg
m	total mass of guest particles, kg
N	impeller speed, s^{-1}
N_R	rotor speed, s^{-1}
N_V	rotational speed of the vessel, s^{-1}
n	number of impellers, dimensionless
n'	number of revolutions of the rotating structure per unit time, s^{-1}

n_a	number of alloy particles, dimensionless
n_c	average number of host particles fed into the minimum clearance region within a single compression, dimensionless
n_r	rotational speed, s^{-1}
P(r)	impact force, N
R_I	radius of an impeller, m
R_V	radius of the mixer vessel, m
r	radius of rotation, m
r_r	radius of rotation, m
S	surface area ratio of alloy particles to media, dimensionless
S_c	contact area of a guest particle with a host particle in indentation, m^2
$S(D_{py})$, S_i	selection functions, s^{-1}
S_V	specific surface area of powder, $m^2 \cdot m^{-3}$
t	time, s
t_d	deformation treatment time, s
t_p	circulating time of particles, s
u	average mass flow rate of host particles, $kg \cdot s^{-1}$
u_r	representative velocity, $m \cdot s^{-1}$
V_0	total solid volume of particles in the mixer, m^3
V_b	volume of the breakage region, m^3
V_C	bulk volume of copper powder, m^3
V_P	bulk volume of particles, m^3
V_v	total volume of the void among media, m^3
v	volume content of guest particles in the composite particles, dimensionless
$v_i(t)$	volume fraction of agglomerates of a diameter D_{px}, dimensionless
v_r	content corresponding to formation of an ideal randomly packed coating layer, dimensionless
Y	Young's modulus of a medium, Pa
z	parameter in Eq. (12), m

Greek Letters

α	constant in Eq. (33), J
β	constant in Eq. (33), J
γ	constant in Eq. (33), J
Δ	extension displacement, m
δ	elastic deformation of a medium, m
δ_h	deformation of a host particle, m
ε	true strain of a copper particle, dimensionless
ε_i	intra-void fraction of an agglomerate, dimensionless
ε_M	void fraction among media, dimensionless
η	degree of attainment to final stage of dispersion, dimensionless
η_d	damping coefficient, $kg \cdot s^{-1}$
Θ	dimensionless parameter defined by Eq. (47), dimensionless
Θ'	dimensionless parameter defined by Eq. (40), dimensionless
θ	parameter in Eq. (12), rad
κ	coordination number of a medium, dimensionless

κ_g distance between centers of guest particles, m
λ time interval of a single compression, s
λ' constant in Eq. (43), dimensionless
ν Poisson's ratio of a medium, dimensionless
Ξ dimensionless parameter defined by Eq. (38), dimensionless
ξ constant in Eq. (58), dimensionless
ρ_A apparent density of an agglomerate, $kg \cdot m^{-3}$
ρ_B bulk density of a powder bed, $kg \cdot m^{-3}$
ρ_C true density of copper, $kg \cdot m^{-3}$
ρ_C' bulk density of copper powder, $kg \cdot m^{-3}$
ρ_{F0} true density of an Fe_2O_3 particle, $kg \cdot m^{-3}$
ρ_f true density of a guest particle, $kg \cdot m^{-3}$
ρ_M true density of a medium, $kg \cdot m^{-3}$
σ stress applied to powder, Pa
σ_{Bg} geometric standard deviation given by Eq. (23), dimensionless
σ_g geometric standard deviation, dimensionless
σ_{st} fracture strength of an agglomerate, Pa
σ_t tensile stress of a Fe_2O_3 powder bed, Pa
σ_{tm} tensile strength of an agglomerate, Pa
φ constant in Eq. (46), dimensionless

REFERENCES

[1] Alonso, M; Satoh, M; Miyanami, K. *Powder Technol.,* 1988, 56, 135.
[2] Satoh, M; Alonso, M; Shigemura, T; Miyanami, K; Suzuki, T. *Trans. Inst. Chem. Eng.,* 1990, 68, 149.
[3] Satoh, M; Yamashita, T; Yoshida, T; Hasegawa, S; Miyanami, K. *J. Soc. Powder Technol., Japan,* 1993, 30, 390.
[4] Murakami, Y; Fujimoto, K; Shimada, T; Yamada, A; Asano, K. *J. Chem. Eng. Japan,* 1972, 5, 297.
[5] Yoshino, H; Taya, Y; Kaga, H; Mukaida, K. *J. Soc. Powder Technol., Japan,* 1995, 32, 866.
[6] Iwasaki, T; Satoh, M. *Adv. Powder Technol.,* 2002, 13, 395.
[7] Iwasaki, T; Satoh, M; Takahashi, T. *Powder Technol.,* 2001, 119, 95.
[8] Iwasaki, T; Kim, JH; Mizuhashi, S; Satoh, M. *J. Electron. Mater.,* 2005, 34, 647.
[9] Iwasaki, T; Kim, JH; Satoh, M. *Chem. Eng. Sci.,* 2006, 61, 1065.
[10] Epstein, B. *Ind. Eng. Chem.,* 1947, 40, 2289.
[11] Bass, L; *Angew, Z. Math. Phys.,* 1954, 5, 283.
[12] Gaudin, AM; Meloy, TP. *Trans. AIME,* 1962, 223, 43.
[13] Thornton, C; Ciomocos, MT; Adams, MJ. *Powder Technol.,* 1999, 105, 74.
[14] Bika, DG; Gentzler, M; Michaels, JN. *Powder Technol.,* 2001, 117, 98.
[15] Hayashi, M; Tanaka, T. *J. Chem. Eng. Japan,* 1972, 5, 308.
[16] Hiramatsu, Y; Oka, Y; Kiyama, H. *J. Min. Inst. Japan,* 1965, 81, 1024.
[17] Hiramatsu, Y; Oka, Y. *Int. J. Rock Mech. Min. Sci.,* 1966, 3, 89.
[18] Ridgeway, K; Tarbuck, KJ. *Brit. Chem. Eng.,* 1967, 12, 384.

[19] Sugai, Y; Satoh, M. *J. Soc. Powder Technol., Japan,* 1999, 36, 10.
[20] Knudsen, FP. *J. Am. Ceram. Soc.,* 1959, 42, 376.
[21] Kuwahara, Y; Yamamoto, H; Ueda, K. *J. Res. Assoc. Powder Technol., Japan,* 1969, 6, 406.
[22] Kuwahara, Y. *J. Chem. Eng. Japan,* 1971, 4, 359.
[23] Sugai, Y; Satoh, M. *J. Soc. Powder Technol., Japan,* 1997, 34, 767.
[24] Satoh, M; Yoshida, T; Miyanami, K; Okudaira, Y. *J. Soc. Powder Technol., Japan,* 1994, 31, 789.
[25] Benjamin, JS. *Metall. Trans.,* 1970, 1, 2943.
[26] Senna, M; Fujiwara, Y; Isobe, T; Tanaka, J. *Solid State Ionics,* 2001, 141-142, 31.
[27] Saito, F; Zhang, QW; Kano, J. *J. Mater. Sci.,* 2004, 39, 5051.
[28] Iwasaki, T; Satoh, M; Koga, T. *Powder Technol.,* 2001, 121, 239.
[29] Fayed, ME; Otten, L. *Handbook of Powder Science and Technology*, 2nd ed., Chapman and Hall, New York, 1997, 108.
[30] Cundall, PA; Strack, ODL. *Geotechnique,* 1979, 29, 47.
[31] Tsuji, Y; Tanaka, T; Ishida, T. *Powder Technol.,* 1992, 71, 239.
[32] Timoshenko, SP; Goodier, JN. *Theory of Elasticity*, 3rd ed., McGraw-Hill, New York, 1970, 409.
[33] Iwasaki, T; Satoh, M; Ito, T. *Powder Technol.,* 2002, 123, 105.
[34] Iwasaki, T; Satoh, M; Ito, T. *J. Mater. Process. Tech.,* 2004, 146, 330.
[35] Reddy, V. *Manuf. Eng.,* 1989, 102, 83.
[36] Alonso, M; Alguacil, FJ. *Rev. Metal. Madrid,* 1999, 35, 315.
[37] Alonso, M; Satoh, M; Miyanami, K; Higashi, K; Ito, T. *Powder Technol.,* 1990, 63, 35.
[38] Tsuji, K; Inada, H; Kojima, K; Satoh, M; Higashi, K; Miyanami, K; Tanimura, S. *J. Mater. Sci.,* 1992, 27, 1179.
[39] Kaga, H; Taya, Y; Katayama, H; Nagasawa, Y; Mukaida, K. *Powder Technol.,* 1997, 93, 143.
[40] Ami, K; Isobe, T; Senna, M. *Powder Technol.,* 1998, 100, 46.
[41] Shinohara, K; Liang, H; Uchiyama, T. *J. Ceram. Soc. Japan,* 2000, 108, 402.
[42] Ramlakhan, M; Wu, CY; Watano, S; Dave, RN; Pfeffer, R. *Powder Technol.,* 2000, 112, 137.
[43] Fukumori, Y; Ichikawa, H; Ueda, M. *Proc. World Cong. Particle Technol.*, 1998, 3, Brighton, UK 1217.
[44] Iwasaki, T; Satoh, M; Masuda, T; Fujita, T. *J. Mater. Sci.,* 2000, 35, 4025.
[45] Watano, S; Imada, Y; Miyanami, K; Wu, CY; Dave, RN; Pfeffer, R; Yoshida, T. *J. Chem. Eng. Japan,* 2000, 33, 848.
[46] Kim, W; Saito, F. *Powder Technol.,* 2000, 113, 109.
[47] Kh. F; Urakaev, VV. Boldyrev, *Powder Technol.,* 2000, 107, 197.
[48] Ueno, K; Shinohara, K. *Kagaku Kogaku Ronbunshu,* 1999, 25, 823.
[49] Otani, M; Minoshima, H; Ura, T; Shinohara, K. *Adv. Powder Technol.,* 1996, 7, 291.
[50] Iwasaki, T; Koga, T; Satoh, M. *J. Chem. Eng. Japan,* 2000, 33, 499.
[51] Endoh, S; Szépvölgyi, J; Hayashi, T; Naito, M; Tanimoto, T. *Kagaku Kogaku Ronbunshu,* 2000, 26, 268.
[52] Alonso, M; Satoh, M; Miyanami, K. *Powder Technol.,* 1990, 62, 35.
[53] Shima, S; Saleh, MAE. *Int. Conf. Mech. Eng. Adv. Tech. Ind. Prod.*, Assiut Univ., Assiut, Egypt, 1994, 323.

[54] Shima, S; Kotera, H; Ujie, Y. *J. Jpn Soc. Powder Powder Metallurgy,* 1995, 42, 341.
[55] Ashby, MF; Jones, DRH. *Engineering Materials* 1, 2nd ed., Butterworth-Heinemann, Oxford, 1996.
[56] Ansell, GS. *Met. Soc. Conf.,* 1968, 47, 61.
[57] Suganuma, K. ed., Lead-Free Soldering in *Electronics: Science, Technology, and Environmental Impact,* Marcel Dekker, New York. 2004.
[58] Shohji, I; Gagg, C; Plumbridge, WJ. *J. Electron. Mater.,* 2004, 33, 923.
[59] Takemoto, T; Miyazaki, M; Matsunawa, A. Proc. 3rd Int. Conf. *On ECOMATERIALS, Tsukuba,* Japan, 1997, 63.
[60] Takemoto, T. Proc. EcoDesign '99, Tokyo, Japan, 1999, 964.
[61] McCormack, M; Jin, S; Chen, HS. *J. Electron. Mater.,* 1994, 23, 687.
[62] Loomans, ME; Vaynman, S; Ghosh, G; Fine, ME. *J. Electron. Mater.* 1994, 23, 741.
[63] Suganuma, K; Murata, T; Noguchi, H; Toyoda, Y. *J. Mater. Res.,* 2000, 15, 884.
[64] Saeda, S; Utsunomiya, M; Murase, N. JP 9–19794 (A), 1997.
[65] Lin, KL; Chen, KI; Shi, PC. *J. Electron. Mater.,* 2003, 32, 1490.
[66] Wu, CML; Yu, DQ; Law, CMT; Wang, L. *J. Electron. Mater.,* 2002, 31, 921.
[67] Wataya, T; Yoshikawa, M; Tanaka, T; Kojima, H. *J. Jpn. Inst. Met.,* 2003, 67, 643.
[68] Kim, JH; Satoh, M; Iwasaki, T. *Mat. Sci. Eng.,* 2002, A342, 258.
[69] Kim, JH; Satoh, M; Iwasaki, T. *Adv. Powder Technol.,* 2005, 16, 61.

In: Powder Engineering, Technology and Applications ISBN: 978-1-61761-216-9
Editor: Jason M. Barker, pp.311-324

Chapter 11

SEMICONDUCTOR CERAMIC MATERIALS PRODUCED FROM AIIBVI NANOPOWDERS*

N. N. Kolesnikov, E. B. Borisenko, V. V. Kveder, D. N. Borisenko, A. V. Timonina and B.A. Gnesin[1]
Institute of Solid State Physics, Russia

ABSTRACT

Based on our experience of conventional crystal growth of AIIBVI, we have developed the techniques for nanocrystal production of some of these compounds.

Further innovations in the technological process appear to significantly reduce the development and production cost of the final product with respect to conventional single crystals, while maintaining a high level of physical properties, which conform to leading standards.

New technology of the vapor phase deposition was developed in our laboratory to produce CdTe nanoparticles of average diameter 8 nm. Further development of the technology allows us to produce 10-nm Cd-Zn-Te nanocrystals and to overcome difficulties in obtaining a ternary solid solution Cd1-xZnxTe (x = 0.04—0.1) with a stable chemical composition. Highly dense CdTe and Cd1-xZnxTe ceramics of high mechanical hardness and durability were produced by the room temperature process without any lubricants or binding materials.

It was found that CdTe ceramics produced from nanocrystals undergo wurtzite—sphalerite transition under pressure. The polymorphic transition from the hexagonal to the cubic phase in Cd1-xZnxTe ceramics caused by annealing is discussed.

The two-component texture composed of <220> axial and {100}<hk0> components was recorded in the as-compressed materials.

The effect of annealing on grain growth and texture in the ceramics is considered. It was found that our compacted ceramics guarantees high transmittance in a wide IR region of 6—25 μm, high specific resistivity on the order of 10^{10} Ωcm, and high microhardness on the order of 10^3 MPa.

---

* A version of this chapter was also published in Nanopowders and Nanocoatings: Production, Properties and Applications, edited by V. F. Cotler, Nova Science Publishers. It was submitted for appropriate modifications in an effort to encourage wider dissemination of research.

The obtained properties make these materials promising for use in IR optics and for ionizing-radiation detectors.

Introduction

$Cd_{1-x}Zn_xTe$ (CZT) single crystals traditionally grown from melt are used in infrared technique and as materials for ionizing radiation detectors.

The development of nanotechnologies provides possibilities for fabrication of materials with new properties and broadens the conventional application fields.

Nanotechnological methods are promising for fabrication of ceramic materials from $A^{II}B^{VI}$ binary and ternary compounds. Among these materials very interesting are CdTe and $Cd_{1-x}Zn_xTe$ compounds which have numerous applications in infrared optics and ionizing radiation detectors. Particularly interesting is $Cd_{0.9}Zn_{0.1}Te$ composition as one, used in ionizing radiation detectors operating at room temperature without cooling.

As compared to the traditional melt growth of $Cd_{1-x}Zn_xTe$ crystals [1], the developed vapor deposition technique is simpler, because no high pressures are required to depress decomposition below the melting point and to decrease loss of volatile components. Furthermore, this technique provides production of a material with a definite uniform stoichiometric composition. This is an advantage, because a composition of a melt-grown bulk, especially $Cd_{1-x}Zn_xTe$ ternary solution, varies along the growth axis [1].

The objective of this study is to investigate the effect of external parameters, such as temperature, pressure of compaction, annealing temperature and time on phase composition, microstructure, and crystallographic texture which determine optical, electrical and mechanical properties of a material.

Experimental

The CdTe and CZT nanopowders were deposited using the technique we have develpoped for vapor deposition of $A^{II}B^{VI}$ nanopowders [2]. To produce a nanopowder, pure CdTe single crystal or synthesized CZT were evaporated from a fused silica boat in helium gas flow. The powders were deposited on a gold substrate (Figure 1). The temperature in the source zone of evaporation was 750—990°C in the case of CdTe and 800—850°C in the case of CZT, the helium flow velocity was 1000—1500 ml/min, the temperature in the deposition zone was 540—610° C. These temperature regimes and helium flow velocity are chosen to provide deposition of 8-nm CdTe and 10-nm CZT particles (Figure 2).

There is a dependence between a composition of $Cd_{1-x}Zn_xTe$ nanopowder and a solid source. It was found experimentally which composition of the evaporation source is required to obtain a definite composition of the nanopowder. For example, in order to obtain $Cd_{0.9}Zn_{0.1}Te$ nanopowder composition, the source composition should be $Cd_{0.5}Zn_{0.5}$ Te. A difference between the source and the nanopowder compositions is due to different partial vapor pressures of the components and diffusion coefficients in helium. The problem of variations in chemical composition under thermal evaporation of $A^{II}B^{VI}$ compounds is discussed elsewhere [3, 4].

A one-die scheme was used to compress CdTe or CZT in an Instron machine. The pressure ranged from 400 to 700 MPa. The temperature of compression varied in the range of T_d = 20-200°C (0.2-0.35 of the melting point T_m). After the required pressure had been achieved, the specimen was kept under pressure for 0.5-120 min. The lower temperatures in the compaction process are preferred for producing large amounts of material, since it ensures the simplest and most economical scheme to obtain material with an acceptable density. The upper temperature limit is restricted due to the tendency of AIIBVI compounds to oxidize and partially decompose, particularly at temperatures above 300°C [2]. The density of the compressed specimens measured by weighing them in water was 92-97% of the calculated density, depending on the compression regime. The relative error of the measurements was ~0.2 %. X-ray phase analysis was carried out using a Siemens D500 diffractometer. A sample was fixed in a ring-shaped holder with a 2-mm accuracy of alignment. Either FeK_α or MoK_α was used as the excitation source.

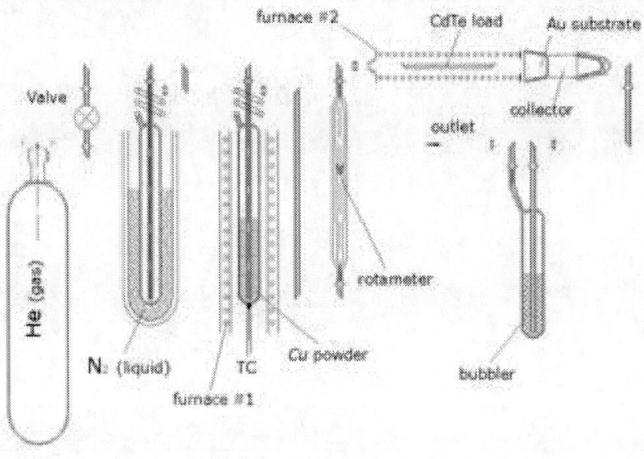

Figure 1. Setup for vapor deposition of $A^{II}B^{VI}$ nanopowders

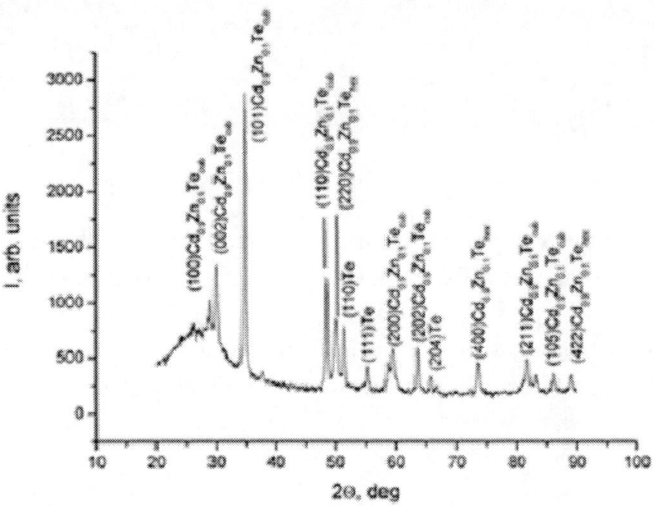

Figure 2. $Cd_{1-x}Zn_xTe$ nanopoweder, TEM image

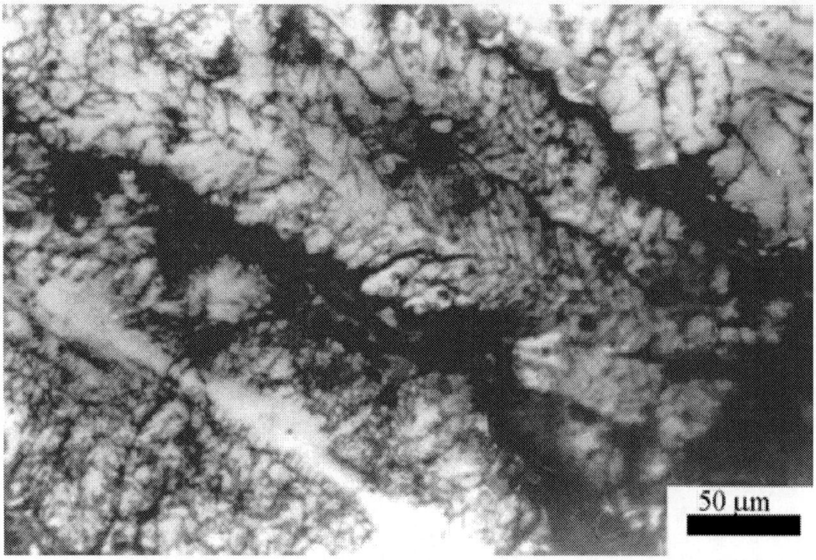

Figure 3. CdTe ceramics produced at T = 20°C, p = 650 MPa, time under pressure t = 10 min

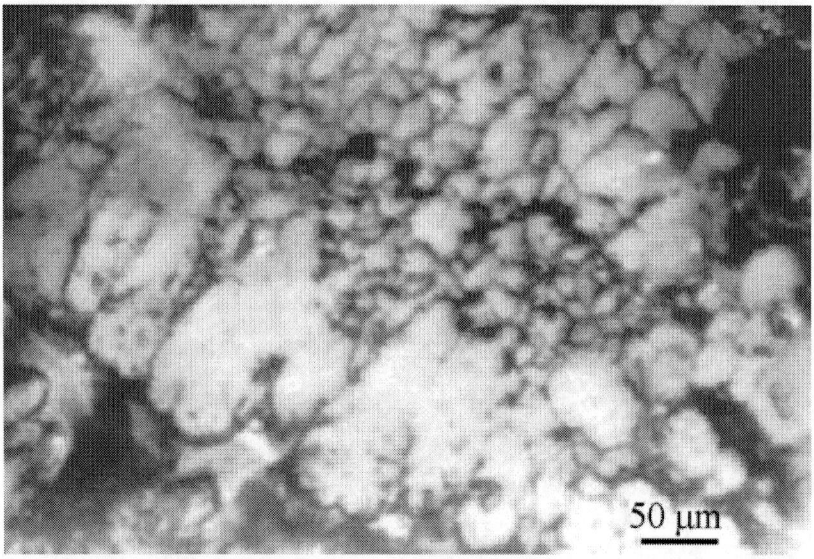

Figure 4. CdTe ceramics produced at T = 200°C, p = 650 MPa, time under pressure t = 10 min

Crystal texture of the compacted samples was studied using the reflection (Schultz) technique. The tilt-angle range was 0-50 °, the azimuth and the polar angle steps were 5 °, and the exposure time at each point was 3 s. To obtain a pole figure (PF), the X-ray detector was set in a certain position based on the previously recorded diffractogram of the compacted CdTe. The background and the defocusing were taken into account when processing the obtained data. The microhardness was measured using a standard Vickers hardness testing technique. The load was 100 g.

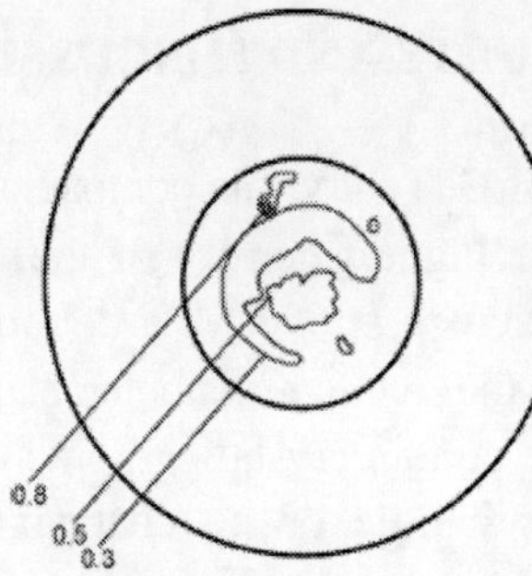

Figure 5. Pole figure {220} of CdTe ceramics produced at T = 20°C, p = 650 MPa

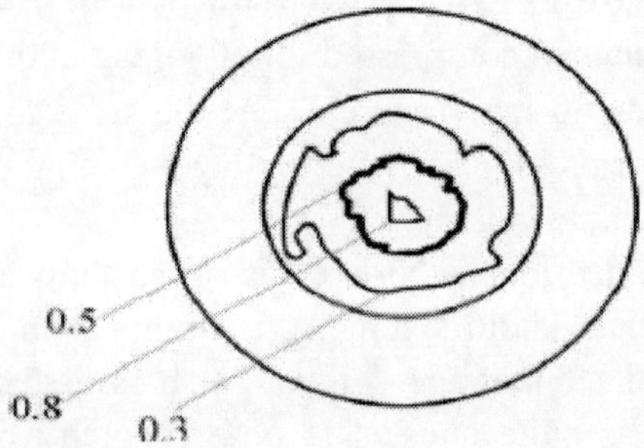

Figure 6. Pole figure {220} of CdTe ceramics produced at T = 200°C, p = 650 MPa

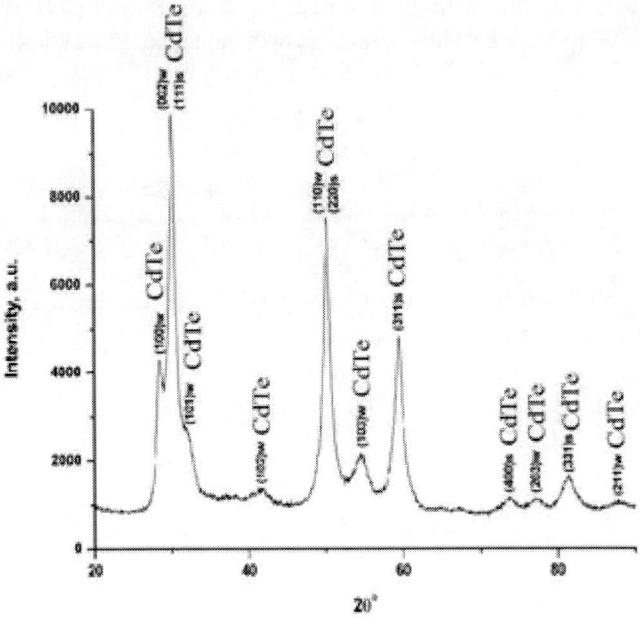

Figure 7. Diffractogram of CdTe nanopowder

The light-transmission spectra were recorded within the wavelength range of 2.5-25 μm using a spectrophotometer Specord 75 IR. The specific electrical resistivity was measured using the four-point-probe technique and direct current. The current-voltage characteristics were obtained using a bias voltage source produced by Ortec.

RESULTS AND DISCUSSION

Microstructures and textures of CdTe crystals compacted at temperatures in the range of 20-200 °C have been studied. The highest density of the ceramics (97% of the calculated density) was achieved under pressure p = 650 MPa at room temperature.. The compacted ceramics consisted of grains with the average diameter 8 μm (Figure 3). In the compacted CdTe, grains are elongated along a certain crystallographic direction (Figure 3). The average grain sizes of the specimens show little or no change under a constant load at room temperature, regardless of time under pressure. The volume fraction of pores was reduced by 2-3 % in a specimen kept for an hour under pressure. An increase in temperature of compaction caused grain growth. The volume fraction of recrystallized grains reached approximately 90 % in a specimen compressed for 60 min at 200 °C under p = 650 MPa. The mean size of new grains was 60 μm (Figure 4). At the same time, micron and sub-micron pores continued to heal under pressure. A layer-by-layer etching of the specimens shows that grains grew throughout the whole volume.

Pole figures {220} obtained for the basic cubic phase show a sharp axial <220> texture. In addition, it was shown that 4-fold symmetry is observed in the {220} PF of the CdTe samples compacted at room temperature (Figure 5). It is associated with another textural component {001}<hk0>. It is quite unusual to observe such a sharp texture in a ceramic, since it is more common for compacted materials to produce rather scattered textures. It was found that though the texture of CdTe specimens compressed at 200 °C remains somewhat axial (Figure 6), it becomes increasingly more scattered than the texture of the specimens compressed at room temperature. Thus, grain growth and textural evolution in the CdTe ceramics compacted at temperatures in the range of 20-200 °C, together with the fact that no strain-hardening was observed in the deformation curves, confirm that dynamic recrystallization takes place during the compression of the nanopowder.

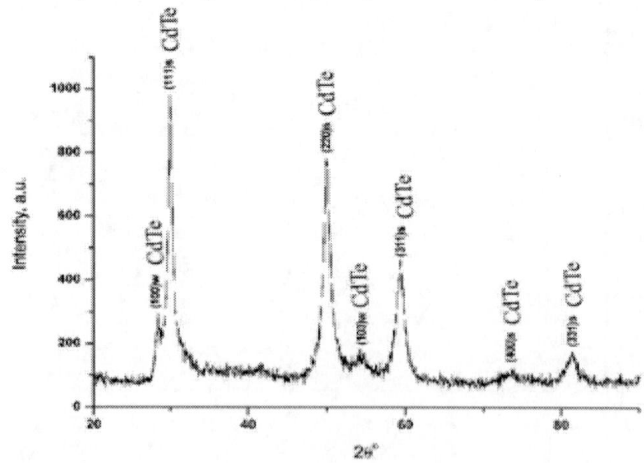

Figure 8. Diffractogram of CdTe ceramics produced at T = 20°C, p = 400 MPa

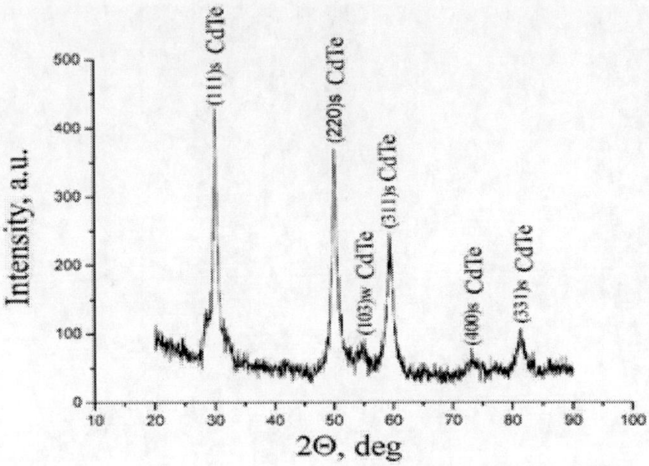

Figure 9. Diffractogram of CdTe ceramics produced at T = 20°C, p = 650 MPa

The CdTe nanopowder consisted of particles of 8 nm in diameter, which was produced through the vapor phase. CdTe usually crystallizes as a cubic phase with sphalerite structure (a = 6.481 Å) [5], which is stable up to near the melting point (T_m = 1092 °C). However, hexagonal structure (wutrtzite) also exists under standard atmospheric pressure. A metastable hexagonal phase with wurtzite structure (a = 4.58 Å, c = 7.50 Å) has been observed in films grown on substrates [6]. The diffractogram in Figure 7 shows that the CdTe nanopowder is a mixture of two phases, wurtzite and sphalerite, and that the wurtzite phase dominates. X-ray diffractograms show that intensities of sphalerite lines increase, while the lines of the hexagonal phase weaken as pressure increases (Figure 8). This means that the hexagonal phase transforms into the cubic phase under pressure. Wurtzite has almost completely converted to sphalerite at a pressure of 650 MPa at T_d = 20°C (Figure 9).

Grain boundaries were revealed by etching the compacted CdTe in a chemical solution containing 7 vol % Br and 93 vol % CH_3OH. In the samples compacted under pressure p = 400 MPa and then subjected to mechanical polishing and etching we observed plates of 20-50 μm in diameter containing twins on the surfaces (Figure 10). The volume fraction of such plates was about 2 % in the specimens that mainly consisted of grains of approximately 8-μm diameter. Twin growth is typical of wurtzite-to-sphalerite phase transition in $A^{II}B^{VI}$ and $A^{III}B^{V}$ crystals [7], so the observed structure gives additional evidence that the phase transition develops under pressure in the compacted CdTe. Further increase in pressure resulted in decrease in volume fraction of the twin plates, which disappear at p = 650°C. They are replaced by grains with high-angle boundaries. The observed changes in the microstructure agree with the data of x-ray phase analysis.

Oppositely, in $Cd_{0.9}Zn_{0.1}Te$ ceramics made of the nanopowder the wurtzite-sphalerite transformation is not completed under pressure.

According to the x-ray phase analysis, 5-10 % of the compacted ceramics is still in the hexagonal phase after compression (Figure 11). The systematic shift of diffraction lines, with respect to pure CdTe, is caused by change in the unit cell parameter from a = 6.481 Á to a = 6.442 Á, which corresponds to the mentioned $Cd_{0.9}Zn_{0.1}Te$ composition.

Figure 10. CdTe ceramics produced at T = 20°C,: (a) p = 400 MPa grains contain twin plates; (b) p = 600 MPa twin plates disappear

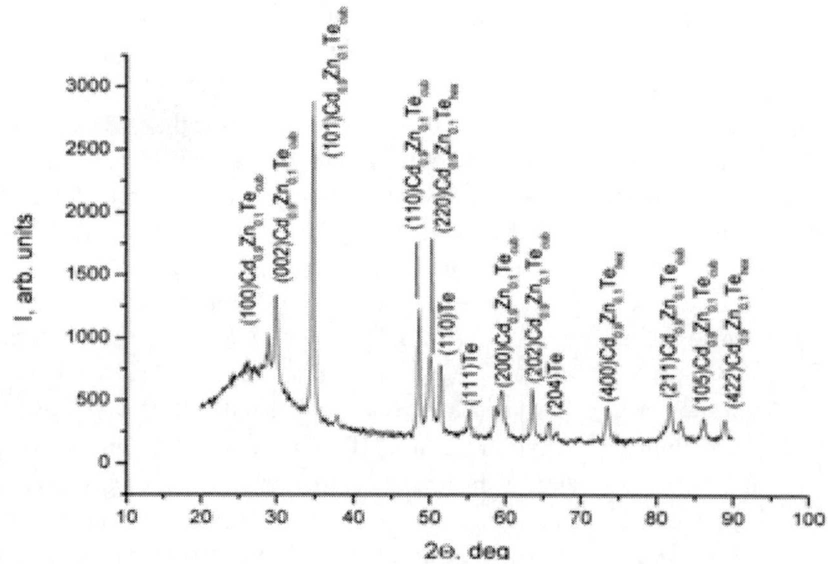

Figure 11. Diffractogram of $Cd_{0.9}Zn_{0.1}Te$ ceramics produced at T = 20°C, p = 400 MPa

The ceramics produced through compaction of CZT nanopowder at room temperature was then annealed in a sealed silica tube at 300ºC for 10—40 min. After 40 min of the annealing, hexagonal lines disappear from diffractograms (Figure 12).

In parallel, some change were also observed in the microstructure. Grains containing twins were observed immediately after compaction (Figure 13). Their volume fraction was about 5%. Twin grains firstly grew during annealing and, as the annealing time increased, the twin grains were replaced by grains with high-angle boundaries (Figure 14).

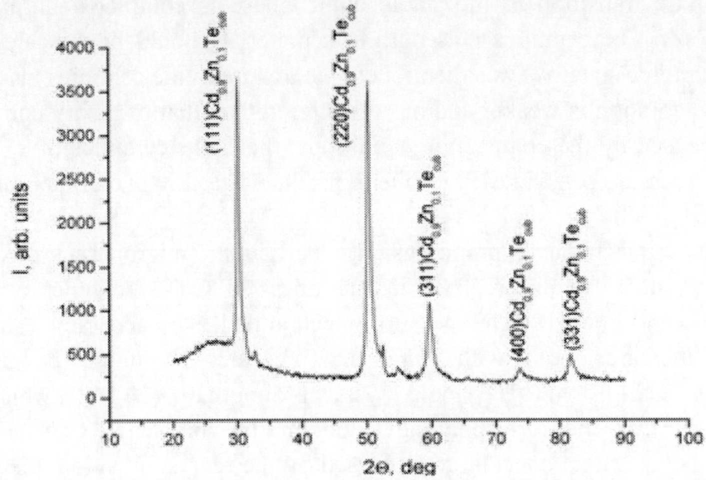

Figure 12. Diffractogram of $Cd_{0.9}Zn_{0.1}Te$ ceramics produced at T = 20°C, p = 400 MPa and annealed at T_{an} = 300°C, t_{an} = 40 min

Figure 13. Grains containing twin plates in $Cd_{0.9}Zn_{0.1}Te$ ceramics produced at T = 20°C, p = 400 MPa

The microstructure of the ceramics compressed under 400 MPa is shown in Figure 15. Grain size distribution in the ceramics is heterogeneous. According to the quantitative estimation, grain size distribution is bimodal with two extremes at 2 and 20 μm, respective volume fractions are 52 and 40%. Fine grains are settled on grain boundaries of more coarse grains. The prevailing grain size after annealing was 40 μm (Figure 16).

It has been found that in the compacted CZT ceramics the <100> axial texture and weak cubic texture present (Figure 17). However, both are less pronounced than in the CdTe ceramics fabricated from nanopowder. As the annealing time increases, the textures become weaker and more scattered. After annealing at 300°C for 40 min no texture was observed (Figure 18). Thus, though the structures of CdTe and $Cd_{0.9}Zn_{0.1}Te$ ceramics made of nanopowders have much in common, the observed differences are substantial. In CdTe ceramics the phase transition to the stable cubic phase is completed during compaction, whereas in $Cd_{0.9}Zn_{0.1}Te$ ceramics it is not. This material should be annealed to attain the stable crystallographic state. As was mentioned, the texture in the ceramics made of the three-component solid solution is weaker and more scattered than that of binary composition. Both results can be caused by the compaction conditions: the CdTe ceramics of 97% density was produced under pressure p = 650 MPa, while to produce $Cd_{0.9}Zn_{0.1}Te$ of the same density, p = 400 MPa was required.

Consequently, mechanical properties, in particular, microhardness of CdTe and $Cd_{0.9}Zn_{0.1}Te$ ceramics also differ. The microhardness of CdTe ceramics as a function of pressure is shown in Figure 19. The measuring technique has an accuracy of about 2 %. The microhardness increases both with time and pressure. It should be noted that the microhardness of melt-grown CdTe single crystals is about H = 430 MPa, which is more than three times lower than the corresponding value for the compacted CdTe H= 1500 MPa. Despite a decrease in defect concentration caused by the recrystallization, the microhardness of the specimens compressed at 200 °C remains rather high H = 1400 MПa. This can probably be explained by the increase in density due to pore healing.

The microhardness of CZT ceramics produced through nanopowder is lower, but as compared to the $Cd_{0.9}Zn_{0.1}Te$ single crystal (H = 450 MPa) it is still rather high, H = 1000 MPa. The described annealing after compaction does not cause deterioration of mechanical hardness.

The specific electrical resistivity (ρ) of the compacted CdTe specimens was tested using the four-point-probe technique, and a value of $5 \cdot 10^9$-$5 \cdot 10^{10}$ Ω·cm was obtained, which is much higher than the typical value of $2 \cdot 10^8$ Ω·cm for CdTe single crystal. This value is in good agreement with the values (approximately 10^{10} Ω·cm) obtained from current-voltage characteristics. Typical current-voltage characteristics are shown in Figure 20.

The infrared transmittance spectra of the compacted CdTe (Figure 21) show significant light scattering in the short wavelength region (wavelengths < 6μm). However, light transmission is rather high in the range of 6-25 μm, but worse than that of CdTe single crystal. This is probably caused by rougher surface of the ceramics, because a proper polishing procedure for these materials has not been developed yet.

Figure 14. The same place on the sample surface as in Figure 13 before annealing of $Cd_{0.9}Zn_{0.1}Te$ ceramics produced at T = 20°C, p = 400 MPa, after annealing at T_{an} = 300°C, t_{an} = 40 min

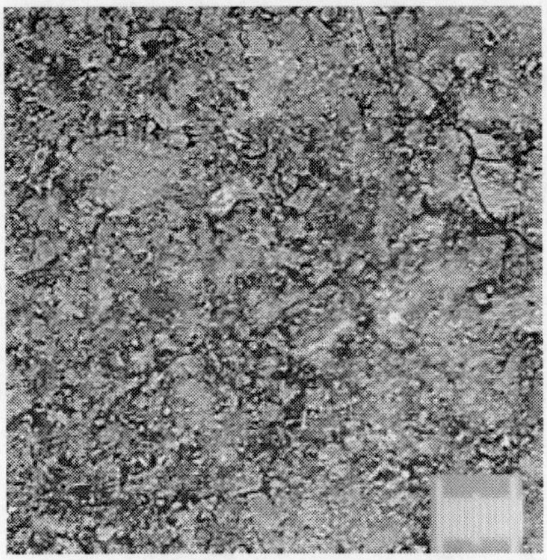

Figure 15. Microstructure of $Cd_{0.9}Zn_{0.1}Te$ ceramics produced at T = 20°C, p = 400 MPa

Figure 16. Microstructure of $Cd_{0.9}Zn_{0.1}Te$ ceramics produced at T = 20°C, p = 400 MPa, annealed at T_{an} = 300°C, t_{an} = 40 min

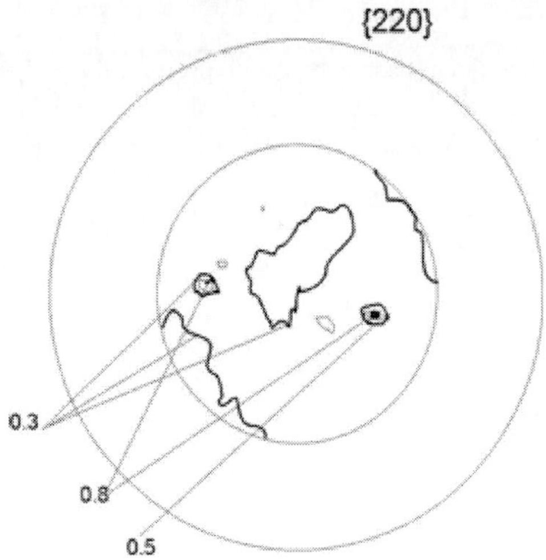

Figure 17. Pole figure {220} of $Cd_{0.9}Zn_{0.1}Te$ ceramics produced at T = 20°C, p = 400 MPa

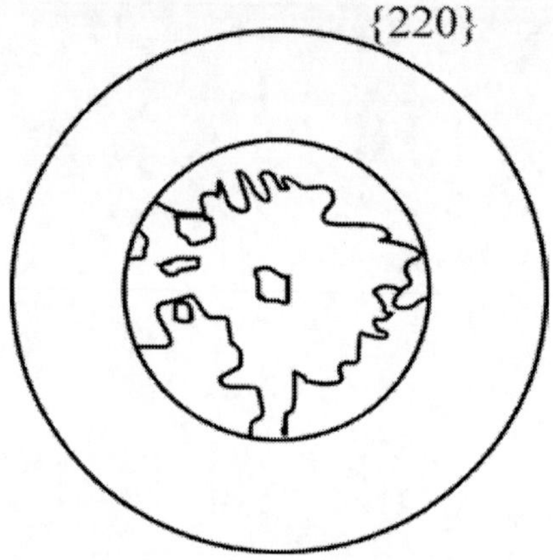

Figure 18. Pole figure {220} of $Cd_{0.9}Zn_{0.1}Te$ ceramics produced at T = 20°C, p = 400 MPa and annealed T_{an} = 300°C, t_{an} = 40 min

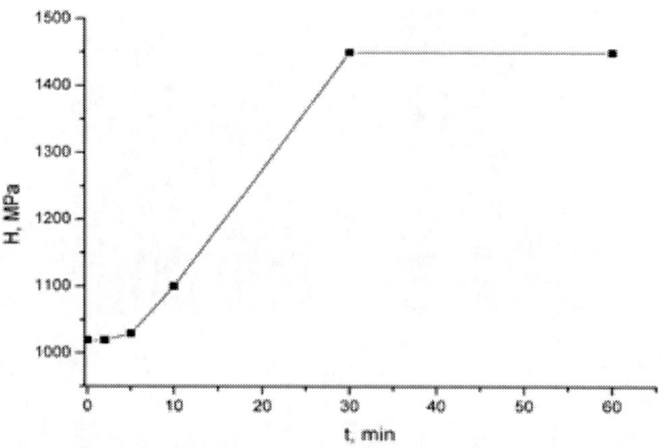

Figure 19. Dependence of microhardness on compaction pressure for CdTe ceramics

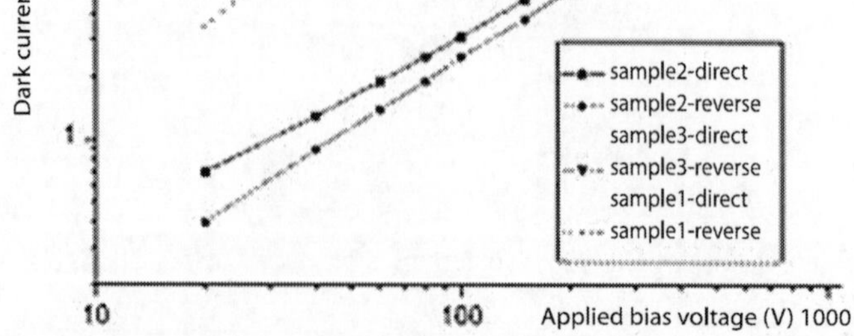

Figure 20. Current-voltage characteristics of CdTe ceramics produced at T = 20°C, p = 650 MPa

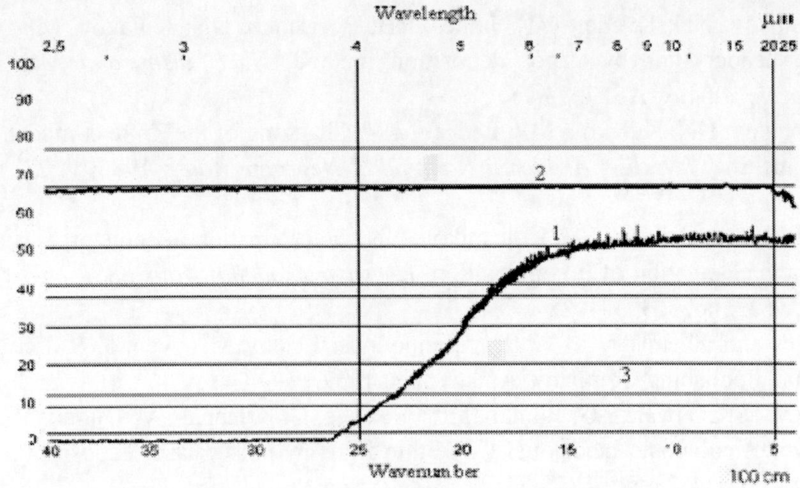

Figure 21. Transmittance spectrum of CdTe in IR range: 1 – single crystal; 2 – ceramics

Thus, the developed technique provides production of materials through the method which is far less complicated and less expensive than traditional melt growth. There is another advantage in production of CZT ceramics as compared to the single crystals: the nanopowder composition is constant and stable and is determined by the source composition, contrary to the ternary composition of Cd-Zn-Te single crystal, which varies along the crystal growth direction.

CONCLUSION

1. A new technique was developed to produce CdTe and $Cd_{1-x}Zn_xTe$ (x = 0.04—0.1) nanopowders consisting of 8-10-nm particles through vapor deposition.
2. Ceramic materials of high 95-97% density were produced from these powders at room temperature
3. It has been shown that recrystallization goes both during compaction of the powders and during annealing after deformation. Grain size can be modified by external parameters, such as temperature and time of compression and annealing.
4. The produced ceramics materials have high mechanical hardness, the specific resistivity is higher than that of the single crystals, its transmission in infrared range is acceptable for optical applications.

REFERENCES

[1] Kolesnikov, NN; James, RB; Berzigiarova, NS; Kulakov, MP. HRB and HPVZM shaped growth of CdZnTe, CdSe and ZnSe crystals. In: James RB, Franks LA, Burger A, Westbrook EM, Durst RD. *Proceedings of SPIE: X-Ray and Gamma-Ray Detectors and Applications IV*; 2002, v. 4784, 93-104.

[2] Kolesnikov, NN; Kveder, VV; James, RB; Borisenko, DN; Kulakov, MP. Growth of CdTe nanocrystals by vapor deposition method, *Nucl. Instr. and Meth. in Phys. Research,* 2004, v. A527, 73-75.

[3] Kolesnikov, NN; Kulakov, MP; Fadeev, AV. Changing of the ZnSe composition during zone melting. *Izvestiya Akademii Nauk SSSR-Neorgan Mater,* 1986, v. 22, no 3, 395-398.

[4] Kulakov, MP; Fadeev, AV; Kolesnikov, NN. Determination of some properties of ZnSe melt and calculation of it composition. *Izvestiya Akademii Nauk SSSR-Neorgan Mater,* 1986, v. 22, no. 3, 399-402.

[5] Physics and Chemistry of $A^{II}B^{VI}$ Compounds, Ed. by M. Aven, J. S. Prener, North-Holland Publishing Company, Amsterdam, 1967.

[6] Pandeya, SK; Tiwaria, U; Ramana, R; Prakasha, C; Krishnab, V; Duttab, V; Zimik, K. Growth of cubic and hexagonal CdTe thin films by pulsed laser deposition, *Thin Solid Films*, 2005, v.473, no. 1, 54-57.

[7] Kulakov MP; Balyakina, IV. Solid State wurtzite-sphalerite transformation and phase boundaries in ZnSe-CdSe, *J. of Crystal Growth*, 1991, 113, 653-657.

INDEX

A

B

C

D

E

M

Q

R

S

T

U

V

W

X

Z